Lecture Notes in Mathematics 2198

More information about this series at http://www.springer.com/series/304

Martin A. Guest • Claus Hertling

Painlevé III: A Case Study in the Geometry of Meromorphic Connections

 Springer

Martin A. Guest
Department of Mathematics
Faculty of Science and Engineering
Waseda University
Tokyo, Japan

Claus Hertling
Lehrstuhl für Mathematik VI
Universität Mannheim
Mannheim, Germany

ISSN 0075-8434   ISSN 1617-9692   (electronic)
Lecture Notes in Mathematics
ISBN 978-3-319-66525-2   ISBN 978-3-319-66526-9   (eBook)
DOI 10.1007/978-3-319-66526-9

Library of Congress Control Number: 2017953455

Mathematics Subject Classification (2010): 34M55, 34Mxx, 32G20

Printed on acid-free paper

This Springer imprint is published by Springer Nature
The registered company is Springer International Publishing AG
The registered company address is: Gewerbestrasse 11, 6330 Cham, Switzerland

Preface

This monograph is about a Painlevé III equation of type $P_{III}(D_6)$ and its relation to isomonodromic families of vector bundles on $\mathbb{P}^1$ with meromorphic connections. This equation is very classical, and indeed many of the results discussed here can be extracted from the literature. But we are motivated by more recent developments, of which this particular equation provides an illuminating and nontrivial example.

The purpose of the monograph is twofold: it offers a conceptual language for the geometrical objects underlying Painlevé equations, and it offers new results on a particular Painlevé III equation of type $P_{III}(D_6)$, which we denote by $P_{III}(0, 0, 4, -4)$. This is equivalent to the radial sine (or sinh) Gordon equation and, as such, it appears very widely in geometry and physics.

The length of the monograph is due to the systematic development of the material in the language of vector bundles with meromorphic connections, together with additional structures which take care of all relevant symmetries. Our motivation is to explain in a leisurely fashion the language of this general theory, by means of a concrete example.

We emphasize that it is not necessary to read this monograph from beginning to end; different parts of the monograph will be of interest to different readers. However, we recommend that Chap. 1 be read from beginning to end, in order to understand how to find those parts.

The relevant geometrical objects are vector bundles with flat meromorphic connections. By choosing local trivializations one obtains scalar (or matrix) differential equations. While the sine-Gordon equation may itself look very natural, certain aspects of its solutions appear quite awkward if the underlying geometrical objects are ignored. This holds even more so for the Painlevé equations, whose explicit formulations have historical but no other significance.

It is well known that the (nonlinear) Painlevé equations describe isomonodromic deformations of associated linear meromorphic differential equations, that is, families of linear equations whose monodromy data remains constant within the family. This theory (like the theory of isospectral deformations) is of great importance as it links nonlinear p.d.e. with algebraic methods of integrable systems theory. It has far-reaching applications to differential geometry (harmonic maps and harmonic bundles), to variations of Hodge structure and their generalizations (such as variations of TERP structures), and to quantum field theory in physics (most recently, mirror symmetry). However, the foundation for all this is the underlying geometry of families of vector bundles with meromorphic connections on $\mathbb{P}^1$, rather than the linear equations themselves, and a description of this geometry forms our starting point.

We emphasize that two variables (or rather two spaces) are involved: the "time variable" $x \in U \subseteq \mathbb{C}$ of the original Painlevé equation, and the deformation variable $z \in \mathbb{P}^1$ of the linear system. Our vector bundles are rank 2 complex vector bundles

$$\pi : E \to U \times \mathbb{P}^1$$

whose restrictions $\pi|_{\{x\} \times \mathbb{P}^1}$ are holomorphic bundles on $\mathbb{P}^1$, or "twistors".

The first half of the article introduces the fundamental moduli space of vector bundles with meromorphic connections on $\mathbb{P}^1$ with specific conditions, then shows how it may be identified with two other spaces. First we have the (slightly less canonical) space M^{mon} of monodromy data, which arises through the Riemann-Hilbert correspondence. Then we have the (much less canonical) space M^{ini} of initial conditions, which arises when local bases are chosen. The familiar Painlevé and sine-Gordon equation appear only at this point.

We obtain complex analytic isomorphisms

$$M^{ini} \cong \text{moduli space of connections} \cong M^{mon}$$

which are highly nontrivial as they encode the solutions of the equations.

The second half of the monograph addresses in this way the complex multi-valued solutions of $P_{III}(0, 0, 4, -4)$ on the "time variable space" $\mathbb{C}^*$. Of particular interest are the poles and zeros of these solutions and the behaviour of the solutions at $x = 0$ and $x = \infty$. So far all data is holomorphic.

In the last four chapters we consider real solutions (with singularities) on $\mathbb{R}_{>0}$. The vector bundles there mix holomorphic and antiholomorphic structures. They are generalizations of variations of Hodge structures, called TERP structures (for Twistor Extension Real Pairing), and are related to tt^* geometry and harmonic bundles. In this language, the monograph gives a complete picture of semisimple rank 2 TERP structures. It also shows what general results on TERP structures tell us about the asymptotics of solutions near 0 and ∞.

In the last chapter, results about the asymptotics near 0 and ∞ are combined with the global geometry of $M_{\mathbb{R}}^{ini}$ and $M_{\mathbb{R}}^{mon}$. This combination leads to a new global picture of all zeros and poles of all real solutions on $\mathbb{R}_{>0}$. Very briefly, this can be stated as follows. For real solutions f of $P_{III}(0, 0, 4, -4)$, there are four types of behaviour at $x = 0$ and four types of behaviour at $x = \infty$:

$$\text{At } x = 0 : \quad \text{At } x = \infty :$$

(i) $f > 0$ (i) $f > 0$

(ii) $f < 0$ (ii) $f < 0$

(iii) f has zeros   (iii) f has zeros with $f'(\text{zero}) > 0$

and poles with $(f^{-1})'(\text{pole}) < 0$

(iv) f^{-1} has zeros (iv) f^{-1} has zeros with $f'(\text{zero}) < 0$

and poles with $(f^{-1})'(\text{pole}) > 0$

We prove that for each solution f the region with the type of behaviour at $x = 0$ and the region with the type of behaviour at $x = \infty$ extend up to some point in $\mathbb{R}_{>0}$ where they meet. Therefore there is no mixed zone with unknown behaviour of zeros and poles. And (i) and (ii) cannot occur together for the same f. Thus only 14 possibilities remain. They correspond to 14 explicit strata of the space $M_{\mathbb{R}}^{mon}$.

We give most proofs in detail, and sometimes alternative proofs, without striving for the shortest one. It would be possible to extract shorter proofs of our new results, depending on the background of the intended audience.

To summarize, let us reiterate the motivation for our approach. The main point is that the bundle point of view deals with intrinsic geometric objects. These become progressively more concrete (but less canonical) as various choices are made. The intrinsic approach leads to a clear understanding of the space of solutions, and of the global (qualitative) properties of the solutions themselves. Moreover, it is an approach which could be adapted for other equations.

We are aware that our approach also has some disadvantages. The length and level of detail may be off-putting for the reader who is interested only in a specific aspect of the solutions. Despite this, the monograph is not entirely self-contained; at several key points we rely on references to the literature, notably for some of the asymptotic results. Another omission is that we have not yet exploited Lie-theoretic language, nor the Hamiltonian or symplectic aspects.

This project started when the authors attempted to reconcile the (algebraic/complex analytic) approach to variations of Hodge structures via meromorphic connections with the (differential geometric) approach to harmonic maps via loop groups. The current article is written primarily from the first viewpoint; a supplementary article from the second viewpoint is planned. However, both approaches are linked by the fundamental idea—the Riemann-Hilbert correspondence—that the monodromy data of the associated linear o.d.e. represents faithfully the solutions of the nonlinear equation. In the first approach, this monodromy data, or rather the meromorphic connection from which it is derived, is explicitly at the forefront. In the second approach, the monodromy data lies behind the so-called generalized Weierstrass or DPW data, and $z \in \mathbb{P}^1$ is the spectral parameter.

Tokyo, Japan   Martin A. Guest
Mannheim, Germany   Claus Hertling

Acknowledgements

This monograph was mainly written during several stays of the second author at Tokyo Metropolitan University. He thanks TMU for hospitality. He also thanks Masa-Hiko Saito for fruitful discussions. The first author thanks Alexander Its and Chang-Shou Lin for explanations of related material.

The first author was partially supported by JSPS grants A25247005 and A21244004, and by Waseda University grant 2013B-083. The second author was partially supported by DFG grant HE 2287/4-1 and JSPS grant S-11023.

Contents

Chapter 1
Introduction

In order to make the results approachable and transparent, this introduction is quite detailed. Unlike the main body of the monograph (Chaps. 2–18), it starts in Sect. 1.1 with the Painlevé III equations, and explains immediately and concretely the space M^{ini} of initial conditions. Although this is quite long, it is just a friendly introduction to essentially well known facts on Painlevé III. Section 1.2 gives, equally concretely, the space M^{mon} of monodromy data (at this point, without explaining where it comes from). Section 1.3 presents the main results on real solutions. No special knowledge is required to understand these statements.

Section 1.4 is intended to orient the reader who is interested primarily in knowing where the above results can be found in the monograph, and which parts of the monograph might be most relevant for applications. The history of the subject is rich, and some references and historical remarks are given in Sect. 1.5.

In Sect. 1.6 we summarize our treatment of vector bundles with meromorphic connections. The terminology is cumbersome, but it is designed to be used, not just looked at. It reflects the properties of these bundles and the additional structures which underlie the symmetries of the equations. Section 1.7 goes on to relate these bundles to the important geometrical concepts which motivated the entire project, and which are (for us) the main applications: TERP structures, variations of Hodge structures, and tt* geometry.

Finally, in Sect. 1.8 we list some problems and questions that we leave for future consideration.

© Springer International Publishing AG 2017
M.A. Guest, C. Hertling, *Painlevé III: A Case Study in the Geometry of Meromorphic Connections*, Lecture Notes in Mathematics 2198,
DOI 10.1007/978-3-319-66526-9_1

1.1   A Painlevé Equation and Its Space of Initial Data M^{ini}

The Painlevé III equation with parameters $(\alpha, \beta, \gamma, \delta) \in \mathbb{C}^4$ is the second order ordinary differential equation

$$P_{III}(\alpha, \beta, \gamma, \delta) : f_{xx} = \frac{f_x^2}{f} - \frac{1}{x}f_x + \frac{1}{x}(\alpha f^2 + \beta) + \gamma f^3 + \delta \frac{1}{f}. \tag{1.1}$$

This is a meromorphic o.d.e. whose coefficients have a simple pole at $x = 0 \in \mathbb{C}$, so we assume from now on that $x \in \mathbb{C}^*$.

It is a basic fact of o.d.e. theory that for any $x_0 \in \mathbb{C}^* = \mathbb{C} \setminus \{0\}$ and any *regular initial datum* $(f_0, \widetilde{f_0}) \in \mathbb{C}^* \times \mathbb{C}$ there exists a unique holomorphic local solution f with $f(x_0) = f_0$ and $f_x(x_0) = \widetilde{f_0}$. The "Painlevé property" is the following nontrivial extension of this:

Theorem 1.1 (Painlevé Property) *Any local solution extends to a global multivalued meromorphic function on* $\mathbb{C}^*$.

Painlevé's proof was incomplete. The Painlevé property follows from the relation to isomonodromic families (and fundamental results on isomonodromic families)— see [FN80, IN86, FIKN06], or Theorem 10.3 below. Other proofs can be found, for example, in [HL01, GLSh02, IS12].

The Painlevé property implies that analytic continuations of solutions with regular initial data give meromorphic functions on the universal covering of $\mathbb{C}^*$, i.e. their singularities are at worst poles (in fact they are all simple poles). However the location of such poles varies drastically with the initial data. In addition, there are other solutions corresponding to "singular initial data", and the same applies to their poles as well. The *global* description of all these solutions and their poles presents a nontrivial problem, which (as far as we know) has not so far been addressed in the existing literature.

If one chooses locally a logarithm

$$\varphi = 2\log f, \text{ i.e. } f = e^{\varphi/2},$$

then φ branches at the poles and zeros of f, but equation (1.1) simplifies to

$$(x\partial_x)^2 \varphi = 2x(\alpha e^{\varphi/2} + \beta e^{-\varphi/2}) + 2x^2(\gamma e^{\varphi} + \delta e^{-\varphi}) \tag{1.2}$$

where ∂_x means d/dx.

One sees immediately the following symmetries (here $k \in \mathbb{Z}$):

$$f \text{ and } \varphi + 4\pi i k \text{ are solutions for } (\alpha, \beta, \gamma, \delta) \iff$$

$$-f \text{ and } \varphi + 2\pi i + 4\pi i k \text{ are solutions for } (-\alpha, -\beta, \gamma, \delta) \iff$$

$$f^{-1} \text{ and } -\varphi + 4\pi i k \text{ are solutions for } (-\beta, -\alpha, -\delta, -\gamma) \iff \tag{1.3}$$

$$-f^{-1} \text{ and } -\varphi + 2\pi i + 4\pi i k \text{ are solutions for } (\beta, \alpha, -\delta, -\gamma).$$

In [OKSK06] four cases are identified:

$$\begin{aligned}
P_{III}(D_6) &\qquad \gamma\delta \neq 0 \\
P_{III}(D_7) &\quad \gamma = 0, \alpha\delta \neq 0 \text{ or } \delta = 0, \beta\gamma \neq 0 \\
P_{III}(D_8) &\qquad \gamma = 0, \delta = 0, \alpha\beta \neq 0 \\
P_{III}(Q) &\qquad \alpha = 0, \gamma = 0 \text{ or } \beta = 0, \delta = 0
\end{aligned}$$

The four parameters can be reduced to 2,1,0,1 parameters (respectively) by rescaling x and f.

In particular, $P_{III}(D_6)$ can be reduced to $P_{III}(\alpha, \beta, 4, -4)$ with $(\alpha, \beta) \in \mathbb{C}^2$. In this monograph we consider only the Painlevé III equation $P_{III}(0, 0, 4, -4)$. Then (1.2) becomes the radial sinh-Gordon equation

$$(x\partial_x)^2 \varphi = 16x^2 \sinh\varphi. \tag{1.4}$$

All four symmetries (1.3) preserve the space of solutions of $P_{III}(0, 0, 4, -4)$.

The following lemma makes precise statements about the zeros and poles of solutions of $P_{III}(0, 0, 4, -4)$ and gives a meaning to *singular initial data* at zeros or poles of solutions. It generalizes to all $P_{III}(D_6)$ equations. The lemma is known, but for lack of a suitable reference we sketch the proof.

Lemma 1.2

(a) Let

$$f(x) = a_1(x - x_0) + a_2(x - x_0)^2 + a_3(x - x_0)^3 + \cdots$$

be a local solution of $P_{III}(0, 0, 4, -4)$ near $x_0 \in \mathbb{C}^*$ with $f(x_0) = 0$. Then

$$a_1 = \pm 2, \quad a_2 = \tfrac{1}{2}a_1/x_0, \tag{1.5}$$

there is no restriction on $a_3 \in \mathbb{C}$, and, for $n \geq 4$, a_n is determined by $a_1, \ldots, a_{n-1}$.

(b) For any $\varepsilon_2 = \pm 1$ and any $\widehat{f}_0 \in \mathbb{C}$, there exists a unique holomorphic solution f of $P_{III}(0, 0, 4, -4)$ with $f(x_0) = 0$ and $\partial_x f(x_0) = -2\varepsilon_2$, $\partial_x^3 f(x_0) = \widehat{f}_0$.

(c) A solution can have only simple zeros and simple poles. There are two types of zeros (according to whether $\partial_x f(x_0) = 2$ or -2) and two types of poles (according to whether $\partial_x(f^{-1})(x_0) = 2$ or -2). There is a 1–1 correspondence between local solutions with zeros or poles at $x_0 \in \mathbb{C}^*$ and the set of singular initial data

$$(\varepsilon_1, \varepsilon_2, \widehat{f}_0) \in \{\pm 1\} \times \{\pm 1\} \times \mathbb{C} \tag{1.6}$$

given by

$$f^{\varepsilon_1}(x_0) = 0, \ \partial_x(f^{\varepsilon_1})(x_0) = -2\varepsilon_2, \ \partial_x^3(f^{\varepsilon_1})(x_0) = \widehat{f}_0. \tag{1.7}$$

Proof Part (a) is obtained by substituting the Taylor series into equation (1.1). In part (b) only the convergence of the resulting formal power series is nontrivial. In this monograph it is proved indirectly via Theorem 10.3 (b). Part (c) follows immediately from the symmetry $f \leftrightarrow f^{-1}$ in (1.3).   □

For any $x_0 \in \mathbb{C}^*$ the space of regular initial data is $\mathbb{C}^* \times \mathbb{C}$ with coordinates $(f_0, \widetilde{f_0})$ and the conditions $f(x_0) = f_0$, $\partial_x f(x_0) = \widetilde{f_0}$. The space of singular initial data is given in (1.6) with the conditions (1.7). It is a rather easy exercise to find new coordinates such that the space of regular initial data glues to any one component of the space of singular initial data to a chart isomorphic to $\mathbb{C} \times \mathbb{C}$. This can be done simultaneously for all $x_0 \in \mathbb{C}^*$. We explain this next, and state the result in Corollary 1.3.

Suppose that x_0 is a zero of f (so $\varepsilon_1 = 1$) with $\varepsilon_2 = 1$ in (1.5). Then for x close to x_0,

$$f = (-2)(x - x_0) + \tfrac{1}{2}(-2/x_0)(x - x_0)^2 + \tfrac{1}{6}\widehat{f}_0(x - x_0)^3 + O((x - x_0)^4)$$
$$= (-2)(x - x_0)(1 + \tfrac{1}{2}(1/x_0)(x - x_0)) + O((x - x_0)^3)$$
$$= (-2)(x - x_0)(1 - \tfrac{1}{4}(f/x_0)) + O((x - x_0)^3),$$

so

$$(-2)(x - x_0) = f(1 + \tfrac{1}{4}f/x_0) + O((x - x_0)^3).$$

Thus

$$\partial_x f = (-2) + \tfrac{-2}{x_0}(x - x_0) + \tfrac{1}{2}\widehat{f}_0(x - x_0)^2 + O((x - x_0)^3)$$
$$= -2 + \tfrac{-2}{x}(x - x_0) + (-2)(\tfrac{1}{x_0} - \tfrac{1}{x})(x - x_0)$$
$$\quad + \tfrac{1}{2}\widehat{f}_0(x - x_0)^2 + O((x - x_0)^3)$$
$$= -2 + \tfrac{-2}{x}(x - x_0) + \tfrac{-2}{x^2}(x - x_0)^2 + \tfrac{1}{2}\widehat{f}_0(x - x_0)^2 + O((x - x_0)^3)$$
$$= -2 + \tfrac{f}{x}(1 + \tfrac{f}{4x_0}) + \tfrac{-1}{2x^2}f^2 + \tfrac{1}{8}\widehat{f}_0 f^2 + O((x - x_0)^3)$$
$$= -2 + \tfrac{1}{x}f - \tfrac{1}{4x^2}f^2 + \tfrac{1}{8}\widehat{f}_0 f^2 + O((x - x_0)^3). \tag{1.8}$$

On the space of regular solutions $\mathbb{C}^* \times \mathbb{C}$ the coordinates $(f_0, \widetilde{f_0})$ can be replaced by the coordinates (f_0, g_0) with

$$\widetilde{f_0} = \frac{2f_0}{x}g_0 \tag{1.9}$$

or by the coordinates $(f_0, \widetilde{g_0})$ with

$$g_0 = -\frac{x}{f_0} + \frac{1}{2} + \frac{f_0}{2}\widetilde{g_0}. \tag{1.10}$$

Then (1.8) shows that $\widetilde{g}_0$ extends to the set of singular initial data in Lemma 1.2 (c) with $\varepsilon_1 = \varepsilon_2 = 1$ and identifies the point $(0, \widetilde{g}_0) \in \{0\} \times \mathbb{C}$ with the point $(1, 1, \widehat{f}_0)$ with

$$\widetilde{g}_0 = \frac{-1}{4x_0} + \frac{x_0}{8}\widehat{f}_0. \tag{1.11}$$

Now Corollary 1.3 follows easily. There and later the four pairs $(\varepsilon_1, \varepsilon_2) \in \{\pm 1\} \times \{\pm 1\}$ are enumerated by $k \in \{0, 1, 2, 3\}$ via the bijection

$$(1, 1) \mapsto 0, \ (-1, 1) \mapsto 1, \ (1, -1) \mapsto 2, \ (-1, -1) \mapsto 3. \tag{1.12}$$

We may now introduce the first "moduli space" that will be of interest to us:

$$M_{3FN}^{ini} = M_{3FN}^{reg} \cup M_{3FN}^{sing}, \quad M_{3FN}^{sing} = \cup_{k=0}^{3} M_{3FN}^{[k]}.$$

The suffix $3FN$ will be explained in Chap. 10 (see also Sect. 1.5).

Corollary 1.3 *The set of regular and singular initial data for local solutions of $P_{III}(0, 0, 4, -4)$ for all $x_0 \in \mathbb{C}^*$ is the 3-dimensional algebraic manifold M_{3FN}^{ini} whose four affine charts $\mathbb{C}^* \times \mathbb{C} \times \mathbb{C}$ have coordinates $(x_0, f_k, \widetilde{g}_k)$ for $k \in \{0, 1, 2, 3\}$.*
Each chart consists of the set M_{3FN}^{reg} of regular initial data and of one of the components $M_{3FN}^{[k]}$ of the set $M_{3FN}^{sing} = \cup_{k=0}^{3} M_{3FN}^{[k]}$ of singular initial data. In any chart $M_{3FN}^{reg} \cong \mathbb{C}^* \times \mathbb{C}^* \times \mathbb{C}$ and $M_{3FN}^{[k]} = \mathbb{C}^* \times \{0\} \times \mathbb{C}$. The charts are related by

$$g_k := -\frac{x_0}{f_k} + \frac{1}{2} + \frac{f_k}{2}\widetilde{g}_k, \tag{1.13}$$

$$(f_0, g_0) = (f_1^{-1}, -g_1) = (-f_2, g_2) = (-f_3^{-1}, -g_3). \tag{1.14}$$

(1.14) comes from the symmetries in (1.3) and from (1.12) and (1.9). Equation (1.13) comes from (1.10).

M_{3FN}^{ini} is in fact the space of initial data of Okamoto [Ok79] for $P_{III}(0, 0, 4, -4)$. He starts with a naive compactification of M_{3FN}^{reg}, blows it up several times and then takes out a certain divisor. The factor f_0^2 before $\widetilde{g}_0$ in

$$\widetilde{f}_0 = \frac{2f_0}{x_0} g_0 = -2 + \frac{f_0}{x_0} + \frac{f_0^2}{2}\widetilde{g}_0$$

reflects the fact that the component $M_{3FN}^{[0]}$ is obtained by two subsequent blowing ups. The complement of M_{3FN}^{ini} in Okamoto's compactification is a divisor of type $\widetilde{D}_6$, which is the origin of the name $P_{III}(D_6)$. This compactification and divisor are not obvious from Corollary 1.3.

Descriptions of M_{3FN}^{ini} by four affine charts (in fact, for the space of initial data of all $P_{III}(D_6)$ equations) can be found in [MMT99] and in [Te07], but they do not make explicit Lemma 1.2 and the notion of singular initial data.

Up to here everything is essentially well known, but the following section contains some new results.

1.2 The Space of Monodromy Data M^{mon}

It is well known that Painlevé equations describe isomonodromic deformations of systems of linear differential equations. This permits a very different and very fruitful point of view on M_{3FN}^{ini} and the solutions of $P_{III}(0, 0, 4, -4)$. It gives rise to a holomorphic isomorphism

$$\Phi_{3FN} : M_{3FN}^{ini} \to M_{3FN}^{mon}$$

where M_{3FN}^{mon} is the space of monodromy data. It is treated in this section as a black box, but will be a major theme in the rest of this monograph.

Let us introduce

$$
\begin{aligned}
V^{mat} &:= \{(s, B) \in \mathbb{C} \times SL(2, \mathbb{C}) \mid B = \left(\begin{smallmatrix} b_1 & b_2 \\ -b_2 & b_1 + s b_2 \end{smallmatrix} \right) \} \\
&\cong V^{mon} := \{(s, b_1, b_2) \in \mathbb{C}^3 \mid b_1^2 + b_2^2 + s b_1 b_2 = 1\}.
\end{aligned}
\tag{1.15}
$$

We have an automorphism

$$
\begin{aligned}
m_{[1]}^2 &: \mathbb{C} \times V^{mat} \to \mathbb{C} \times V^{mat} \\
(\beta, s, B) &\mapsto (\beta + 4\pi i, s, \mathrm{Mon}_0^{mat}(s)^{-2} B)
\end{aligned}
\tag{1.16}
$$

where

$$\mathrm{Mon}_0^{mat}(s) := S^t S^{-1} = \begin{pmatrix} 1 & -s \\ s & 1 - s^2 \end{pmatrix}, \quad S := \begin{pmatrix} 1 & s \\ 0 & 1 \end{pmatrix}. \tag{1.17}$$

(The significance of these matrices will be explained later.) Define

$$M_{3FN}^{mon} := \mathbb{C} \times V^{mat} / \langle m_{[1]}^2 \rangle, \tag{1.18}$$

where $\langle m_{[1]}^2 \rangle$ is the group (isomorphic to $\mathbb{Z}$) generated by $m_{[1]}^2$. While V^{mat} and $\mathbb{C} \times V^{mat}$ are algebraic manifolds, M_{3FN}^{mon} is only an analytic manifold. But the fibres $M_{3FN}^{mon}(x)$ of the natural projection

$$pr_{3FN}^{mon} : M_{3FN}^{mon} \to \mathbb{C}^*, \quad [(\beta, s, B)] \mapsto \tfrac{1}{2} e^{-\beta/2} = x \tag{1.19}$$

are algebraic, as any choice of β with $\frac{1}{2}e^{-\beta/2} = x$ yields an isomorphism

$$pr_{mat} : M^{mon}_{3FN}(x) \to V^{mat}, \tag{1.20}$$

$$[(\widetilde{\beta}, s, \widetilde{B})] \mapsto \text{ the pair } (s, B) \text{ with } [(\widetilde{\beta}, s, \widetilde{B})] = [(\beta, s, B)].$$

The trivial foliation on $\mathbb{C} \times V^{mat}$ over $\mathbb{C}$ with leaves $\mathbb{C} \times \{(s, B)\}$ induces a foliation of M^{mon}_{3FN} over $\mathbb{C}^*$. A leaf through $[(\beta, s, B)]$ has finitely many branches if and only if $\text{Mon}_0^{mat}(s)^2$ has finite order, and then the number of branches is this order.

Theorem 1.4 below is a key result of this monograph. It follows from Theorem 10.3, which in turn depends on the Riemann-Hilbert correspondence of Chap. 2. Its origin and relation to results in the literature are discussed in Sect. 1.6.

Theorem 1.4 *The relation of $P_{III}(0, 0, 4, -4)$ to isomonodromic families gives rise to a holomorphic isomorphism*

$$\Phi_{3FN} : M^{ini}_{3FN} \to M^{mon}_{3FN} \tag{1.21}$$

such that any local solution of $P_{III}(0, 0, 4, -4)$ is mapped to a local leaf in M^{mon}_{3FN}.

This means that the restriction of the meromorphic function f_0 on M^{ini}_{3FN} to one leaf in M^{mon}_{3FN} (composed with the projection of this leaf to $\mathbb{C}^*$) gives a multi-valued solution of $P_{III}(0, 0, 4, -4)$ and that all solutions arise in this way. If a local solution f is defined in a simply connected subset U of $\mathbb{C}^*$ then, for any choice of a lift $U' \subset \mathbb{C}$ of U to the universal covering $\mathbb{C} \to \mathbb{C}^*$, $\beta \mapsto \frac{1}{2}e^{-\beta/2}$, we obtain an element (s, B) of V^{mat}. Here s is unique, and B depends on the choice of U'.

This (s, B) encodes the geometry of the solution f in a very nice way. One instance is the following obvious corollary.

Corollary 1.5 *A global multi-valued solution has finitely many branches if and only if $\text{Mon}_0^{mat}(s)^2$ has finite order, and then the number of branches is this order.*

Less obvious are the following statements, which are amongst our main results:

(1) The pair (s, B) and the asymptotics of f for $x \to 0$ determine one another (see Sect. 1.4).
(2) In the case of a real solution of $P_{III}(0, 0, 4, -4)$ on $\mathbb{R}_{>0}$, one can read off the sequence of zeros and poles of f from the pair (s, B) with $\beta \in \mathbb{R}$ (Sect. 1.3 and Chap. 18).

We shall deal with multi-valued solutions f on $\mathbb{C}^*$ of $P_{III}(0, 0, 4, -4)$ where one branch over $\mathbb{C} - \mathbb{R}_{\leq 0}$ is distinguished. The choice of β with $\beta \in (-2\pi, 2\pi)$ for this branch yields a unique $(s, B) \in V^{mat}$, and the solution is denoted by $f_{mult}(., s, B)$. Then V^{mat} parametrizes all multi-valued solutions with a distinguished branch.

1.3 Real Solutions of $P_{III}(0, 0, 4, -4)$ on $\mathbb{R}_{>0}$

In this section we sketch the new results concerning real solutions of $P_{III}(0, 0, 4, -4)$ on $\mathbb{R}_{>0}$. Before this is formulated in Theorem 1.6 below, we discuss the subspaces $M^{ini}_{3FN,\mathbb{R}}$ and $M^{mon}_{3FN,\mathbb{R}}$ corresponding to initial conditions and monodromy data of real solutions.

$M^{ini}_{3FN,\mathbb{R}}$ is the real semialgebraic submanifold of M^{ini}_{3FN} which is obtained from Corollary 1.3 simply by replacing $x_0 \in \mathbb{C}^*$ by $x_0 \in \mathbb{R}_{>0}$ and the spaces $\mathbb{C}, \mathbb{C}^*$ by $\mathbb{R}$, $\mathbb{R}^*$. Thus

$$M^{ini}_{3FN,\mathbb{R}} = M^{reg}_{3FN,\mathbb{R}} \cup \left(\cup^3_{k=0} M^{[k]}_{3FN,\mathbb{R}}\right) = M^{reg}_{3FN,\mathbb{R}} \cup M^{sing}_{3FN,\mathbb{R}},$$

$$M^{reg}_{3FN,\mathbb{R}} \cong \mathbb{R}_{>0} \times \mathbb{R}^* \times \mathbb{R},$$

$$M^{[k]}_{3FN,\mathbb{R}} \cong \mathbb{R}_{>0} \times \{0\} \times \mathbb{R}, \qquad (1.22)$$

$$M^{reg}_{3FN,\mathbb{R}} \cup M^{[k]}_{3FN,\mathbb{R}} \cong \mathbb{R}_{>0} \times \mathbb{R} \times \mathbb{R} \quad \text{for any } k \in \{0, 1, 2, 3\},$$

and these last four spaces are charts which cover $M^{ini}_{3FN,\mathbb{R}}$. The space $M^{reg}_{3FN,\mathbb{R}}$ has two connected components, one for $f(x_0) > 0$ and one for $f(x_0) < 0$. They are supplemented by the four real hypersurfaces of singular initial data, which give the two types of zeros and poles.

A real solution of $P_{III}(0, 0, 4, -4)$ on $\mathbb{R}_{>0}$ is the restriction to $\mathbb{R}_{>0}$ of the distinguished branch of a multi-valued solution $f_{mult}(., s, B)$. The space of pairs (s, B) corresponding to real solutions on $\mathbb{R}_{>0}$ is (see Theorem 15.5 (a))

$$V^{mat,\mathbb{R}} := \{(s, B) \in V^{mat} \mid (s, B) = (\bar{s}, \overline{B}^{-1})\}$$

$$= \{(s, B) \in V^{mat} \mid s, b_5, b_6 \in \mathbb{R}\} \qquad (1.23)$$

$$\cong V^{mon,\mathbb{R}} := \{(s, b_5, b_6) \in \mathbb{R}^3 \mid b_5^2 + (\tfrac{1}{4}s^2 - 1)b_6^2 = 1\}$$

where $b_5 := b_1 + \tfrac{1}{2}sb_2$, $b_6 := ib_2$. This surface is illustrated in Fig. 1.1.

The isomorphism $\Phi_{3FN} : M^{ini}_{3FN} \to M^{mon}_{3FN}$ restricts to a real analytic isomorphism

$$\Phi_{3FN,\mathbb{R}} : M^{ini}_{3FN,\mathbb{R}} \to M^{mon}_{3FN,\mathbb{R}}. \qquad (1.24)$$

Restricted to real numbers, the covering $\mathbb{R} \to \mathbb{R}_{>0}$, $\beta \mapsto \tfrac{1}{2}e^{-\beta/2}$ is bijective, so we have a natural isomorphism

$$\psi_{mat} : M^{mon}_{3FN,\mathbb{R}} \xrightarrow{\cong} \mathbb{R}_{>0} \times V^{mat,\mathbb{R}}. \qquad (1.25)$$

At the end of the monograph, in Remarks 18.5 and 18.7, pictures are presented which represent the images in $M^{mon}_{3FN,\mathbb{R}}$ of the hypersurfaces $M^{[k]}_{3FN,\mathbb{R}}$. These pictures (the one in Remark 18.7 is conjectural) are consistent with, and are derived from, Theorem 1.6.

Fig. 1.1 The space $V^{mat,\mathbb{R}}$

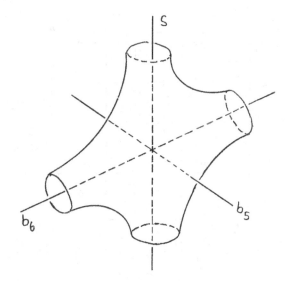

The four strata

$$
\begin{array}{ll}
\text{(i)} & |s| \le 2, \ b_5 \ge 1 \\
\text{(ii)} & |s| \le 2, \ b_5 \le -1 \\
\text{(iii)} & s > 2 \\
\text{(iv)} & s < -2
\end{array}
\tag{1.26}
$$

of $V^{mat,\mathbb{R}}$ will be important for the behaviour of solutions near 0. The four strata

$$
\begin{array}{ll}
\text{(i)} & B = 1_2 \\
\text{(ii)} & B = -1_2 \\
\text{(iii)} & b_6 < 0 \\
\text{(iv)} & b_6 > 0
\end{array}
\tag{1.27}
$$

will be important for the behaviour of solutions near ∞. Taking all possible intersections we obtain 14 strata, as illustrated in Fig. 1.2.

The following notation allows a precise formulation of the behaviour of the solutions near 0 and ∞. Let $f = f_{mult}(., s, B)|_{\mathbb{R}_{>0}}$ be a real solution of $P_{III}(0, 0, 4, -4)$ on $\mathbb{R}_{>0}$. A zero $x_0 \in \mathbb{R}_{>0}$ of f with $\partial_x f(x_0) = \pm 2$ is denoted $[0\pm]$, a pole $x_1 \in \mathbb{R}_{>0}$ of f with $\partial_x(f^{-1})(x_1) = \pm 2$ is denoted $[\infty\pm]$.

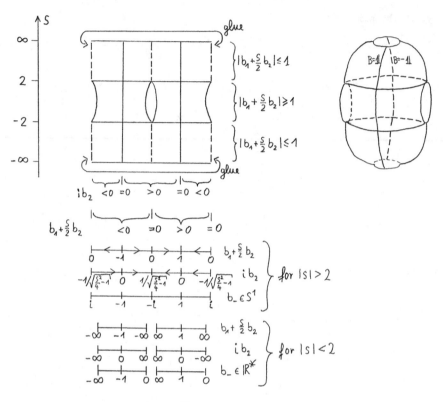

Fig. 1.2 Stratification of the space $V^{mat,\mathbb{R}}$

If a $y_0 \in \mathbb{R}_{>0}$ exists such that $f|_{(0,y_0]}$ is positive then f and $f|_{(0,y_0]}$ have type $\overleftarrow{> 0}$. If a $y_0 \in \mathbb{R}_{>0}$ exists such that $f(y_0) < 0$ and $f|_{(0,y_0]}$ has infinitely many zeros $x_1, x_2, x_3, \ldots$ with $x_1 > x_2 > x_3 > \ldots$ and such that x_k is of type $[0-]$ for odd k and of type $[0+]$ for even k, then f and $f|_{(0,y_0]}$ have type $\overleftarrow{[0+][0-]}$. The types $\overleftarrow{< 0}$, $\overleftarrow{[0-][0+]}$, $\overleftarrow{[\infty+][\infty-]}$ and $\overleftarrow{[\infty-][\infty+]}$ for f and $f|_{(0,y_0]}$ are defined analogously.

The types $\overrightarrow{> 0}$, $\overrightarrow{< 0}$, $\overrightarrow{[0+][\infty-]}$, $\overrightarrow{[\infty-][0+]}$, $\overrightarrow{[0-][\infty+]}$ and $\overrightarrow{[\infty+][0-]}$ are defined analogously. Of course f has type $\overrightarrow{[0+][\infty-]}$ if and only if it has type $\overrightarrow{[\infty-][0+]}$. But for $f|_{[y_0,\infty)}$ the smallest zero or pole is important.

Parts (a) and (b) of Theorem 1.6 below give local information near 0 and near ∞. Part (c) shows that there is no intermediate mixed zone: the types near 0 and near ∞ determine the sequence of zeros and poles completely.

Theorem 1.6 (Theorems 18.2 and 18.4) *Fix a real solution* $f = f_{mult}(., s, B)|_{\mathbb{R}_{>0}}$ *of* $P_{III}(0, 0, 4, -4)$ *on* $\mathbb{R}_{>0}$ *and its monodromy data* $(s, B) \in V^{mat,\mathbb{R}}$.

(a) *The type of f near ∞ depends only on the stratum in (1.27) in which (s, B) lies. The following table lists the types.*

$\begin{aligned}B &= 1_2\\ B &= -1_2\end{aligned}$	$\begin{aligned}&\overrightarrow{>0}\\ &\overrightarrow{<0}\end{aligned}$
$b_6 < 0$	$\overrightarrow{[\infty+][0-]}$ and $\overrightarrow{[0-][\infty+]}$
$b_6 > 0$	$\overrightarrow{[\infty-][0+]}$ and $\overrightarrow{[0+][\infty-]}$

$$(1.28)$$

(b) *The type of f near 0 depends only on the stratum in (1.26) in which (s, B) lies. The following table lists the types.*

$\begin{aligned}|s| &\leq 2,\ b_5 \geq 1\\ |s| &\leq 2,\ b_5 \leq -1\end{aligned}$	$\begin{aligned}&\overleftarrow{>0}\\ &\overleftarrow{<0}\end{aligned}$
$s > 2$	$\overleftarrow{[0+][0-]}$ and $\overleftarrow{[0-][0+]}$
$s < -2$	$\overleftarrow{[\infty+][\infty-]}$ and $\overleftarrow{[\infty-][\infty+]}$

$$(1.29)$$

(c) *There exists some $y_0 \in \mathbb{R}_{>0}$ with $f(y_0) \neq 0$ such that $f|_{(0,y_0]}$ is of type (i), (ii), (iii), or (iv) near 0 and $f|_{[y_0,\infty)}$ is of type (i), (ii), (iii), or (iv) near ∞. Thus the sequence of zeros and/or poles on $\mathbb{R}_{>0}$ of f is completely determined by the 14 strata.*

Parts (a) and (b) are known, but part (c) is new.

For part (a) there are several sources. Reference [MTW77] studies the solutions of type $\overrightarrow{>0}$ or $\overrightarrow{<0}$, but does not identify (s, B). Reference [IN86, ch. 11] obtains $B = \pm 1_2$ for these solutions in the case $|s| \leq 2$, and it shows that all solutions with $B \neq \pm 1_2$ and $|s| \leq 2$ have infinitely many zeros or poles near ∞. Probably the arguments in [IN86, ch. 11] do not require $|s| \leq 2$. Anyway, the same without the restriction to $|s| \leq 2$ follows from the characterization of nilpotent orbits of TERP structures by mixed TERP structures in [HS07, Theorem 9.3] and [Mo11b, Corollary 8.15] (see Sect. 1.7 and Chap. 17).

Part (b) follows from the asymptotic formulae for $x \to 0$ in [Ni09] which are rewritten in Theorem 12.4.

Part (c) follows from a combination of (a) and (b) with basic properties of the space $M^{ini}_{3FN,\mathbb{R}}$ and the isomorphism $\Phi_{3FN,\mathbb{R}} : M^{ini}_{3FN,\mathbb{R}} \to \mathbb{R}_{>0} \times V^{mat,\mathbb{R}}$. One needs essentially only that $M^{ini}_{3FN,\mathbb{R}}$ has the four smooth hypersurfaces of singular initial data and that their images in $\mathbb{R}_{>0} \times V^{mat,\mathbb{R}}$ are transversal to the fibres of the projection $pr_{mat} : \mathbb{R}_{>0} \times V^{mat,\mathbb{R}} \to V^{mat,\mathbb{R}}$, which follows from the zeros and poles being simple. It is also convenient to use the existence of the two smooth solutions $f_{mult}(., 0, \pm 1_2) = \pm 1$.

The argument for (c), which combines local asymptotic information near 0 and ∞ with the global geometry of the moduli spaces, seems to be new in the theory of Painlevé III equations.

1.4   User's Guide to the Results

This section is for the reader who wishes to see concrete results on solutions of $P_{III}(0,0,4,-4)$ before (or instead of) the general theory of meromorphic connections and the proofs of those results. Such readers can restrict attention to Chaps. 1, 12, 15 and 18.

Section 1.1 contains basic elementary facts about the equation $P_{III}(0,0,4,-4)$, as well as our formulation of the "space of initial conditions". Section 1.2 gives the monodromy data (the derivation of which can be found in Chap. 2).

Theorem 12.4 in Chap. 12 rewrites the asymptotic formulae for $x \to 0$ from [IN86, FIKN06, Ni09]. They are remarkable for several reasons.

First, they yield a direct connection between solutions f and their monodromy data (s, B). The leading term is determined by and determines the pair (s, B), so it determines the solution. This is independent of the moduli space viewpoint, although it cannot be formulated as a global statement. However, if one regards the leading term as an initial datum for a solution "at $x = 0$", then the space of such data can be identified with our space of monodromy data V^{mat}.

Second, the concrete form of the leading term is important. Theorem 12.4 describes which leading terms are possible. Consider the stratification $V^{mat} = V^{mat,a} \cup V^{mat,b+} \cup V^{mat,b-} \cup V^{mat,c+} \cup V^{mat,c-}$, where

$$V^{mat,a} := \{(s, B) \in V^{mat} \mid s \in \mathbb{C} - (\mathbb{R}_{\leq -2} \cup \mathbb{R}_{\geq 2})\},$$

$$V^{mat,b\pm} := \{(s, B) \in V^{mat} \mid \pm s \in \mathbb{R}_{>2}\}, \tag{1.30}$$

$$V^{mat,c\pm} := \{(s, B) \in V^{mat} \mid s = \pm 2\}.$$

If $(s, B) \in V^{mat,a}$ then f has leading term

$$\frac{\Gamma(\frac{1}{2} - \alpha_-)}{\Gamma(\frac{1}{2} + \alpha_-)} \, b_- \cdot \left(\frac{x}{2}\right)^{2\alpha_-} \tag{1.31}$$

where α_- is defined by

$$e^{-2\pi i \alpha_-} = \lambda_- := (1 - \tfrac{1}{2}s^2) - s\sqrt{\tfrac{1}{4}s^2 - 1} \text{ and } \Re(\alpha_-) \in (-\tfrac{1}{2}, \tfrac{1}{2}) \tag{1.32}$$

and

$$b_- = (b_1 + \tfrac{1}{2}sb_2) + \sqrt{\tfrac{1}{4}s^2 - 1} \, b_2. \tag{1.33}$$

Here, λ_- is an eigenvalue of $\mathrm{Mon}_0^{mat}(s)$, and b_- is an eigenvalue of B with the same eigenvector (B and $\mathrm{Mon}_0^{mat}(s)$ commute). The function $(x/2)^{2\alpha_-}$ is a multivalued function with a distinguished branch. Equation (1.31) implies that on any multisector a neighbourhood of 0 exists on which f has no zeros or poles.

If $(s, B) \in V^{mat,b+}$ then f has leading term

$$- \frac{x}{t^{NI}} \sin\left(2t^{NI} \log \frac{1}{2}x - 2 \arg \Gamma(1 + it^{NI}) + \delta^{NI}\right) \tag{1.34}$$

where

$$t^{NI} := \frac{1}{2\pi} \log|\lambda_-|, \tag{1.35}$$

$$\delta^{NI} \in \mathbb{C} \text{ with } \Re(\delta^{NI}) \in [0, 2\pi] \text{ and } e^{i\delta^{NI}} = b_- \tag{1.36}$$

with λ_- and b_- as above. In particular, f has in any sufficiently large multisector no poles near 0 and exactly one sequence of zeros converging to 0. They have all approximately the same argument $\frac{\log|b_-|}{2t^{NI}}$; see formulae (12.25) and (12.26).

If $(s, B) \in V^{mat,c+}$ then f has leading term

$$- 2x\left(b_1 + \tfrac{1}{2}sb_2\right)\left(\log \frac{x}{2} - \frac{i\pi}{2}(b_1 + \tfrac{1}{2}sb_2)b_2 + \gamma_{Euler}\right). \tag{1.37}$$

In particular, on any multisector a neighbourhood of 0 exists on which f has no zeros or poles.

If $(s, B) \in V^{mat,b-\cup c-}$ then f^{-1} has the monodromy data

$$\left(-s, \left(\begin{smallmatrix} 1 & 0 \\ 0 & -1 \end{smallmatrix}\right) B \left(\begin{smallmatrix} 1 & 0 \\ 0 & -1 \end{smallmatrix}\right)\right) \in V^{mat,b+\cup c+},$$

and one can derive the leading term of f from the leading term of f^{-1}.

The proof in [Ni09] treats the three cases $(s, B) \in V^{mat,a}, V^{mat,b+}$ and $V^{mat,c+}$ separately. In Chap. 12 the formulae for $(s, B) \in V^{mat,b+}$ and for $(s, B) \in V^{mat,c+}$ are derived from the formula (with the leading term *and* the next term) for $(s, B) \in V^{mat,a}$. Also in other aspects the formulae are made more transparent here.

Chapter 13 offers a proof in the language of this monograph. But the heart of the proof is close to that in [Ni09].

Chapters 12 and 13 explain the shape of the formulae, the only surprise being the constant factors, in particular $\Gamma(\frac{1}{2} - \alpha_-)/\Gamma(\frac{1}{2} + \alpha_-)$ in (1.31), which come from Hankel functions.

Chapter 18 gives a complete picture of the real solutions on $\mathbb{R}_{>0}$ and their zeros and poles. This is summarized in Sect. 1.3—see Theorem 1.6.

The real solutions discussed so far correspond to the real solutions of the radial sinh-Gordon equation

$$(x\partial_x)^2\varphi = 16x^2 \sinh \varphi.$$

Chapter 15 discusses also solutions on $\mathbb{R}_{>0}$ with values in $i\mathbb{R}_{>0}$ or in S^1. They correspond to real solutions of the radial sinh-Gordon equation

$$(x\partial_x)^2\varphi = -16x^2 \sinh \varphi$$

with negative sign, and the radial sine-Gordon equation

$$(x\partial_x)^2\varphi = 16x^2 \sin\varphi.$$

These solutions are much easier to deal with: they are all smooth on $\mathbb{R}_{>0}$, because the condition $\partial_x f(x_0) = \pm 2$ in Lemma 1.2 is impossible when the values are in $i\mathbb{R}_{>0}$ or in S^1.

1.5   Related Work on Painlevé III and Meromorphic Connections

This section summarizes some history, and some relations between this monograph and the work of other authors. More information is given in Chap. 9.

The history of the relation between Painlevé III and meromorphic connections is long. Already shortly after the discovery of the Painlevé equations Garnier found a family of second order linear differential equations with rational coefficients such that isomonodromic subfamilies are governed by solutions of $P_{III}(D_6)$. Okamoto [Ok86] and Ohyama-Okumura [OO06] considered this more systematically.

At the beginning of the 1980s Flaschka-Newell [FN80] and Jimbo-Miwa [JM81] wrote down first order linear systems of differential equations in 2×2 matrices whose isomonodromic families are governed by solutions of $P_{III}(D_6)$. But the appearance of the solutions in the matrices was different, and the framework in [FN80] works only for $P_{III}(0,0,4,-4)$. We use the suffix 3FN in M_{3FN}^{ini}, M_{3FN}^{mon} to indicate that we are working with this framework (the 3 indicates the III in $P_{III}(D_6)$).

Unpublished calculations of the second author show the following. An isomonodromic family as in [FN80] gives four solutions of $P_{III}(0,0,4,-4)$ which are related by the symmetries in (1.3). On the other hand, the recipe in [JM81] gives for the same isomonodromic family one solution of $P_{III}(0,4,4,-4)$. The four solutions are mapped to the one solution by the 4:1 folding transformation in [TOS05] and [Wi04] which is called $\psi^{[4]}_{III(D_6^{(1)})}$ in [TOS05]. With hindsight, this folding transformation could have been found in the early 1980s by comparing [FN80] and [JM81].

All of [IN86], [FIKN06, ch. 13–16] and [Ni09] work with the framework in [FN80]. We also work with this framework, but lift it to vector bundles with additional structures. This is important in order to obtain the full space M_{3FN}^{mon} of monodromy data. Reference [IN86] works with a Zariski open set. References [Ni09, FIKN06] treat also the complement, but keep it separate from the Zariski open set (separatrix solutions and generic solutions).

Lifting the framework to vector bundles is also important in order to write down the bundles which correspond to the singular initial data and to obtain the isomorphism $\Phi_{3FN} : M_{3FN}^{ini} \to M_{3FN}^{mon}$. Our M_{3FN}^{mon} is the space of monodromy data of such bundles.

The isomorphism $\Phi_{3FN} : M_{3FN}^{ini} \to M_{3FN}^{mon}$ is a *Riemann-Hilbert isomorphism*
in the notation of [FIKN06, Mo11a] and other papers and a *monodromy map*
in the notation of [Bo01]. These references contain general results assuring the
holomorphicity of such Riemann-Hilbert maps. In order to obtain an isomorphism
one has to be careful about the objects included. In many situations, a set of
vector bundles with meromorphic connections and naively chosen conditions is not
Hausdorff; one needs the right stability conditions.

The framework of [JM81] for all $P_{III}(D_6)$ equations was taken up in the language
of vector bundles with connections first in [Bo01], and later in [PS09]. There on both
sides of the Riemann-Hilbert correspondence only Zariski open subset were considered. [PT14] builds on [PS09] and constructs a Riemann-Hilbert isomorphism. It has
some similarities, but also some differences from our isomorphism. All reducible
vector bundles with connections are completely reducible in our case, and are not
completely reducible in their case. They have more objects, whereas our objects (the
P_{3D6}-TEJPA bundles) are richer.

1.6 P_{3D6}-TEJPA Bundles

Our fundamental ingredient is the concept of P_{3D6} bundle, a geometrical object
which (on making certain choices) gives rise to the $P_{III}(D_6)$ equations and their
solutions. In this section we summarize how this concept will be developed and
used.

The isomonodromic families for $P_{III}(0,0,4,-4)$ in [FN80], as well as the
isomonodromic families for all $P_{III}(D_6)$ equations in [JM81], are, in the terminology
of Chaps. 2 and 4, isomonodromic families of trace free P_{3D6} bundles. The
appearance of the solutions of the P_{III} equations in the normal forms differs in
[FN80, JM81]; we shall work with the recipe in [FN80]. Such P_{3D6} bundles have
additional symmetries, giving rise to the notion of P_{3D6}-TEJPA bundles.

A P_{3D6} bundle is a 4-tuple $(H, \nabla, u_0^1, u_\infty^1)$ (Definition 2.1). Here $H \to \mathbb{P}^1$ is
a holomorphic vector bundle of rank 2 on $\mathbb{P}^1$, or *twistor*. ∇ is a meromorphic
connection whose only poles are at 0 and ∞, and both have order 2. Both poles
are semisimple. We denote by $u_0^1 \neq u_0^2$ the eigenvalues of the endomorphism
$[z\nabla_{z\partial_z}] : H_0 \to H_0$ and by $u_\infty^1 \neq u_\infty^2$ those of $[-\nabla_{\partial_z}] : H_\infty \to H_\infty$. The ordering of
the eigenvalues is part of the data.

Chapter 2 introduces also P_{3D6} monodromy tuples (intrinsic formulation of
Stokes data) and P_{3D6} numerical tuples (classical matrix formulation of Stokes
data), and states a Riemann-Hilbert correspondence between P_{3D6} bundles and
P_{3D6} monodromy tuples (Theorem 2.3). This goes back to Sibuya [Si67, Si90]. A
local Riemann-Hilbert correspondence for one singular point is given in [Ma83a]
and in much more detail in [BV89]. Modern versions of a Riemann-Hilbert
correspondence with parameters which contain Theorem 2.3 as very special case
are in [Bo01] and in [Mo11a, Theorem 4.3.1]. The latter is the most general version
we know. See also Remark 2.4.

Chapter 3 analyzes reducibility of P_{3D6} bundles in terms of their monodromy tuples. This makes statements in [PS09] more explicit.

Chapter 4 considers isomonodromic families of P_{3D6} bundles. They have four parameters, the eigenvalues $u_0^1, u_0^2, u_\infty^1, u_\infty^2$, but in fact they reduce to just one essential parameter.

A P_{3D6} bundle is *trace free* if its determinant bundle $\det(H, \nabla)$ with connection is the trivial rank 1 bundle with trivial flat connection. Then $u_0^2 = -u_0^1, u_\infty^2 = -u_\infty^1$.

Chapter 4 discusses a solution in [Heu09] of the *inverse monodromy problem*, in the case of irreducible trace free P_{3D6} bundles. This problem asks whether, for a given monodromy group, there is an isomonodromic family of holomorphic vector bundles on $\mathbb{P}^1$ with meromorphic connections, such that each member has the given monodromy group and such that a generic member is holomorphically trivial. By [Heu09] (see Theorem 4.2), the generic members of any universal isomonodromic family of trace free P_{3D6} bundles are trivial holomorphic bundles (they are *pure twistors*), and all others, which form a (possibly empty) hypersurface, are isomorphic to $\mathcal{O}_{\mathbb{P}^1}(1) \oplus \mathcal{O}_{\mathbb{P}^1}(-1)$ (they are $(1, -1)$-*twistors*; this notation is established in Remark 4.1 (iv)). For the case of trace free P_{3D6}-TEP bundles we shall give a short proof of this in Theorem 8.2 (a) and Lemma 8.5. A more involved proof can be found in [Ni09].

The P_{3D6} bundles which are relevant to [FN80] can be equipped with three types of additional structure: a pairing P, an automorphism A and an automorphism J. Chapter 6 introduces the pairing P and P_{3D6}-TEP bundles (T = *twistor* $\sim$ holomorphic vector bundle, E = *extension* $\sim$ meromorphic connection, P = pairing) and studies their monodromy tuples and isomorphism classes. A P_{3D6} bundle can be equipped with a pairing P if and only if it is irreducible or completely reducible and the "exponents of formal monodromy" are all zero.

In Chap. 7 we shall see that P_{3D6}-TEP bundles can be equipped with automorphisms A and J if and only if they are trace free. Then A and J are each unique up to the sign. The choice of A and J can be considered as a marking which fixes these signs. The choice distinguishes also two out of eight 4-tuples of certain bases of the monodromy tuple. The two 4-tuples of bases differ only by a global sign. They are useful for the moduli spaces and normals forms. With the choice of A and J, the trace free P_{3D6}-TEP bundle becomes a P_{3D6}-TEJPA bundle.

Next we introduce moduli spaces M_{3TJ}^{mon} (in Chap. 7) and M_{3TJ}^{ini} (in Chap. 8), which are analogous to M_{3FN}^{mon} and M_{3FN}^{ini}. Both can be identified with the moduli space M_{3TJ} of all isomorphism classes of P_{3D6}-TEJPA bundles. We obtain a natural analytic isomorphism

$$\Phi_{3TJ} : M_{3TJ}^{mon} \to M_{3TJ}^{ini}. \tag{1.38}$$

Like Φ_{3FN}, it is a Riemann-Hilbert correspondence. There are natural projections

$$pr_{3TJ}^{mon} : M_{3TJ}^{mon} \to \mathbb{C}^* \times \mathbb{C}^*, \quad pr_{3TJ}^{ini} : M_{3TJ}^{ini} \to \mathbb{C}^* \times \mathbb{C}^* \tag{1.39}$$

given by $(H, \nabla, u_0^1, u_\infty^1, P, A, J) \to (u_0^1, u_\infty^1)$. The space M_{3FN}^{ini} is defined at this point using a choice of normal form of the flat connection; only later, in Theorem 10.3, is M_{3FN}^{ini} identified with the space of initial data of the solutions of $P_{III}(0, 0, 4, -4)$. The derivation of $P_{III}(0, 0, 4, -4)$ from the normal form of the connection is carried out in Chap. 10, after some preparation and general remarks on Painlévé equations in Chap. 9.

In fact we define M_{3TJ}^{mon} and M_{3TJ}^{ini} first, then construct the spaces M_{3FN}^{mon} and M_{3FN}^{ini} (in Chap. 10) as pull-backs of M_{3TJ}^{mon} and M_{3TJ}^{ini} via the diagonal embedding

$$c^{diag} : \mathbb{C}^* \to \mathbb{C}^* \times \mathbb{C}^*, \quad x \mapsto (x, x). \tag{1.40}$$

Like M_{3FN}^{mon}, the space M_{3TJ}^{mon} is the quotient of the algebraic manifold $\mathbb{C} \times \mathbb{C}^* \times V^{mat}$ by a group action of $\langle m_{[1]} \rangle \cong \mathbb{Z}$. It inherits a foliation above pr_{3TJ}^{mon} from this, and the fibres $M_{3TJ}^{mon}(u_0^1, u_\infty^1)$ of pr_{3TJ}^{mon} are isomorphic as algebraic manifolds to one another and to V^{mat}.

As in the case of M_{3FN}^{ini}, the fibres $M_{3TJ}^{ini}(u_0^1, u_\infty^1)$ of pr_{3TJ}^{ini} are algebraic surfaces with four affine charts, and M_{3TJ}^{ini} is an algebraic manifold of dimension 4. It has three strata (unlike M_{3FN}^{ini}, which has five): the open stratum of normal forms of P_{3D6}-TEJPA bundles which are pure twistors, and two smooth hypersurfaces of P_{3D6}-TEJPA bundles which are $(1, -1)$-twistors. In the 1-parameter isomonodromic families of P_{3D6}-TEJPA bundles which correspond to solutions of $P_{III}(0, 0, 4, -4)$, the $(1, -1)$-twistors correspond to the zeros and poles of the solutions (Theorem 10.3).

Matrices which amount to normal forms (but without identifying P, A and J) for the pure twistor P_{3D6}-TEJPA bundles are also given in [FN80, IN86, FIKN06, Ni09]. But the normal forms in Theorem 8.2 for the $(1, -1)$ twistor P_{3D6}-TEJPA bundles are new. The relations between our approach and [IN86, FIKN06, Ni09] is discussed in detail in Chaps. 11–13.

The three symmetries R_1, R_2, R_3 of P_{3D6}-TEJPA bundles introduced in Chap. 7 are supplemented by further symmetries R_4, R_5 in Chap. 14; R_5 is a "reality condition". These are used in Chap. 15 to study real solutions of $P_{III}(0, 0, 4, -4)$ on $\mathbb{R}_{>0}$. Chapter 18 brings together the results obtained so far to describe the space of all real solutions.

1.7 TERP(0) Bundles, Generalizations of Hodge Structures

One motivation for the second author to study real solutions (possibly with zeros and/or poles) of $P_{III}(0, 0, 4, -4)$ on $\mathbb{R}_{>0}$ is that they are equivalent to Euler orbits of semisimple rank 2 TERP(0) bundles.

A pure and polarized TERP(0) bundle generalizes a pure and polarized Hodge structure. An isomonodromic family of them is called a pure and polarized TERP structure and generalizes a variation of Hodge structures. TERP structures were defined in [He03] in order to give a framework for the data in [CV91, CV93, Du93]. The semisimple rank 2 case is the simplest essentially new case beyond variations of Hodge structures.

In [CV91, CV93, Du93] it was noticed that this case is related to real solutions of $P_{III}(0, 0, 4, -4)$ on $\mathbb{R}_{>0}$. In [CV91, CV93] some of the globally smooth and positive solutions (namely, some of the solutions $f_{mult}(., s, \mathbf{1}_2)|_{\mathbb{R}_{>0}}$ with $|s| \leq 2$) are obtained from the data there.

The discussion in Chap. 16 will show that semisimple rank 2 TERP structures (not necessarily pure and polarized) correspond to solutions of $P_{III}(0, 0, 4, -4)$ on $\mathbb{R}_{>0}$ with values in $\mathbb{R}$ or in S^1. The solutions with values in S^1 are all globally smooth, and their TERP structures are all pure, but not polarized.

A real solution f on $\mathbb{R}_{>0}$ corresponds to an Euler orbit $\cup_{x>0} \text{TERP}_f(x)$, a certain 1-parameter isomonodromic family of TERP(0) bundles. $\text{TERP}_f(x)$ is a $(1, -1)$ twistor if and only if x is a zero or a pole of f. If $f(x) > 0$ then $\text{TERP}_f(x)$ is pure and polarized; if $f(x) < 0$ then $\text{TERP}_f(x)$ is pure, but not polarized.

In [Sch73] and [CKS86] nilpotent orbits of Hodge structures are studied, and a beautiful correspondence with polarized mixed Hodge structures is established. In [CV91, CV93] a renormalization group flow leads to an *infrared limit* of the Euler orbits of those TERP(0) bundles which are studied there implicitly.

This motivated the definition in [HS07] of a *nilpotent orbit* of TERP(0) bundles (Definition 17.3): An Euler orbit $\cup_{x>0} G(x)$ of TERP(0) bundles (Definition 17.1) is a *nilpotent orbit* if for large x the TERP(0) bundle $G(x)$ is pure and polarized.

It also motivated the conjecture 9.2 in [HS07] that (in the case when the formal decomposition of the pole at 0 is valid without ramification) an Euler orbit $\cup_{x>0} G(x)$ is a nilpotent orbit if and only if one (equivalently: any) TERP(0) bundle $G(x)$ is a *mixed TERP structure*. The general definition of a mixed TERP structure is given in [HS07], and the semisimple case is in Definition 17.6. Roughly, $G(x)$ is a mixed TERP structure if the real structure and Stokes structure are compatible and if the regular singular pieces of the formal decomposition of the pole at 0 induce certain polarized mixed Hodge structures.

The conjecture was proved in [HS07] (the direction $\Leftarrow$ and the regular singular case of $\Rightarrow$, building on [Mo03]) and in [Mo11b] (the general case of the more difficult direction $\Rightarrow$, building on [Mo11a]). The semisimple case is formulated in Theorem 17.9.

In our context here it shows that a real solution $f_{mult}(., s, B)|_{\mathbb{R}_{>0}}$ of $P_{III}(0, 0, 4, -4)$ on $\mathbb{R}_{>0}$ is smooth and positive for large x if and only if $B = \mathbf{1}_2$ (Corollary 17.10).

In [HS07] the concept of *Sabbah orbit* is defined. An Euler orbit $\cup_{x>0} G(x)$ of TERP(0) bundles $G(x)$ is a Sabbah orbit if for small x the TERP(0) bundle is pure and polarized. And [HS07, Theorem 7.3] characterizes Sabbah orbits by the property that they induce (in a different way than above) polarized mixed Hodge structures. This result gives a characterization of those real solutions $f_{mult}(., s, B)|_{\mathbb{R}_{>0}}$ which are smooth and positive for small x. But we do not need it here, as the asymptotic formulae in [Ni09] (see Theorem 12.4 and Sect. 1.4) give information which is more precise and easier to handle.

Finally, a result in [HS11] gives certain Euler orbits $\cup_{>0}G(x)$ of TERP(0) bundles such that all $G(x)$ are pure and polarized (see Theorem 17.11 for the semisimple case). In our context here, this result applies and gives in fact all globally smooth and positive solutions of $P_{III}(0,0,4,-4)$ on $\mathbb{R}_{>0}$ (Corollary 17.12, Remark 17.13 (i), Theorem 18.2).

Chapters 16 and 18 give by the correspondence with solutions on $\mathbb{R}_{>0}$ (with values in $\mathbb{R}$ or in S^1) a rather complete picture of the semisimple rank 2 TERP structures and their Euler orbits. The applications here of the general results on TERP structures illustrate what these general results give and how they work.

1.8 Open Problems

We collect here some questions and conjectures.

(I) (Remarks 9.1 and 9.2 and Sect. 1.5) Reference [JM81] proposes a recipe which relates all solutions of $P_{III}(D_6)$ to isomonodromic families of trace free P_{3D6} bundles (see also [FIKN06, ch. 5], [PS09, PT14]). A different recipe, which applies only for the solutions of $P_{III}(0,0,4,-4)$, is proposed in [FN80] (see also [IN86], [FIKN06, ch. 7–16], [Ni09]) and this monograph. The P_{3D6} bundles in [FN80] are related in [JM81] to solutions of $P_{III}(0,4,4,-4)$. The second author made calculations which show that these recipes are related by the 4:1 folding transformation in [Wi04] and [TOS05] which is called $\psi^{[4]}_{III(D_6^{(1)})}$ in [TOS05] and which maps almost all solutions of $P_{III}(0,0,4,-4)$ to almost all solutions of $P_{III}(0,4,4,-4)$.

It seems interesting to study what one obtains from the other two folding transformations $\psi^{[4]}_{IV}$ and $\psi^{[2]}_{II}$ in [TOS05] and known recipes for P_{IV} and P_{II}.

(II) (Remark 10.4 (i)) By [Ok79] the algebraic surface $M^{ini}_{3FN}(x)$ of initial data at $x \in \mathbb{C}^*$ of solutions of $P_{III}(0,0,4,-4)$ is the complement of a divisor Y of type $\widetilde{D}_6$ in a natural compactification S. Can the points in Y be given a meaning in terms of suitable generalizations of P_{3D6} bundles? If so, can this be applied to a better understanding of the solutions of $P_{III}(0,0,4,-4)$?

(III) (Remark 10.4 (vi)) The multi-valued function f_{mult} on M^{ini}_{3FN} in Remark 10.2 unites all solutions of $P_{III}(0,0,4,-4)$ and depends holomorphically on the parameters $(s,B) \in V^{mat}$. Can this dependence be controlled (beyond the asymptotic formulae for small or large x)? In particular, are there differential equations governing the dependence on s and B? We do not see any, and wonder why not.

For constant s and varying B, the monodromy of a family of P_{3D6}-TEP bundles is constant, but the poles along 0 and ∞ of the covariant derivatives along B of sections have infinite order. Therefore it is not isomonodromic, and it does not give differential equations in an obvious way.

(IV) (Conjecture 18.6 (a)) We expect that the hypersurfaces in $\mathbb{R}_{>0} \times V^{mat,a\cup c,\mathbb{R}}$ and in $\mathbb{R}_{>0} \times V^{mat,b\cup c,\mathbb{R}}$ which give the $(1, -1)$-twistors (the sheets of zeros and poles of solutions) are convex.

(V) Conjectures 18.6 (b) and (c) formulate our expectations on the behaviour of the limits of the sheets of zeros and poles of solutions when (s, B) approach the holes $s = \pm\infty$ or $ib_2 = \pm\infty$ in $V^{mat,\mathbb{R}}$.

(VI) In Chap. 12 we have analyzed and rewritten the asymptotic formulae as $x \to 0$ for solutions $f_{mult}(., s, B)$ from [Ni09]. Similar general formulae for $x \to \infty$ are missing. Special cases are in [IN86], for example Chapter 8 contains asymptotic formulae as $x \to \infty$ for solutions on $\mathbb{R}_{>0}$ which are smooth near ∞. Chapter 11 contains asymptotic formulae as $x \to \infty$ for the zeros and poles of real solutions on $\mathbb{R}_{>0}$. This should be extended to asymptotic formulae as $x \to \infty$ for the multi-valued solutions $f_{mult}(., s, B)$ on $\mathbb{C}^*$ and their zeros and poles.

(VII) The matrix $T(s) = \left(\begin{smallmatrix} 0 & 1 \\ -1 & s \end{smallmatrix}\right)$ in (5.26) is equivalent to the parameter $s = \operatorname{tr} T$. The parameters s and $2b_5 = \operatorname{tr} B$ in V^{mat} have completely different roles, yet there is a symmetry between them. The defining equation for V^{mat} is

$$
\begin{aligned}
0 = 4 \det B - 4 &= 4b_1^2 + 4b_2^2 + 4sb_1b_2 - 4 \\
&= (2b_1 + sb_2)^2 - (s^2 - 4)b_2^2 - 4 \\
&= ((\operatorname{tr} B)^2 - 4) + (ib_2)^2((\operatorname{tr} T)^2) - 4). \quad (1.41)
\end{aligned}
$$

The map

$$
V^{mat} \dashrightarrow V^{mat}, \quad (\operatorname{tr} T, \operatorname{tr} B, ib_2) \mapsto (\operatorname{tr} B, \operatorname{tr} T, \frac{1}{ib_2}) \quad (1.42)
$$

is birational and exchanges the traces of T and of B, and it restricts to $V^{mat,\mathbb{R}}$.

Within $V^{mat,\mathbb{R}}$, either $|\operatorname{tr} T| \le 2$ and $|\operatorname{tr} B| \ge 2$ (in $V^{mat,a\cup c,\mathbb{R}}$) or $|\operatorname{tr} T| \ge 2$ and $|\operatorname{tr} B| \le 2$ (in $V^{mat,b\cup c,\mathbb{R}}$). We wonder whether there is more behind this symmetry.

Further, we wonder about the apparent rotational symmetry by $\frac{\pi}{2}$ in the picture in Remark 18.7. Of course, because of the sizes of the spirals for small x_0 and large x_0, one should invert small and large x_0. Are the spirals, i.e. the four hypersurfaces in $\mathbb{R}_{>0} \times V^{mat,\mathbb{R}}$, exchanged by a suitable automorphism which is roughly a rotation by $\frac{\pi}{2}$ in the picture of $V^{mat,\mathbb{R}}$ and which exchanges small and large x_0? If so, this might relate asymptotic formulae for $x \to 0$ with asymptotic formulae for $x \to \infty$.

Chapter 2
The Riemann-Hilbert Correspondence for P_{3D6} Bundles

This chapter will formulate the Riemann-Hilbert correspondence for those holomorphic vector bundles on $\mathbb{P}^1$ with meromorphic connections which are central for the Painlévé III(D_6) equations. Everything in this chapter is classical, though presented in the language of bundles. We shall give references after Theorem 2.3.

We define first P_{3D6} bundles, then P_{3D6} monodromy tuples, then we state the Riemann-Hilbert correspondence in Theorem 2.3. After that, the correspondence will be explained. Finally we introduce P_{3D6} numerical tuples, a more concrete version of monodromy tuples.

Definition 2.1 A P_{3D6} *bundle* is a 4-tuple $(H, \nabla, u_0^1, u_\infty^1)$ consisting of the following ingredients. First, $H \to \mathbb{P}^1$ is a holomorphic vector bundle of rank 2 on $\mathbb{P}^1$ (a "twistor"), and ∇ is a meromorphic (hence flat) connection on H, which is holomorphic on $H|_{\mathbb{C}^*}$ but has poles at 0 and ∞ of order 2. Writing z for the coordinate on $\mathbb{C} \subset \mathbb{P}^1$, the eigenvalues of the endomorphism $[z\nabla_{z\partial_z}] : H_0 \to H_0$ (i.e. the coefficient of z^{-2} in a matrix representation of ∇_{∂_z}) are assumed distinct, and will be denoted $u_0^1, u_0^2 \in \mathbb{C}$. In particular this endomorphism is semisimple. Similarly, the eigenvalues of the endomorphism $[-\nabla_{\partial_z}] : H_\infty \to H_\infty$ are assumed distinct, and will be denoted u_∞^1, u_∞^2.

Distinguishing one eigenvalue at 0 and one at ∞ is part of the definition of P_{3D6} bundle. For the structure of P_{3D6}-bundles this choice is inessential, but it will facilitate our treatment of normal forms of P_{3D6} bundles and families with varying eigenvalues.

Before defining P_{3D6} monodromy tuples, it is necessary to introduce sectors in $\mathbb{C}^*$ which will be used for the Stokes structures. They depend on $u_0^1 \neq u_0^2, u_\infty^1 \neq$

© Springer International Publishing AG 2017

M.A. Guest, C. Hertling, *Painlevé III: A Case Study in the Geometry of Meromorphic Connections*, Lecture Notes in Mathematics 2198, DOI 10.1007/978-3-319-66526-9_2

$u_\infty^2 \in \mathbb{C}$. At $z = 0$ we define

$$\zeta_0 := i(u_0^1 - u_0^2)/|u_0^1 - u_0^2| \in S^1,$$
$$I_0^a := \{z \in S^1 \mid \Re(\tfrac{1}{z}(u_0^1 - u_0^2)) < 0\},$$
$$I_0^b := -I_0^a = \{z \in S^1 \mid \Re(\tfrac{1}{z}(u_0^1 - u_0^2)) > 0\}, \qquad (2.1)$$
$$I_0^+ := I_0^a \cup I_0^b \cup \{\zeta_0\} = S^1 - \{-\zeta_0\},$$
$$I_0^- := -I_0^+ = I_0^a \cup I_0^b \cup \{-\zeta_0\} = S^1 - \{\zeta_0\}.$$

At $z = \infty$ we use the coordinate $\widetilde{z} := \tfrac{1}{z}$ on $\mathbb{P}^1 - \{0\}$, so we define

$$\zeta_\infty := |u_\infty^1 - u_\infty^2|/i(u_\infty^1 - u_\infty^2) \in S^1,$$
$$I_\infty^a := \{z \in S^1 \mid \Re(z(u_\infty^1 - u_\infty^2)) < 0\},$$
$$I_\infty^b := -I_\infty^a = \{z \in S^1 \mid \Re(z(u_\infty^1 - u_\infty^2)) > 0\}, \qquad (2.2)$$
$$I_\infty^+ := I_\infty^a \cup I_\infty^b \cup \{\zeta_\infty\} = S^1 - \{-\zeta_\infty\},$$
$$I_\infty^- := -I_\infty^+ = I_\infty^a \cup I_\infty^b \cup \{-\zeta_\infty\} = S^1 - \{\zeta_\infty\}.$$

For any (connected and open) subset $I \subset S^1$, we denote by

$$\widehat{I} := \mathbb{R}_{>0} I := \{z \in \mathbb{C}^* \mid z/|z| \in I\}$$

the corresponding sector in $\mathbb{C}^*$. The following pictures show the sectors for some choice of $u_0^1, u_0^2, u_\infty^1, u_\infty^2$ with $u_0^1 = u_\infty^1 = -u_0^2 = -u_\infty^2$.

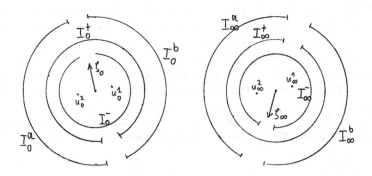

Definition 2.2 A P_{3D6} *monodromy tuple* is a 17-tuple

$$(u_0^j, u_\infty^j, \alpha_0^j, \alpha_\infty^j, L, L_0^{+j}, L_0^{-j}, L_\infty^{+j}, L_\infty^{-j}(j = 1, 2)). \qquad (2.3)$$

Here $u_0^j, u_\infty^j, \alpha_0^j, \alpha_\infty^j \in \mathbb{C}$, and u_0^j, u_∞^j satisfy $u_0^1 \neq u_0^2, u_\infty^1 \neq u_\infty^2$; they will represent the eigenvalues of the "pole parts" of a connection. The various bundles will represent the decompositions at irregular singularities given by the classical formal theory, and $\alpha_0^j, \alpha_\infty^j$ will represent the exponents of formal monodromy. More precisely, they are assumed to satisfy the following conditions. L is a flat rank 2 vector bundle on $\mathbb{C}^*$. $L_0^{\pm j}$ are flat rank 1 subbundles of $L|_{\widehat{I_0^\pm}}$, and $L_\infty^{\pm j}$ are flat rank 1 subbundles of $L|_{\widehat{I_\infty^\pm}}$, with

$$L|_{\widehat{I_0^\pm}} = L_0^{\pm 1} \oplus L_0^{\pm 2}, \quad L|_{\widehat{I_\infty^\pm}} = L_\infty^{\pm 1} \oplus L_\infty^{\pm 2}, \tag{2.4}$$

$$L_0^{+1}|_{\widehat{I_0^a}} = L_0^{-1}|_{\widehat{I_0^a}}, \quad L_0^{+2}|_{\widehat{I_0^b}} = L_0^{-2}|_{\widehat{I_0^b}}, \tag{2.5}$$

$$L_\infty^{+1}|_{\widehat{I_\infty^a}} = L_\infty^{-1}|_{\widehat{I_\infty^a}}, \quad L_\infty^{+2}|_{\widehat{I_\infty^b}} = L_\infty^{-2}|_{\widehat{I_\infty^b}}. \tag{2.6}$$

Denote (for 0 and ∞, this is abbreviated by $0/\infty$) by

$$t_{0/\infty}^{aij} : L_{0/\infty}^{+i}|_{\widehat{I_{0/\infty}^a}} \to L_{0/\infty}^{-j}|_{\widehat{I_{0/\infty}^a}}$$

the projection induced by the restriction to $\widehat{I_{0/\infty}^a}$ of the splitting $L|_{\widehat{I_{0/\infty}^-}} = L_{0/\infty}^{-1} \oplus L_{0/\infty}^{-2}$, and similarly $t_{0/\infty}^{bij}$. Then $t_{0/\infty}^{a11} = \mathrm{id}, t_{0/\infty}^{a12} = 0$, thus $t_{0/\infty}^{a22}$ is an isomorphism, and $t_{0/\infty}^{b22} = \mathrm{id}, t_{0/\infty}^{b21} = 0$, thus $t_{0/\infty}^{b11}$ is an isomorphism. Gluing by $t_{0/\infty}^{aij}$ and $t_{0/\infty}^{bij}$ (for fixed j) gives flat rank 1 bundles $L_{0,\infty}^{+j} \cup_{t's} L_{0/\infty}^{-j}$ on $\mathbb{C}^*$. They and the numbers $\alpha_{0/\infty}^j$ are linked by the compatibility conditions

$$\text{eigenvalue of monodromy of } L_0^{+j} \cup_{t's} L_0^{-j} = e^{-2\pi i \alpha_0^j},$$

$$\text{eigenvalue of monodromy of } L_\infty^{+j} \cup_{t's} L_\infty^{-j} = e^{2\pi i \alpha_\infty^j}. \tag{2.7}$$

Theorem 2.3 (Riemann-Hilbert Correspondence)

(a) *There is a canonical 1:1 correspondence (described below) between P_{3D6} bundles and P_{3D6} monodromy tuples.*

(b) *It extends to a canonical 1:1 correspondence between holomorphic families of P_{3D6} bundles and holomorphic families of P_{3D6} monodromy tuples (over the same complex base manifold).*

Remarks 2.4

(i) More commonly, the correspondence (a) is formulated for meromorphic bundles with meromorphic connections, not for holomorphic bundles, see [Ma83a] or the much more detailed account [BV89] (which works with holomorphic bundles, but meromorphic bundle maps). But in our semisimple case one can extract from [Si90] the version for holomorphic bundles. Below we shall follow [Sa02, II 5] for the treatment of holomorphic bundles.

(ii) Several strong generalizations of the classical correspondence for a meromorphic bundle with a meromorphic connection are given in [Sa13]: for (not
necessarily regular) holonomic D-modules on curves [Sa13, Theorem 5.14],
and for good meromorphic connections in higher dimensions [Sa13, Theorem 10.8 and Theorem 12.16].

(iii) It is also more common to formulate (a) alone, i.e. for single bundles, rather
than families. Indeed, in the case of families, one has to be careful with the
regular singular pieces which have to be encoded in the monodromy tuple. But
in our semisimple case this is not a problem and can also be extracted from
[Si90]. A remark on families is also made in [Ma83a], but for meromorphic
bundles.

(iv) A correspondence for families of holomorphic vector bundles is also important
in [FIKN06], but there it is discussed in detail only in an example.

Reference [Bo01] gives a beautiful description of a Riemann-Hilbert correspondence (he calls it the monodromy map) which covers Theorem 2.3. He
restricts to the semisimple case, and considers two versions, one with and one
without framings on both sides. The version with framings has the advantage
that all spaces are manifolds. The central statements are in [Bo01, ch. 3,7] after
proposition 3.7 and after definition 7.3.

(v) The most general version of the Riemann-Hilbert correspondence for holomorphic vector bundles with meromorphic connections that we know is given
in [Mo11a, Theorem 4.3.1]. Theorem 2.3 is a very special case of this.
Mochizuki considers families of vector bundles on complex manifolds of
arbitrary dimension with meromorphic poles along normal crossing divisors
such that a formal decomposition exists locally (without ramification) and
such that the regular singular pieces have logarithmic poles. Our u_0^j, u_∞^j
($j = 1, 2$) correspond to the *good sets of irregular values* there, and the
rest of our monodromy data (2.3) is called the *full Stokes data*. Theorem
4.3.1 in [Mo11a] includes the possibility of subfamilies on which the family
of connections extends to a flat connection on (z, t)-space, where t is the
deformation parameter, and which are thus isomonodromic subfamilies. We
did not include that in Theorem 2.3. We shall discuss isomonodromic families
in our situation separately in Chap. 3.

(vi) We are assuming here an obvious formulation of the notion of a holomorphic
family of P_{3D6} bundles. One has a holomorphic vector bundle $H^{(T)}$ on $\mathbb{P}^1 \times T$,
where T is a (usually connected) complex manifold, and a relative connection
$\nabla^{(T)}$ which gives on each bundle $H^{(t)} = H^{(T)}|_{\mathbb{P}^1 \times \{t\}}$ for $t \in T$ a flat
meromorphic connection $\nabla^{(t)}$, such that $(H^{(t)}, \nabla^{(t)})$ is a P_{3D6} bundle. Then,
automatically, the numbers $u_0^j(t), u_\infty^j(t)$ ($j = 1, 2$) vary holomorphically.

The sectors in (2.1) and (2.2) also vary holomorphically. The notion of a
holomorphic family of P_{3D6} monodromy tuples is now clear. $L^{(T)}$ is a holomorphic
vector bundle on $\mathbb{C}^* \times T$, with a relative connection which gives on each bundle $L^{(t)}$,
$t \in T$, a flat structure. The flat subbundles $L_0^{\pm j}, L_\infty^{\pm j}$ vary (within the varying sectors)
holomorphically. The numbers $\alpha_0^j, \alpha_\infty^j$ depend holomorphically on $t \in T$.

From a P_{3D6} bundle to a P_{3D6} monodromy tuple: This will make one direction in Theorem 2.3 (a) precise. The following is well-known—references are, for example, [Ma83a, BV89, Si90, Sa02, PS03, FIKN06].

Let $(H, \nabla, u_0^1, u_\infty^1)$ be a P_{3D6} bundle. Then u_0^j, u_∞^j are given by Definition 2.1, and L is the restriction $H|_{\mathbb{C}^*}$ with its flat structure. The subbundles and the numbers $\alpha_0^j, \alpha_\infty^j$ will come from sectorial decompositions which "lift" the formal decompositions of Hukuhara, Turrittin and others at the poles.

We discuss first the pole at 0. The pole at ∞ will (with the new coordinate $\tilde{z} = \frac{1}{z}$) be entirely analogous. The formal decomposition at the pole at 0 exists without ramification in the case of P_{3D6} bundles, because the pole is semisimple with distinct eigenvalues (e.g. [Sa02, II 5.7 Theorem]). Usually the formal decomposition is written for meromorphic bundles. But again because the pole is semisimple with distinct eigenvalues, it is valid also for holomorphic P_{3D6} bundles (e.g. [Sa02, II 5.8 Remark]).

Writing $\mathbb{C}[[z]]$ for the algebra of formal power series and $\mathbb{C}\{z\}$ for the subalgebra of convergent power series, one can express the formal decomposition at 0 as

$$\Psi_0^{for} : (\mathcal{O}(H)_0, \nabla) \otimes_{\mathbb{C}\{z\}} \mathbb{C}[[z]] \xrightarrow{\cong} \oplus_{j=1}^2 (\mathcal{O}(H_0^j)_0, \nabla_0^j) \otimes_{\mathbb{C}\{z\}} \mathbb{C}[[z]] \qquad (2.8)$$

where $\mathcal{O}(H)$ is the sheaf of holomorphic sections of H, (H_0^j, ∇_0^j) is a rank 1 holomorphic vector bundle on $\mathbb{C}$, with a meromorphic (flat) connection with a pole at 0 of order ≤ 2 (2 if $u_0^j \neq 0$; 1 if $u_0^j = 0$), with eigenvalue u_0^j of the pole part $[z\nabla_{z\partial_z}] : (H_0^j)_0 \to (H_0^j)_0$. Then $(H_0^j, \nabla_0^j - d(\frac{-u_0^j}{z}))$ has a logarithmic pole at 0. That means that there is a unique $\alpha_0^j \in \mathbb{C}$ such that

$$\mathcal{O}(H_0^j) = \mathcal{O}_{\mathbb{C}} z^{\alpha_0^j} e^{-u_0^j/z} e_0^j$$

where e_0^j is a multi-valued flat global section of $H_0^j|_{\mathbb{C}^*}$. Then the monodromy of (H_0^j, ∇_0^j) has eigenvalue $e^{-2\pi i \alpha_0^j}$, and

$$(\nabla_0^j)_{z\partial_z}(z^{\alpha_0^j} e^{-u_0^j/z} e_0^j) = z\partial_z(z^{\alpha_0^j} e^{-u_0^j/z}) e_0^j$$

$$= (\alpha_0^j + u_0^j/z) z^{\alpha_0^j} e^{-u_0^j/z} e_0^j.$$

This formal decomposition lifts to a holomorphic decomposition in sectors which contain at most one of the two Stokes directions $\mathbb{R}_{>0} (\pm\zeta_0)$. In sectors which contain exactly one of the two Stokes directions, it lifts uniquely. So a unique lift of the formal decomposition exists in the sectors $\widehat{I}_0^\pm$, and they are maximal with this property.

To formulate the sectorial decomposition, we need the sheaf $\mathcal{A}$ on S^1 of holomorphic functions in (intersections with neighbourhoods of 0 of) sectors which have an asymptotic expansion in $\mathbb{C}[[z]]$. For the definition we refer to [Ma83a] (it is instructive to compare the definitions in [Sa02, PS03, Mo11a]). For $f \in \mathcal{A}_\xi$, $\xi \in S^1$, we denote by $\hat{f} \in \mathbb{C}[[z]]$ its asymptotic expansion.

Then there exist two unique flat isomorphisms

$$\Psi_0^\pm : H|_{\widehat{I_0^\pm}} \xrightarrow{\cong} H_0^1|_{\widehat{I_0^\pm}} \oplus H_0^2|_{\widehat{I_0^\pm}} \tag{2.9}$$

which extend to isomorphisms

$$\Psi_0^\pm : (\mathcal{O}(H)_0, \nabla) \otimes_{\mathbb{C}\{z\}} \mathcal{A}|_{I_0^\pm} \xrightarrow{\cong} \oplus_{j=1}^2 (\mathcal{O}(H_0^j)_0, \nabla_0^j) \otimes_{\mathbb{C}\{z\}} \mathcal{A}|_{I_0^\pm}, \tag{2.10}$$

lifting Ψ_0^{for}. The flat rank 1 subbundles

$$L_0^{\pm j} := (\Psi_0^\pm)^{-1}(H_0^j|_{\widehat{I_0^\pm}}) \tag{2.11}$$

satisfy $L|_{\widehat{I_0^\pm}} = L_0^{\pm 1} \oplus L_0^{\pm 2}$ and (2.5). Let us explain the part $L_0^{+1}|_{\widehat{I_0^a}} = L_0^{-1}|_{\widehat{I_0^a}}$ of (2.5). If $e_0^{\pm 1}$ and $e_0^\pm$ are flat generating sections of $L_0^{\pm 1}$ and $L_0^{\pm 2}$, then $z^{\alpha_0^1} e^{-u_0^1/z} e_0^{\pm 1}$ and $z^{\alpha_0^2} e^{-u_0^2} e_0^{\pm 2}$ are generating sections of $\mathcal{O}(H)|_0 \otimes_{\mathbb{C}\{z\}} \mathcal{A}|_{I_0^\pm}$. If $e_{0|I_0^a}^{-1} = \varepsilon_1 e_{0|I_0^a}^{+1} + \varepsilon_2 e_{0|I_0^a}^{+2}$ with $\varepsilon_2 \neq 0$, then also $z^{\alpha_0^1} e^{-u_0^1} e_0^2$ would be a section of $\mathcal{O}(H)|_0 \otimes_{\mathbb{C}\{z\}} \mathcal{A}|_{I_0^a}$. Then $z^{\alpha_0^1} e^{-u_0^1}$ would be in $z^{\alpha_0^2} e^{-u_0^2} \mathcal{A}|_{I_0^a}$. But that is impossible as, for $z \in \widehat{I_0^a}$ close to 0, $e^{-u_2/z}$ is exponentially much smaller than $e^{-u_0^1/z}$.

The discussion of the pole at ∞ is entirely analogous, with the new coordinate $\widetilde{z} = \frac{1}{z}$ on $\mathbb{P}^1 - \{0\}$. We have $\widetilde{z} \partial_{\widetilde{z}} = -z \partial_z$, $\widetilde{z}^2 \partial_{\widetilde{z}} = -\partial_z$, and

$$\mathcal{O}(H_\infty^j) = \mathcal{O}_{\mathbb{P}^1 - \{0\}} \widetilde{z}^{\alpha_\infty^j} e^{-u_\infty^j/\widetilde{z}} = \mathcal{O}_{\mathbb{P}^1 - \{0\}} z^{-\alpha_\infty^j} e^{-u_\infty^j z}.$$

One has to observe that the monodromy with respect to $\widetilde{z}$ is the inverse of the monodromy (with respect to z). This is the reason for the different signs in the two compatibility conditions in (2.7).

All data of the P_{3D6} monodromy tuple have now been defined.

From a P_{3D6} monodromy tuple to a P_{3D6} bundle: This will make the other direction in Theorem 2.3 (a) precise. It is as classical as the first direction.

Let $(u_0^j, u_\infty^j, \alpha_0^j, \alpha_\infty^j, L, L_0^{\pm j}, L_\infty^{\pm j} (j = 1, 2))$ be a P_{3D6} monodromy tuple. L will be the restriction to $\mathbb{C}^*$ of a bundle $H \to \mathbb{P}^1$, whose extensions to 0 and ∞ still have to be constructed. We shall discuss the extension to 0 (the extension to ∞ will be analogous). Choose flat generating sections $e_0^{\pm j}$ of the flat rank 1 subbundles $L_0^{\pm j}$ of $L|_{\widehat{I_0^\pm}}$ with

$$e_0^{-1}|_{\widehat{I_0^a}} = e_0^{+1}|_{\widehat{I_0^a}} \text{ and } e_0^{-2}|_{\widehat{I_0^a}} = e^{-2\pi i \alpha_0^2} e_0^{+2}|_{\widehat{I_0^a}}. \tag{2.12}$$

Consider the holomorphic sections

$$z_0^{\alpha^j} e^{-u_0^j/z} e_0^{\pm j} \quad \text{on } \widehat{I}_0^{\pm}. \tag{2.13}$$

Then there exist coefficients $a_0^{\pm ij} \in \mathcal{A}|_{I_0^{\pm}}$ with $\widehat{a}_0^{+ij} = \widehat{a}_0^{-ij}$ and $\widehat{a}_0^{\pm ij}(0) = \delta_{ij}$ and such that for both $j = 1, 2$ the sections on the (intersections with a neighbourhood of 0 of the) sectors $\widehat{I}_0^+$ and $\widehat{I}_0^-$

$$\sum_{i=1}^{2} z_0^{\alpha^i} e^{-u_0^i/z} e_0^{\pm i} a_0^{\pm ij} \tag{2.14}$$

glue to a holomorphic section on (a neighbourhood of 0 in) $\mathbb{C}^*$, and the two sections form a basis. Here $z_0^{\alpha^i_0}$ is defined on $\widehat{I}_0^+$ using an arbitrary branch of $\log z$ on $\widehat{I}_0^+$. It is defined on $\widehat{I}_0^-$ by extending the branch on $\widehat{I}_0^+$ counterclockwise (that means, over $\widehat{I}_0^a$) to $\widehat{I}_0^-$. Then the extension $\mathcal{O}(H)_0$ to 0 of $L = H|_{\mathbb{C}^*}$ is defined by these sections. The sections and the coefficients are not at all unique, but the extension is unique and has a pole of order 2 at 0 with eigenvalues u_0^j of the pole part.

The same procedure works at ∞, with the new coordinate $\widetilde{z} = 1/z$. This completes the construction of the P_{3D6} bundle.   □

Remarks 2.5

(i) Let $e_0^{\pm j}$ be as above. Then $\underline{e}_0^{\pm} := (e_0^{\pm 1}, e_0^{\pm 2})$ is a flat basis of $L|_{\widehat{I}_0^{\pm}}$. The construction above can be rephrased by saying that there exist two matrices $A_0^{\pm} \in GL(2, \mathcal{A}|_{I_0^{\pm}})$ with $\widehat{A}_0^+ = \widehat{A}_0^-$ and $\widehat{A}_0^{\pm}(0) = \mathbb{1}_2$ such that the two holomorphic bases

$$\underline{e}_0^{\pm} \begin{pmatrix} z_0^{\alpha^1} e^{-u_0^1/z} & 0 \\ 0 & z_0^{\alpha^2} e^{-u_0^2/z} \end{pmatrix} A_0^{\pm}(z) \tag{2.15}$$

glue to a basis φ on (a neigborhood of 0 in) $\mathbb{C}^*$, with which the extension of $H|_{\mathbb{C}^*}$ to 0 is defined.

Here we use (for $n = m = 2$) the convention that the matrix multiplication extends in the following way to a product of an n-tuple (= a row vector) with entries $v_1, \ldots .v_n$ in a K-vector space V and a matrix $(a_{ij}) \in M(n \times m, K)$,

$$(v_1, \ldots, v_n)(a_{ij}) = (\sum_{i=1}^{n} a_{i1} v_i, \ldots, \sum_{i=1}^{n} a_{im} v_i).$$

So, the product is an m-tuple with entries in V.

(ii) Given a basis $\underline{\varphi}$ of $\mathcal{O}(H)_0$ the following conditions are equivalent.

(α) $[z\nabla_{z\partial_z}](\underline{\varphi}(0)) = \underline{\varphi}(0) \begin{pmatrix} u_0^1 & 0 \\ 0 & u_0^2 \end{pmatrix}$, $\qquad$ (2.16)

(β) $\nabla_{z\partial_z}\underline{\varphi}(z) = \underline{\varphi}(z) \left[\dfrac{1}{z}\begin{pmatrix} u_0^1 & 0 \\ 0 & u_0^2 \end{pmatrix} + B(z) \right]$ $\qquad$ (2.17)

$\quad$ with $B(z) \in M(2 \times 2, \mathbb{C}\{z\})$,

(γ) flat generating sections $e_0^{\pm j}$ of $L_0^{\pm}$ and matrices $A_0^{\pm}$ as $\qquad$ (2.18)

$\quad$ in (i) exist such that the basis $\underline{\varphi}$ takes the form (2.15) on $\widehat{I}_0^{\pm}$.

The first equivalence is trivial, the second is at the heart of the story. Here the correspondence

$$\varphi(0) \longleftarrow \underline{\varphi} \longrightarrow e_0^{\pm}$$

is 1:many:1. It induces a canonical 1:1 correspondence

$$\underline{\varphi}(0) \longleftrightarrow e_0^{\pm} \qquad (2.19)$$

between bases $\underline{\varphi}(0)$ of H_0 of eigenvectors of $[z\nabla_{z\partial_z}]$ and bases $e_0^{\pm}$ of $L|_{\widehat{I}_0^{\pm}}$ which respect the splitting $L|_{\widehat{I}_0^{\pm}} = L_0^{\pm 1} \oplus L_0^{\pm 2}$ and satisfy (2.12).

The P_{3D6} monodromy tuples are good for conceptual arguments, e.g. regarding irreducibility in Chap. 3, but less well suited for calculations such as those in Chap. 6. For this reason we shall construct P_{3D6} *numerical tuples* from P_{3D6} monodromy tuples, although this involves making further choices.

Let a P_{3D6} monodromy tuple be given. Choose bases $e_0^+ = (e_0^{+1}, e_0^{+2})$ and $e_\infty^+ = (e_\infty^{+1}, e_\infty^{+2})$ of flat generating sections of the bundles $L_0^{+1}, L_0^{+2}, L_\infty^{+1}, L_\infty^{+2}$. Then there are unique bases $e_0^- = (e_0^{-1}, e_0^{-2})$ and $e_\infty^- = (e_\infty^{-1}, e_\infty^{-2})$ of flat generating sections of the bundles $L_0^{-1}, L_0^{-2}, L_\infty^{-1}, L_\infty^{-2}$ such that (2.12) and the analogous condition at ∞ hold. Furthermore, there are unique numbers $s_0^a, s_0^b, s_\infty^a, s_\infty^b \in \mathbb{C}$ such that the matrices

$$S_0^a := \begin{pmatrix} 1 & s_0^a \\ 0 & 1 \end{pmatrix}, \quad S_0^b := \begin{pmatrix} e^{-2\pi i\alpha_0^1} & 0 \\ 0 & e^{-2\pi i\alpha_0^2} \end{pmatrix}\begin{pmatrix} 1 & 0 \\ s_0^b & 1 \end{pmatrix},$$

$$S_\infty^a := \begin{pmatrix} 1 & s_\infty^a \\ 0 & 1 \end{pmatrix}, \quad S_\infty^b := \begin{pmatrix} e^{-2\pi i\alpha_\infty^1} & 0 \\ 0 & e^{-2\pi i\alpha_\infty^2} \end{pmatrix}\begin{pmatrix} 1 & 0 \\ s_\infty^b & 1 \end{pmatrix}$$

$\qquad$ (2.20)

and the bases $\underline{e}_0^\pm, \underline{e}_\infty^\pm$ satisfy

$$\underline{e}_0^- |_{\widehat{I_0^a}} = \underline{e}_0^+ |_{\widehat{I_0^a}} S_0^a, \text{ i.e. } \underline{e}_0^-(-z) = \underline{e}_0^+(ze^{\pi i}) S_0^a \text{ for } z \in \widehat{I_0^+},$$

$$\underline{e}_0^- |_{\widehat{I_0^b}} = \underline{e}_0^+ |_{\widehat{I_0^b}} S_0^b, \text{ i.e. } \underline{e}_0^-(-z) = \underline{e}_0^+(ze^{-\pi i}) S_0^b \text{ for } z \in \widehat{I_0^+},$$

$$\underline{e}_\infty^- |_{\widehat{I_\infty^a}} = \underline{e}_\infty^+ |_{\widehat{I_\infty^a}} S_\infty^a, \text{ i.e. } \underline{e}_\infty^-(-z) = \underline{e}_\infty^+(ze^{-\pi i}) S_\infty^a \text{ for } z \in \widehat{I_\infty^+},$$

$$\underline{e}_\infty^- |_{\widehat{I_\infty^b}} = \underline{e}_\infty^+ |_{\widehat{I_\infty^b}} S_\infty^b, \text{ i.e. } \underline{e}_\infty^-(-z) = \underline{e}_\infty^+(ze^{\pi i}) S_\infty^b \text{ for } z \in \widehat{I_\infty^+}.$$

(2.21)

Here the meaning of $\underline{e}_0^+(ze^{\pi i})$ (or $\underline{e}_0^+(ze^{-\pi i})$) is that the basis $\underline{e}_0^+$ of flat sections is extended (flatly) counterclockwise (or clockwise) to $\widehat{I_0^-}$, and similarly for $e_\infty^+(ze^{\pm\pi i})$.

Writing

$$\text{Mon} = \text{monodromy of } L$$

we have

$$\text{Mon}(\underline{e}_0^+)(z) = \underline{e}_0^+(ze^{2\pi i}) = \underline{e}_0^-(-ze^{\pi i})(S_0^a)^{-1}$$

$$= \underline{e}_0^+(z) S_0^b (S_0^a)^{-1},$$

$$\text{Mon}(\underline{e}_\infty^-)(z) = \underline{e}_\infty^-(ze^{2\pi i}) = \underline{e}_\infty^+(-ze^{\pi i}) S_\infty^a$$

$$= \underline{e}_\infty^-(z) (S_\infty^b)^{-1} S_\infty^a.$$

(2.22)

Let us choose a number $\beta \in \mathbb{C}^*$ such that

$$e^{-\beta} = (u_0^1 - u_0^2)(u_\infty^1 - u_\infty^2).$$

(2.23)

Then

$$e^\beta (-\zeta_0) \mathbb{R}_{>0} = \zeta_\infty \mathbb{R}_{>0},$$

$$e^\beta \widehat{I_0^+} = \widehat{I_\infty^-},$$

$$e^\beta \widehat{I_0^a} = \widehat{I_\infty^a}.$$

For an arbitrary point $e^\xi \in \widehat{I_0^+}$, this β encodes a (homotopy class) of a path from this point e^ξ to the point $e^{\xi+\beta}$, namely a path whose lift to the universal covering $\exp : \mathbb{C} \to \mathbb{C}^*$ with starting point ξ ends at $\xi + \beta$. The homotopy class of the path is unique. The path (or its homotopy class) is called $[\beta]$.

Using this path, the connection matrix

$$B(\beta) = \begin{pmatrix} b_1 & b_2 \\ b_3 & b_4 \end{pmatrix} \in GL(2, \mathbb{C})$$

can be defined by

$$\underline{e}_\infty^- = (\underline{e}_0^+ \text{ extended flatly along } [\beta] \text{ from } \widehat{I_0^+} \text{ to } \widehat{I_\infty^-}) \, B(\beta). \tag{2.24}$$

From this definition we have

$$b_1 b_4 - b_2 b_3 \neq 0 \tag{2.25}$$

and

$$(\underline{e}_0^+ \text{ extended along } [\beta] \text{ to } \widehat{I_\infty^-})$$
$$= \text{Mon}\left[(\underline{e}_0^+ \text{ extended along } [\beta - 2\pi i] \text{ to } \widehat{I_\infty^-}) \right].$$

This implies

$$(\underline{e}_0^+ \text{ extended along } [\beta] \text{ to } \widehat{I_\infty^-}) \, B(\beta - 2\pi i)$$
$$= \text{Mon}\left[(\underline{e}_\infty^-) \right] = \underline{e}_\infty^- \, (S_\infty^b)^{-1} S_\infty^a$$
$$= (\underline{e}_0^+ \text{ extended along } [\beta] \text{ to } \widehat{I_\infty^-}) \, B(\beta) \, (S_\infty^b)^{-1} S_\infty^a$$
$$\text{and also } = \text{Mon}\left[(\underline{e}_0^+ \text{ extended along } [\beta] \text{ to } \widehat{I_\infty^-}) \, B(\beta) \right]$$
$$= (\underline{e}_0^+ \text{ extended along } [\beta] \text{ to } \widehat{I_\infty^-}) \, S_0^b \, (S_0^a)^{-1} \, B(\beta).$$

This shows

$$B(\beta) \, (S_\infty^b)^{-1} S_\infty^a = S_0^b \, (S_0^a)^{-1} B(\beta) \tag{2.26}$$
$$= B(\beta - 2\pi i). \tag{2.27}$$

The group $\mathbb{Z}$ acts on the pairs $(\beta, B(\beta))$; its generator $[1]$ acts via

$$[1] \cdot (\beta, B(\beta)) = (\beta + 2\pi i, B(\beta + 2\pi i)) \tag{2.28}$$
$$= (\beta + 2\pi i, (S_0^b \, (S_0^a)^{-1})^{-1} \, B(\beta)).$$

The bases $\underline{e}_0^\pm$ and $\underline{e}_\infty^\pm$ can be changed by diagonal base changes

$$\widetilde{\underline{e}}_0^\pm = \underline{e}_0^\pm \begin{pmatrix} \lambda_1 & 0 \\ 0 & \lambda_2 \end{pmatrix}, \quad \widetilde{\underline{e}}_\infty^\pm = \underline{e}_\infty^\pm \begin{pmatrix} \lambda_3 & 0 \\ 0 & \lambda_4 \end{pmatrix}$$

with $(\lambda_1, \lambda_2, \lambda_3, \lambda_4) \in (\mathbb{C}^*)^4$. Then $u_0^j, u_\infty^j, \alpha_0^j, \alpha_\infty^j, \beta$ are unchanged, but

$$\widetilde{S}_0^{a/b} = \begin{pmatrix} \lambda_1 & 0 \\ 0 & \lambda_2 \end{pmatrix}^{-1} S_0^{a/b} \begin{pmatrix} \lambda_1 & 0 \\ 0 & \lambda_2 \end{pmatrix},$$

$$\widetilde{S}_\infty^{a/b} = \begin{pmatrix} \lambda_3 & 0 \\ 0 & \lambda_4 \end{pmatrix}^{-1} S_\infty^{a/b} \begin{pmatrix} \lambda_3 & 0 \\ 0 & \lambda_4 \end{pmatrix},$$

$$\widetilde{B}(\beta) = \begin{pmatrix} \lambda_1 & 0 \\ 0 & \lambda_2 \end{pmatrix}^{-1} B(\beta) \begin{pmatrix} \lambda_3 & 0 \\ 0 & \lambda_4 \end{pmatrix}, \qquad (2.29)$$

$$(\widetilde{s}_0^a, \widetilde{s}_0^b, \widetilde{s}_\infty^a, \widetilde{s}_\infty^b, \widetilde{b}_1, \widetilde{b}_2, \widetilde{b}_3, \widetilde{b}_4)$$

$$= \left(\tfrac{\lambda_2}{\lambda_1} s_0^a, \tfrac{\lambda_1}{\lambda_2} s_0^b, \tfrac{\lambda_4}{\lambda_3} s_\infty^a, \tfrac{\lambda_3}{\lambda_4} s_\infty^b, \tfrac{\lambda_3}{\lambda_1} b_1, \tfrac{\lambda_4}{\lambda_1} b_2, \tfrac{\lambda_3}{\lambda_2} b_3, \tfrac{\lambda_4}{\lambda_2} b_4 \right).$$

The actions of $\mathbb{Z}$ in (2.28) and of $(\mathbb{C}^*)^4$ in (2.29) commute.

Definition 2.6 A P_{3D6} *numerical tuple* is a 17-tuple of complex numbers

$$(u_0^j, u_\infty^j, \alpha_0^j, \alpha_\infty^j (j = 1, 2), s_0^a, s_0^b, s_\infty^a, s_\infty^b, \beta, b_1, b_2, b_3, b_4) \qquad (2.30)$$

such that $u_0^1 \neq u_0^2, u_\infty^1 \neq u_\infty^2$, and (2.23), (2.25) and (2.26) are satisfied.

The group $(\mathbb{C}^*)^4 \times \mathbb{Z}$ acts via (2.28) and (2.29) on the set of P_{3D6} numerical tuples.

Lemma 2.7

(a) *There is a canonical 1:1 correspondence between P_{3D6} monodromy tuples and $(\mathbb{C}^*)^4 \times \mathbb{Z}$-orbits of P_{3D6} numerical tuples.*

(b) *Any holomorphic family of P_{3D6} numerical tuples induces a holomorphic family of P_{3D6} bundles. Any holomorphic family of P_{3D6} bundles can locally be represented by a holomorphic family of P_{3D6} numerical tuples.*

Proof Starting from a P_{3D6} monodromy tuple, the $(\mathbb{C}^*)^4 \times \mathbb{Z}$-orbit of P_{3D6} numerical tuples has been constructed above. Going backwards through the construction, one can construct from a P_{3D6} numerical tuple a P_{3D6} monodromy tuple. The statements regarding holomorphic families are obvious.   □

Remarks 2.8

(i) Let us fix the 9-tuple $(u_0^j, u_\infty^j, \alpha_0^j, \alpha_\infty^j (j = 1, 2), \beta)$. The quotient set

$$\{(s_0^a, s_0^b, s_\infty^a, s_\infty^b, b_1, b_2, b_3, b_4)\}/(\mathbb{C}^*)^4$$

of those 8-tuples which, together with the fixed part, form a P_{3D6} numerical tuple, parametrizes the set of isomorphism classes of P_{3D6} bundles with the given fixed part.

In the case of trace free P_{3D6} bundles (see Remark 4.1 (ii) and (iii)) this quotient was considered in [PS09]. The categorical quotient was calculated there. See Remark 3.3 (i) and (ii).

(ii) In a P_{3D6} monodromy tuple, the link between the data at 0 and at ∞ is built in, but the compatibility condition (2.7) has to be formulated. In a P_{3D6} numerical tuple, the compatibility condition (2.7) is built in, but the link between the data at 0 and at ∞ has to be formulated—it is the connection matrix $B(\beta)$.

Chapter 3
(Ir)Reducibility

A pair (H, ∇), where $H \to M$ is a holomorphic vector bundle on a complex manifold M, and ∇ is a (flat) meromorphic connection, is said to be reducible if there exists a subbundle $G \subset H$ with $0 < \operatorname{rank} G < \operatorname{rank} H$ which is (at all nonsingular points of the connection) a flat subbundle. Such a G will simply be called a flat subbundle. A pair (H, ∇) is completely reducible if it decomposes into a sum of flat rank 1 subbundles.

The (ir)reducibility of a P_{3D6} bundle is easily characterized in terms of properties of its P_{3D6} monodromy tuple, and the reducible P_{3D6} bundles can be described explicitly. This is the content of Lemma 3.1 and Corollary 3.2 below. They are essentially contained in [PS09], though not in this detail.

Lemma 3.1 (cf. [PS09])

(a) *A P_{3D6} bundle is reducible but not completely reducible if and only if exactly one of the following four conditions (C_{ij}), $i, j = 1, 2$, holds.*

> *Condition (C_{ij}):*
> $$\alpha_0^i + \alpha_\infty^j \in \mathbb{Z}$$
> L_0^{+i} *and* L_0^{-i} *glue to a (flat, rank 1) subbundle* L_0^i *of* L
> L_∞^{+j} *and* L_∞^{-j} *glue to a (flat, rank 1) subbundle* L_∞^j *of* L
> $L_0^i = L_\infty^j.$

If (C_{ij}) holds, let e^{ij} be a multi-valued flat global generating section of $L_0^i = L_\infty^j$. Then the sections

$$\sigma^{ij} := z^{\alpha_0^i} e^{-u_0^i/z - u_\infty^j z} \, e^{ij} \text{ and } z^{-\alpha_\infty^j - \alpha_0^i} \, \sigma^{ij}$$

generate the restriction to $\mathbb{C}$ and the restriction to $\mathbb{P}^1 - \{0\}$ of a flat subbundle W of (H, ∇), and $\deg W = -\alpha_0^i - \alpha_\infty^j$.

© Springer International Publishing AG 2017
M.A. Guest, C. Hertling, *Painlevé III: A Case Study in the Geometry of Meromorphic Connections*, Lecture Notes in Mathematics 2198, DOI 10.1007/978-3-319-66526-9_3

(b) A P_{3D6} bundle is completely reducible if and only if condition (C_{ij}) (for some i, j) and also condition $(C_{3-i,3-j})$ hold.

Proof If $W \subset H$ is a flat rank 1 subbundle, then $W|_{\mathbb{C}^*} \subset H|_{\mathbb{C}^*} = L$ must be compatible with all four splittings $L|_{\widehat{I}_{0/\infty}^{\pm}} = L_{0/\infty}^{\pm 1} \oplus L_{0/\infty}^{\pm 2}$ of L (on $\widehat{I}_0^{\pm}$ and $\widehat{I}_\infty^{\pm}$). Then $W|_{\mathbb{C}^*}$ must coincide with one subbundle of the splitting on each of the four sets $\widehat{I}_0^{\pm}$ and $\widehat{I}_\infty^{\pm}$. This shows that one of the four conditions (C_{ij}) holds.

Conversely, if (C_{ij}) holds, then a flat rank 1 subbundle W can be constructed as in part (a) of the lemma.

If a second condition holds, it must be $(C_{3-i,3-j})$, and then one can construct a second flat rank 1 subbundle, so that (H, ∇) is completely reducible.

Conversely, if (H, ∇) is completely reducible, two conditions (C_{ij}) and $(C_{3-i,3-j})$ hold. This establishes (a) and (b).   □

Corollary 3.2 (cf. [PS09])   Fix $(u_0^j, u_\infty^j, \alpha_0^j, \alpha_\infty^j (j = 1, 2), \beta)$ with $e^{-\beta} = (u_0^1 - u_0^2)(u_\infty^1 - u_\infty^2)$. Consider the set $M(u_0^j, u_\infty^j, \alpha_0^j, \alpha_\infty^j)$ of isomorphism classes of P_{3D6} bundles whose P_{3D6} monodromy tuples contain $u_0^j, u_\infty^j, \alpha_0^j, \alpha_\infty^j$.

(a) Then

$$\deg H = \deg \det H = -\alpha_0^1 - \alpha_0^2 - \alpha_\infty^1 - \alpha_\infty^2 \in \mathbb{Z}, \qquad (3.1)$$

$$\text{so } \alpha_0^1 + \alpha_\infty^1 \in \mathbb{Z} \iff \alpha_0^2 + \alpha_\infty^2 \in \mathbb{Z}, \qquad (3.2)$$

$$\alpha_0^1 + \alpha_\infty^2 \in \mathbb{Z} \iff \alpha_0^2 + \alpha_\infty^1 \in \mathbb{Z}. \qquad (3.3)$$

(b) If neither (3.2) nor (3.3) hold, then $M(u_0^j, u_\infty^j, \alpha_0^j, \alpha_\infty^j)$ does not contain reducible P_{3D6} bundles.

(c) If conditions (3.2) or conditions (3.3) hold, $M(u_0^j, u_\infty^j, \alpha_0^j, \alpha_\infty^j)$ contains three families of reducible bundles, and these families map to one point in the categorical quotient

$$\{P_{3D6} \text{ numerical tuples containing } (u_0^j, u_\infty^j, \alpha_0^j, \alpha_\infty^j, \beta)\}//(\mathbb{C}^*)^4 \qquad (3.4)$$

where $(\mathbb{C}^*)^4$ acts as in (2.29). One family is parametrized by a point; it is a completely reducible P_{3D6} bundle. The other two families are parametrized by $\mathbb{P}^1$ and contain reducible, but not completely reducible, P_{3D6} bundles.

(d) If both conditions (3.2) and (3.3) hold, the categorical quotient has two special points, and above each of them there are three families as in (c).

Proof

(a) By definition, $\deg H = \deg \det H$. The bundle $\det(H, \nabla)$ is a product of a factor $e^{-(u_0^1 + u_0^2)/z - (u_\infty^1 + u_\infty^2)z} \times (trivial flat bundle)$ and a bundle with logarithmic poles at 0 and ∞, with eigenvalues $\alpha_0^1 + \alpha_0^2$ and $-(\alpha_\infty^1 + \alpha_\infty^2)$ of $[\nabla_{z\partial_z}]$ on $(\det H)_0$ and $(\det H)_\infty$. This establishes (3.1). Now (3.2) and (3.3) follow trivially.

(b) If none of the conditions in (3.2) and (3.3) hold, then none of the conditions (C_{ij}) hold for a bundle in $M(u_0^j, u_\infty^j, \alpha_0^j, \alpha_\infty^j)$. Hence, by Lemma 3.1, none of these bundles can be reducible.

(c)+(d) First we consider condition (C_{11}). A P_{3D6} monodromy tuple satisfies (C_{11}) if and only if any associated P_{3D6} numerical tuple satisfies $s_0^b = s_\infty^b = b_3 = 0$. As u_0^j and u_∞^j are fixed, we can also fix some β with (2.23) and restrict to the connection matrix $B = B(\beta)$. Then the set of isomorphism classes of P_{3D6} monodromy tuples corresponds to the quotient (as a set) of the 5-tuples $(s_0^a, s_\infty^a, b_1, b_2, b_4)$ by the action of $(\mathbb{C}^*)^4$ as in (2.29) where $(\lambda_1, \lambda_2, \lambda_3, \lambda_4)$ maps $(s_0^a, s_\infty^a, b_1, b_2, b_4)$ to

$$\left(\tfrac{\lambda_2}{\lambda_1} s_0^a, \ \tfrac{\lambda_4}{\lambda_3} s_\infty^a, \ \tfrac{\lambda_3}{\lambda_1} b_1, \ \tfrac{\lambda_4}{\lambda_1} b_2, \ \tfrac{\lambda_4}{\lambda_2} b_4 \right). \tag{3.5}$$

As $b_1 \neq 0$ and $b_4 \neq 0$, they can be normalized to $b_1 = b_4 = 1$. Then the action reduces to the action of $(\mathbb{C}^*)^2$ on triples (s_0^a, s_∞^a, b_2), where (λ_1, λ_2) maps this triple to

$$\left(\tfrac{\lambda_2}{\lambda_1} s_0^a, \ \tfrac{\lambda_2}{\lambda_1} s_\infty^a, \ \tfrac{\lambda_2}{\lambda_1} b_2 \right).$$

Equation (2.26) connects s_0^a, s_∞^a and b_2 linearly. It is equivalent to

$$\begin{pmatrix} 1 & b_2 \\ 0 & 1 \end{pmatrix} \begin{pmatrix} e^{2\pi i \alpha_\infty^1} & 0 \\ 0 & e^{2\pi i \alpha_\infty^2} \end{pmatrix} \begin{pmatrix} 1 & s_\infty^a \\ 0 & 1 \end{pmatrix}$$

$$= \begin{pmatrix} e^{-2\pi i \alpha_0^1} & 0 \\ 0 & e^{-2\pi i \alpha_0^2} \end{pmatrix} \begin{pmatrix} 1 & -s_\infty^a \\ 0 & 1 \end{pmatrix} \begin{pmatrix} 1 & b_2 \\ 0 & 1 \end{pmatrix}$$

and thus equivalent to (3.2) and $s_\infty^a = -s_0^a + (1 - e^{2\pi i (\alpha_0^1 - \alpha_0^2)}) b_2$. This means that the isomorphism classes are parametrized by the orbits of the standard $\mathbb{C}^*$ action on $\mathbb{C}^2$ with coordinates (s_0^a, b_2). The point $(s_0^a, b_2) = (0, 0)$ corresponds to the completely reducible bundle which satisfies (C_{11}) and (C_{22}). The other orbits form a $\mathbb{P}^1$ and correspond to the reducible but not completely reducible P_{3D6} bundles which satisfy only (C_{11}). Obviously these orbits cannot be separated from the orbit $\{(0,0)\}$ by invariant functions and are thus mapped to the same point in the categorical quotient in (3.4).

The P_{3D6} bundles which satisfy (C_{22}) are represented in exactly the same way by $\mathbb{C}^2$ with the standard $\mathbb{C}^*$ action. The orbit $\{(0,0)\}$ corresponds again to the completely reducible bundle which satisfies (C_{11}) and (C_{22}), and the other orbits form a $\mathbb{P}^1$ and correspond to the reducible but not completely reducible P_{3D6} bundles which satisfy only (C_{22}). Therefore these are also mapped to the same point in the categorical quotient in (3.4).

The P_{3D6} bundles which satisfy either (C_{12}) or (C_{21}) or both, behave in exactly the same way. But they are mapped to another point in the categorical quotient in (3.4), as their P_{3D6} numerical tuples satisfy $s_0^a = s_\infty^a = 0$ and either $b_4 = 0$ or $b_1 = 0$ or both. This proves (c) and (d).   □

Remarks 3.3

(i)   In the case of trace free P_{3D6} bundles, the categorical quotient in (3.4) had been calculated in [PS09]. It is an affine cubic surface in $\mathbb{C}^3$. If neither (3.2) nor (3.3) hold, it is smooth and a geometric quotient and equips $M(u_0^j, u_\infty^j, \alpha_0^j, \alpha_\infty^j)$ with a canonical structure of an affine algebraic manifold.

If either (3.2) or (3.3) hold, it is not a geometric quotient. Then it has an A_1 singularity, and above the singular point, one has the three families of reducible bundles. Then $M(u_0^j, u_\infty^j, \alpha_0^j, \alpha_\infty^j)$ with the natural quotient topology is not Hausdorff.

One obtains a moduli space if one either restricts to the completely reducible P_{3D6} bundle or to one of the two families of reducible, but not completely reducible P_{3D6} bundles. In the latter two cases, the moduli space is obtained from the categorical quotient by blowing up the A_1 singularity with an exceptional divisor $\mathbb{P}^1$.

If (3.2) and (3.3) hold, the categorical quotient has two A_1 singularities and above each of the two singular points one has three families of reducible bundles. Again $M(u_0^j, u_\infty^j, \alpha_0^j, \alpha_\infty^j)$ with the quotient topology is not Hausdorff. Again one can obtain a moduli space by choosing at either point one of the three families of reducible bundles.

(ii) The eigenvalues of the monodromy of a reducible bundle are the eigenvalues $e^{-2\pi i \alpha_0^1}$ and $e^{-2\pi i \alpha_0^2}$ of the formal monodromy.

If at least one of the conditions (3.2) or (3.3) holds, then the eigenvalues $e^{-2\pi i \alpha_0^1}$ and $e^{-2\pi i \alpha_0^2}$ coincide if and only if (3.2) and (3.3) hold both. In this case a family parametrized by $\mathbb{P}^1$ of reducible but not completely reducible P_{3D6} bundles, splits into a family parametrized by $\mathbb{C}$, where the monodromy has a 2×2 Jordan block and a single bundle with semisimple monodromy.

(iii) The case of interest to us is the case with $\alpha_0^j = \alpha_\infty^j = 0$. There the categorical quotient has two A_1-singularities. But we shall consider at each of the two points only the corresponding completely reducible connection. Therefore we obtain a geometric quotient. See Theorem 7.6 (b).

Chapter 4
Isomonodromic Families

Isomonodromic families of P_{3D6} bundles can be characterized easily in terms of the P_{3D6} monodromy tuples. We shall see that a universal isomonodromic family in four parameters exists, but that only one parameter is essential.

A result of Heu [Heu09] applies and shows that generic members of isomonodromic families of irreducible P_{3D6} bundles are trivial as holomorphic vector bundles, and that the other members (if such exist) have sheaves of holomorphic sections isomorphic to $\mathcal{O}_{\mathbb{P}^1}(1) \oplus \mathcal{O}_{\mathbb{P}^1}(-1)$. This implies solvability of the inverse monodromy problem—which asks whether some given monodromy data can be realized by a pure twistor with meromorphic connection—in this case.

Consider a holomorphic family $(H^{(t)}, \nabla^{(t)}, u_0^{1,(t)}, u_\infty^{1,(t)})_{t \in T}$ of P_{3D6} bundles over a complex base manifold T and the corresponding family of P_{3D6} monodromy tuples $(u_0^{j,(t)}, u_\infty^{j,(t)}, \alpha_0^{j,(t)}, \alpha_\infty^{j,(t)}, L^{(t)}, L_0^{\pm j,(t)}, L_\infty^{\pm j,(t)})_{t \in T}$. Suppose that all $L^{(t)}$ have isomorphic monodromy. Then one can locally (over $S \subset T$, say) extend the family of connections on the bundles $L^{(t)}$ to a flat connection on $\cup_{t \in S} L^{(t)}$. But in general the covariant derivatives of holomorphic sections in the T direction will have essential singularities along $\{0\} \times S$ and $\{\infty\} \times S$.

The family $\cup_{t \in S} H^{(t)}$ with extended flat connection is considered *isomonodromic* only if this extended connection has poles of Poincaré rank ≤ 1 along $\{0\} \times S$ and $\{\infty\} \times S$. In other words the covariant derivatives of holomorphic sections by logarithmic vector fields along $\{0\} \times S$ and $\{\infty\} \times S$ have at most simple poles along $\{0\} \times S$ and $\{\infty\} \times S$.

It is well known (see [Ma83b]) that this holds if the splittings

$$L|_{\widehat{T}_{0/\infty}^{\pm,(t)}} = L_{0/\infty}^{\pm 1,(t)} \oplus L_{0,\infty}^{\pm 2,(t)}$$

vary in a flat way and if the $\alpha_0^{j,(t)}, \alpha_\infty^{j,(t)}$ are constant. Thus, the only freedom is the holomorphic variation of $u_0^{j,(t)}, u_\infty^{j,(t)}$. This gives four parameters for a universal

© Springer International Publishing AG 2017

M.A. Guest, C. Hertling, *Painlevé III: A Case Study in the Geometry of Meromorphic Connections*, Lecture Notes in Mathematics 2198, DOI 10.1007/978-3-319-66526-9_4

isomonodromic family over the universal covering U^{univ} of

$$U := \{(u_0^1, u_0^2, u_\infty^1, u_\infty^2) \in \mathbb{C}^4 \mid u_0^1 \neq u_0^2, u_\infty^1 \neq u_\infty^2\},$$

with covering map $c^{univ} : U^{univ} \to U$.

However, three of the four parameters are inessential in the following sense. Define c^{uc*} by

$$c^{uc*} : U \to \mathbb{C}^*, \quad (u_0^1, u_0^2, u_\infty^1, u_\infty^2) \mapsto (u_0^1 - u_0^2)(u_\infty^1 - u_\infty^2). \qquad (4.1)$$

Then the P_{3D6} bundles in one fibre of c^{uc*} can be obtained by the following operations from one another. If $(\lambda_0, \lambda_\infty) \in \mathbb{C}^2$ then one can define for any P_{3D6} bundle the following isomonodromic family in two parameters

$$(H^{(\lambda_0, \lambda_\infty)}, \nabla^{(\lambda_0, \lambda_\infty)}, u_0^{j,(\lambda_0, \lambda_\infty)}, u_\infty^{j,(\lambda_0, \lambda_\infty)}) \qquad (4.2)$$

$$\text{with} \quad \mathcal{O}(H^{(\lambda_0, \lambda_\infty)}) := e^{\lambda_0/z + \lambda_\infty z} \mathcal{O}(H),$$

$$\nabla^{(\lambda_0, \lambda_\infty)} := \nabla,$$

$$\text{thus} \quad u_0^{j,(\lambda_0, \lambda_\infty)} = u_0^j - \lambda_0, \; u_\infty^{j,(\lambda_0, \lambda_\infty)} = u_\infty^j - \lambda_\infty.$$

If $\lambda_1 \in \mathbb{C}^*$ and $m_{\lambda_1} : \mathbb{C} \to \mathbb{C}, z \to \lambda_1 z$, then one can define the following isomonodromic family in one parameter,

$$(H^{(\lambda_1)}, \nabla^{(\lambda_1)}) := m_{\lambda_1}^*(H, \nabla), \qquad (4.3)$$

$$\text{thus} \quad u_0^{j,(\lambda_1)} = u_0^j / \lambda_1, \; u_\infty^{j,(\lambda_1)} = \lambda_1 u_\infty^j.$$

Combining these operations, one obtains from one P_{3D6} bundle an isomonodromic family with trivial transversal monodromy over one fibre of c^{uc*}. This isomonodromic family is considered *inessential*.

The restriction of the universal family over U^{univ} to one fibre of $c^{uc*} \circ c^{univ}$ pulls down to a family over the fibre of c^{uc*}. The whole universal family pulls down to a family over the fibre product U^{beta}, which is defined by c^{uc*} and e, where $e : \mathbb{C} \to \mathbb{C}^*, \zeta \mapsto e^\zeta$, is the universal covering of $\mathbb{C}^*$:

$$
\begin{array}{ccc}
U^{beta} & \xrightarrow{\;c^{uc}\;} & \mathbb{C} \\
\downarrow{\scriptstyle c^{beta}} & & \downarrow{\scriptstyle e} \\
U & \xrightarrow[c^{uc*}]{} & \mathbb{C}^*.
\end{array}
$$

$$(4.4)$$

The covering $c^{beta} : U^{beta} \to U$ is a smaller covering of U than the universal covering $U^{univ} \to U$. Finally, by the operations above, all bundles in one c^{uc}-fibre of U^{beta} are isomorphic as holomorphic vector bundles because the isomonodromic family in one c^{uc}-fibre is inessential.

Remarks 4.1

(i) The properties of being

> irreducible
>
> completely reducible
>
> reducible, but not completely reducible

are independent of the values u_0^j, u_∞^j. Therefore all bundles in one isomonodromic family have the same property. Furthermore, if they are all reducible, a flat rank 1 subbundle of one P_{3D6} bundle extends at least locally to a flat family of flat subbundles (on $\mathbb{C}^* \times T$).

If one and hence all bundles in an isomonodromic family are completely reducible, they are all isomorphic as holomorphic bundles, because the family splits locally into two holomorphic families of rank 1 subbundles, and the degrees of these rank 1 subbundles are constant.

(ii) Definition: Let $(H \to \mathbb{P}^1, \nabla)$ be a holomorphic vector bundle with a meromorphic connection. It is called *trace free* if its determinant bundle $\det(H, \nabla)$ with connection is the trivial bundle with trivial flat connection.

(iii) The description of the determinant bundle $\det(H, \nabla)$ of a P_{3D6} bundle in the proof of Corollary 3.2 (a) shows that the P_{3D6} bundle is trace free if and only if

$$\sum_{j=1}^{2} u_0^j = \sum_{j=1}^{2} u_\infty^j = \sum_{j=1}^{2} \alpha_0^j = \sum_{j=1}^{2} \alpha_\infty^j = 0. \tag{4.5}$$

(iv) The following terminology will be useful here and in Chap. 7. Given a holomorphic vector bundle $H \to \mathbb{P}^1$ of some rank $n \geq 1$, by Birkhoff-Grothendieck there exist unique integers $k_1, \ldots, k_n$ with $k_1 \geq \cdots \geq k_n$ such that

$$\mathcal{O}(H) \cong \mathcal{O}_{\mathbb{P}^1}(k_1) \oplus \cdots \oplus \mathcal{O}_{\mathbb{P}^1}(k_n).$$

The vector bundle H is then called a $(k_1, \ldots, k_n)$-twistor. A $(0, \ldots, 0)$-twistor is also called *pure*.

Theorem 4.2 ([Heu09]) *Let $(H, \nabla, u_0^1, u_\infty^1)$ be a trace free irreducible P_{3D6} bundle. Then there exists a discrete (possibly empty) subset $\Sigma \subset \mathbb{C}$ such that, in the universal isomonodromic family above U^{beta}, all bundles in $(c^{uc})^{-1}(\mathbb{C} - \Sigma)$ are pure (i.e. trivial as holomorphic vector bundles), and the bundles in $(c^{uc})^{-1}(\Sigma)$ are $(1, -1)$-twistors.*

Proof The restriction of the isomonodromic family over U^{beta} to the 2-dimensional subset $U^{beta,tr} \subset U^{beta}$ of trace free P_{3D6} bundles still has two parameters: the parameter λ_1 from above is inessential, and the parameter in the base of the map $c^{uc} : U^{beta,tr} \to \mathbb{C}$ is essential.

In a much more general situation of trace free rank 2 bundles with meromorphic connections on closed complex curves, in [Heu10] universal isomonodromic families are constructed. In [Heu09] the subsets of non-pure twistors are estimated. In our situation the result of [Heu09] is that the subset

$$T_k := \{t \in U^{beta,tr} \mid \kappa(H^{(t)}) \leq k\} \quad \text{with} \tag{4.6}$$

$$\kappa(H^{(t)}) := \min(-2\deg W^{(t)} \mid W^{(t)} \subset H^{(t)} \text{ is a rank 1 subbundle})$$

is a closed analytic subset of codimension at least $-1 - k$. Because of the one inessential parameter in the 2-dimensional manifold $U^{beta,tr}$, T_k consists of fibres of $c^{uc} : U^{beta,tr} \to \mathbb{C}$. Therefore it is either of codimension 0 or 1 or is empty. Thus

$$T_{-4} = \{t \in U^{beta,tr} \mid H^{(t)} \text{ is an } (l, -l)\text{-twistor for some } l \geq 2\}$$

has codimension $\geq -1 - (-4)$, so it is empty. Similarly,

$$T_{-2} = \{t \in U^{beta,tr} \mid H^{(t)} \text{ is an } (1, -1)\text{-twistor}\}$$

has codimension $\geq -1 - (-2) = 1$ or is empty. So it is empty, or $T_{-2} = (c^{uc})^{-1}(\Sigma)$ for a discrete set $\Sigma \subset \mathbb{C}$.   □

Remarks 4.4

(i) For the case of P_{3D6}-TEP bundles (Definition 6.1), which is our main interest, we shall reprove Theorem 4.2 in Chap. 8 (in Theorem 8.2 (a) and in Lemma 8.5) by elementary calculations using normal forms of such bundles. For this case Theorem 4.2 is also proved in [Ni09]. However, Heu's result is much more general, and the elementary proof in Chap. 8 seems simpler.

(ii) Theorem 4.2 will allow us to restrict to special cases of the following favorable situation.

Let $V \to \mathbb{P}^1 \times T$ be a holomorphic rank 2 vector bundle on $\mathbb{P}^1$ times a (connected complex analytic) manifold T, such that any bundle $V^{(t)} := V|_{\mathbb{P}^1 \times \{t\}}$ for $t \in T$ is either pure or a $(1, -1)$-twistor.

First, the set

$$\Theta := \{t \in T \mid V^{(t)} \text{ is a } (1, -1)\text{-twistor}\}$$

is either empty or an analytic hypersurface or $\Theta = T$ [Sa02, I 5.3 Theorem].

Second, (with $\pi : \mathbb{P}^1 \times T \to T$ the projection)

$$\pi_* \mathcal{O}(V) \text{ is locally free of rank 2 and} \tag{4.7}$$

$$\pi_* \mathcal{O}(V)|_t := \pi_* \mathcal{O}(V)_t / \mathbf{m}_{t,T} \cong \pi_* \mathcal{O}(V^{(t)}) \text{ for any } t \in T, \tag{4.8}$$

which implies that any global section of some $V^{(t)}$ extends (non-uniquely) to a global section of $V|_{\mathbb{P}^1 \times S}$ for some small neighbourhood $S \subset T$ of t.

Moreover, (4.7) and (4.8) imply that for any small open set $S \subset T$ one can choose two global sections on $V|_{\mathbb{P}^1 \times S}$ such that they restrict for any $t \in S$ to a basis of the 2-dimensional space $\pi_* \mathcal{O}(V^{(t)})$.

The proof of (4.7) and (4.8) is an application of fundamental results on coherent sheaves in [BS76, ch. III] and basic facts on the cohomology groups $H^l(\mathbb{P}^1, \mathcal{O}_{\mathbb{P}^1}(k))$. It is well known [Sa02, I 2] that

$$\dim H^0(\mathbb{P}^1, \mathcal{O}_{\mathbb{P}^1}(k)) = k + 1 \text{ for } k \geq -1,$$

$$\dim H^l(\mathbb{P}^1, \mathcal{O}_{\mathbb{P}^1}(k)) = 0 \text{ for } l \geq 1, k \geq -1.$$

This implies that

$$\dim H^0(\mathbb{P}^1, \mathcal{O}(V^{(t)})) = 2,$$

$$\dim H^l(\mathbb{P}^1, \mathcal{O}(V^{(t)})) = 0 \text{ for } l \geq 1.$$

The latter implies that $R^l \pi_* \mathcal{O}(V) = 0$ for $l \geq 1$ [BS76, III Cor. 3.11]. Now the base change theorem [BS76, III Cor. 3.5] applies and gives (4.8). Finally, $\dim H^0(\mathbb{P}^1, \mathcal{O}(V^{(t)})) = 2$ for all $t \in T$ and [BS76, III Lemma 1.6] give (4.7).

(iii) All statements and arguments in (ii) generalize to holomorphic vector bundles $V \to \mathbb{P}^1 \times T$ of rank $n \geq 2$ such that for any $t \in T$ the bundle $V^{(t)}$ is a $(k_1, \ldots, k_n)$-twistor with $k_n \geq -1$ and $\sum_{j=1}^n k_j = 0$.

Chapter 5
Useful Formulae: Three 2 × 2 Matrices

In this chapter we collect together some elementary formulae concerning three matrices which appear in the monodromy data: a Stokes matrix S, a monodromy matrix Mon, and a connection matrix B.

Let us fix $s \in \mathbb{C}$ and define two matrices

$$S := \begin{pmatrix} 1 & s \\ 0 & 1 \end{pmatrix}, \quad \text{Mon}_0^{mat} := S^t S^{-1} = \begin{pmatrix} 1 & -s \\ s & 1-s^2 \end{pmatrix} \in SL(2, \mathbb{C}). \tag{5.1}$$

In this chapter (and later) the notation $\sqrt{\frac{1}{4}s^2 - 1}$ indicates the square root with argument in $[0, \pi)$ if $s \neq \pm 2$ (and 0 if $s = \pm 2$).

First we consider the case $s = \pm 2$. Then Mon_0^{mat} has a 2×2 Jordan block with eigenvalues $\lambda_+ = \lambda_- = -1$. A suitable basis of $M(2 \times 1, \mathbb{C})$ is

$$v_1 = \begin{pmatrix} 1 \\ s/2 \end{pmatrix}, \quad v_2 = \begin{pmatrix} 0 \\ 1 \end{pmatrix} \tag{5.2}$$

with $\text{Mon}_0^{mat} v_1 = -v_1$, $\text{Mon}_0^{mat} v_2 = -v_2 - sv_1$. We define $\alpha_\pm := \text{sign}(s)\,(\mp\frac{1}{2})$. Then $e^{-2\pi i \alpha_\pm} = -1 = \lambda_\pm$.

Now we consider the case $s \in \mathbb{C} - \{\pm 2\}$. Then Mon_0^{mat} is semisimple with two distinct eigenvalues

$$\lambda_\pm = (1 - \tfrac{1}{2}s^2) \pm s\sqrt{\tfrac{1}{4}s^2 - 1}. \tag{5.3}$$

Thus $\lambda_+ + \lambda_- = 2 - s^2$, $\lambda_+ \lambda_- = 1$. For eigenvectors we take

$$v_\pm = \begin{pmatrix} 1 \\ \mp\sqrt{\frac{1}{4}s^2 - 1} + \frac{1}{2}s \end{pmatrix} \tag{5.4}$$

© Springer International Publishing AG 2017
M.A. Guest, C. Hertling, *Painlevé III: A Case Study in the Geometry of Meromorphic Connections*, Lecture Notes in Mathematics 2198, DOI 10.1007/978-3-319-66526-9_5

with $\mathrm{Mon}_0^{mat} \, v_\pm = \lambda_\pm \, v_\pm$.

For $s \in \mathbb{C} - (\mathbb{R}_{\geq 2} \cup \mathbb{R}_{\leq -2})$ there exist unique $\alpha_\pm \in \mathbb{C}$ with

$$\Re(\alpha_\pm) \in (-\tfrac{1}{2}, \tfrac{1}{2}) \text{ and } e^{-2\pi i \alpha_\pm} = \lambda_\pm. \tag{5.5}$$

For $s \in \mathbb{R}_{>2} \cup \mathbb{R}_{<-2}$ there exist unique $\alpha_\pm$ with

$$e^{-2\pi i \alpha_\pm} = \lambda_\pm \text{ and } \Re(\alpha_\pm) = \mathrm{sign}(s)\,(\mp\tfrac{1}{2}). \tag{5.6}$$

Lemma 5.1 *For any $s \in \mathbb{C}$:*

(a)

$$\lambda_+(s)\,\lambda_-(s) = 1, \quad \alpha_+(s) + \alpha_-(s) = 0, \tag{5.7}$$

$$\lambda_\pm(s) = \lambda_\mp(-s), \quad \alpha_\pm(s) = \alpha_\mp(-s), \tag{5.8}$$

$$\lambda_+, \lambda_- \in S^1 \iff s \in [-2,2] \iff \alpha_\pm \in [-\tfrac{1}{2}, \tfrac{1}{2}], \tag{5.9}$$

$$|\lambda_+| < 1 \iff \Im(\alpha_+) < 0 \iff \Im(s) > 0 \text{ or } s \in \mathbb{R}_{>2}, \tag{5.10}$$

$$|\lambda_+| > 1 \iff \Im(\alpha_+) > 0 \iff \Im(s) < 0 \text{ or } s \in \mathbb{R}_{<-2}, \tag{5.11}$$

$$\lambda_+, \lambda_- \in \mathbb{R}_{>0} \iff s \in i\mathbb{R}, \iff \alpha_+, \alpha_- \in i\mathbb{R}, \tag{5.12}$$

$$\Im(\lambda_+), \Im(\lambda_-) > 0 \iff \Re(\alpha_+) \in (-\tfrac{1}{2}, 0)$$
$$\iff \Re(s) > 0, s \notin \mathbb{R}_{\geq 2}. \tag{5.13}$$

The map

$$\mathbb{C} \to \{z \in \mathbb{C} \mid -\tfrac{1}{2} < \Re(z) < \tfrac{1}{2}\} \cup (-\tfrac{1}{2} + i\mathbb{R}_{\leq 0}) \cup (\tfrac{1}{2} + i\mathbb{R}_{\geq 0})$$
$$s \mapsto \alpha_+(s) \tag{5.14}$$

is bijective. It is continuous in each of the three regions $\mathbb{C} - (\mathbb{R}_{>2} \cup \mathbb{R}_{<-2})$, $\{z \in \mathbb{C} \mid \Im(z) > 0\} \cup \mathbb{R}_{\geq 2}$, $\{z \in \mathbb{C} \mid \Im(z) < 0\} \cup \mathbb{R}_{\leq -2}$. *The maps* $s \mapsto \sqrt{\tfrac{1}{4}s^2 - 1}$ *and* $s \mapsto \lambda_+(s)$ *are also continuous in these three regions.*

(b)

$$(-i)e^{-\pi i \alpha_\pm} = -\sqrt{\tfrac{1}{4}s^2 - 1} \pm \tfrac{1}{2}s \text{ for all } s \in \mathbb{C},$$
$$\text{and } = \tfrac{1}{2}(1 + \lambda_\pm)/\sqrt{\tfrac{1}{4}s^2 - 1} \text{ for } s \neq \pm 2, \tag{5.15}$$

$$\cos(\pi \alpha_\pm) = (-i)\sqrt{\tfrac{1}{4}s^2 - 1}, \quad \sin(\pi \alpha_\pm) = \mp\tfrac{1}{2}s. \tag{5.16}$$

(c) *In the special case $s \in \mathbb{R}_{>2}$, define*

$$t^{NI} = \frac{1}{2\pi} \log |\lambda_-| = \frac{1}{\pi} \log \sqrt{|\lambda_-|} \in \mathbb{R}_{>0}. \tag{5.17}$$

Then

$$\alpha_\pm = \mp \left(\tfrac{1}{2} + it^{NI} \right), \tag{5.18}$$

$$\sqrt{|\lambda_\pm|} = |e^{-\pi i \alpha_\pm}| = |e^{\mp \pi t^{NI}}| = \mp\sqrt{\tfrac{1}{4}s^2 - 1} + \tfrac{s}{2}, \tag{5.19}$$

$$\cosh(\pi t^{NI}) = \tfrac{1}{2}s, \ \sinh(\pi t^{NI}) = \sqrt{\tfrac{1}{4}s^2 - 1}. \tag{5.20}$$

Proof

(a) Equations (5.7), (5.8) and (5.12) are obvious. For (5.9) observe that $\lambda_+ \lambda_- = 1$. Now (5.10) and (5.11) follow easily as follows. They can be verified at some special values, and then (5.9) and the bijectivity and continuity of the map in (5.14) give (5.10) and (5.11) for all values. Similarly, (5.13) follows from (5.12). The map (5.14) is bijective because the map $\mathbb{C} - \{\pm 2\} \rightarrow \mathbb{C}^*$, $s \mapsto \lambda_+(s)$, is injective. The continuity in the first region is trivial. The continuity in the other two regions follows easily from definitions (5.5) and (5.6). The continuity of the maps $s \mapsto \sqrt{\tfrac{1}{4}s^2 - 1}$ and $s \mapsto \lambda_+(s)$ in the three regions follows from the choice arg $\sqrt{\tfrac{1}{4}s^2 - 1} \in [0, \pi)$.

(b) The case $s = \pm 2$ is trivial. In the case $s \neq \pm 2$, we have

$$\left(\frac{1 + \lambda_\pm}{2\sqrt{\tfrac{1}{4}s^2 - 1}} \right)^2 = \frac{1 + 2\lambda_\pm + \lambda_\pm^2}{s^2 - 4} = \frac{(4 - s^2)\lambda_\pm}{s^2 - 4} \tag{5.21}$$

$$= -\lambda_\pm = -e^{-2\pi i \alpha_\pm}.$$

This and (5.3) give the formulae in (5.15) up to a sign. At $s = 0$ ($\lambda_\pm = 1, \alpha_\pm = 0$) the sign is correct. Because of the continuity of the three maps in (a) in the three regions, the sign in (5.15) is correct for all $s \neq \pm 2$. Equation (5.16) follows from (5.15) and from $\alpha_+ + \alpha_- = 0$.

(c) These are easy consequences of part (b). □

Lemma 5.2

(a) *For any matrix $B \in SL(2, \mathbb{C})$ the following equivalences hold.*

$$BS = S(B^t)^{-1} \iff S^t B^t = B^{-1} S^t$$

$$\iff BS^t = S^t(B^t)^{-1} \tag{5.22}$$

$$\Longleftrightarrow B = \begin{pmatrix} b_1 & b_2 \\ -b_2 & b_1 + sb_2 \end{pmatrix}.$$

If $s \neq 0$ then these conditions are equivalent to

$$Mon_0^{mat} B = B Mon_0^{mat}. \tag{5.23}$$

(b) If the conditions in (a) hold, and $s \neq \pm 2$, then

$$B v_\pm = b_\pm v_\pm$$

$$\text{with } b_\pm = (b_1 + \tfrac{1}{2}sb_2) \mp \sqrt{\tfrac{1}{4}s^2 - 1}\, b_2 = b_1 \mp ie^{-\pi i\alpha_\pm} b_2 \tag{5.24}$$

and $b_+ b_- = \det B = 1$. If $s = \pm 2$, then

$$B v_1 = \widetilde{b}_1 v_1, \; B v_2 = \widetilde{b}_1 v_2 + b_2 v_1$$

$$\text{with } \widetilde{b}_1 = b_1 + \tfrac{1}{2}sb_2 \in \{\pm 1\}. \tag{5.25}$$

If $s \neq \pm 2$, B as in (a) is determined by its eigenvalue b_-, and any value $b_- \in \mathbb{C}^*$ is realizable by a matrix B as in (a). If $s = \pm 2$, B is determined by its eigenvalue $\widetilde{b}_1 \in \{\pm 1\}$ and by b_2, and any pair $(\widetilde{b}_1, b_2) \in \{\pm 1\} \times \mathbb{C}$ is realizable by a matrix as in (a).

The proof consists of elementary calculations and is omitted. The eigenvalues $b_\pm$ of B will be important in Chap. 12.

The following matrix $T(s)$ is a square root of $-Mon_0^{mat}(s)$. Its appearance looks surprising. Lemma 5.3 formulates its properties. We shall use it in the proofs of Lemma 13.1 (a) and Theorem 14.1 (a).

Lemma 5.3 Fix $s \in \mathbb{C}$. The matrix

$$T(s) := \begin{pmatrix} 0 & 1 \\ -1 & s \end{pmatrix} \tag{5.26}$$

satisfies

$$T(s)^{-1} = \begin{pmatrix} s & -1 \\ 1 & 0 \end{pmatrix}, \; T(s)^2 = -Mon_0^{mat}(s), \tag{5.27}$$

$$T(s) B = B T(s) \quad \text{for } B \text{ as in (5.22).} \tag{5.28}$$

If $s \neq \pm 2$, $T(s)$ is semisimple with eigenvectors $v_\pm$ and eigenvalues $\mp ie^{-\pi i\alpha_\pm}$, i.e.

$$T(s)(v_+ \; v_-) = (v_+ \; v_-) \begin{pmatrix} -ie^{-\pi i\alpha_+} & 0 \\ 0 & ie^{-\pi i\alpha_-} \end{pmatrix}. \tag{5.29}$$

If $s = \pm 2$, $T(s)$ has the eigenvalue $\frac{1}{2}s$ and a 2×2 Jordan block and satisfies

$$T(s)(v_1 \ v_2) = (v_1 \ v_2) \begin{pmatrix} \frac{1}{2}s & 1 \\ 0 & \frac{1}{2}s \end{pmatrix}. \tag{5.30}$$

The proof consists of elementary calculations and is omitted.

Chapter 6
P_{3D6}-TEP Bundles

In this monograph we are interested in the Painlevé III(D_6) equation of type $(\alpha, \beta, \gamma, \delta) = (0, 0, 4, -4)$ and in the isomonodromic families of P_{3D6} bundles which are associated to it in [FN80, IN86, FIKN06, Ni09]. These P_{3D6} bundles are special and can be equipped with rich additional structure. We shall develop this structure in two steps in Chaps. 6 and 7. The most important part is the TEP structure below. In Chap. 7 it will be further enriched to a TEJPA structure. Isomonodromic families of P_{3D6}-TEJPA bundles will correspond to solutions of the equation $P_{III}(0, 0, 4, -4)$ (Theorem 10.3).

First, some notation will be fixed. The following holomorphic or antiholomorphic maps most of which are involutions of $\mathbb{P}^1$ will be used here and later:

$$j : \mathbb{P}^1 \to \mathbb{P}^1, \quad z \mapsto -z, \tag{6.1}$$

$$\rho_c : \mathbb{P}^1 \to \mathbb{P}^1, \quad z \mapsto \tfrac{1}{c}z \quad \text{for some } c \in \mathbb{C}^*, \tag{6.2}$$

$$\gamma : \mathbb{P}^1 \to \mathbb{P}^1, \quad z \mapsto 1/\overline{z}, \tag{6.3}$$

$$\sigma : \mathbb{P}^1 \to \mathbb{P}^1, \quad z \mapsto -1/\overline{z}. \tag{6.4}$$

Apart from ρ_c, these were also used in [He03, HS07, HS11, Mo11b, Sa02, Sa05b].

Definition 6.1

(a) A TEP bundle is a holomorphic vector bundle $H \to \mathbb{P}^1$ of rank $n \geq 1$, with a (flat) meromorphic connection ∇ and a $\mathbb{C}$-bilinear pairing P. The connection is holomorphic on $\mathbb{C}^*$ and the pole at 0 has order ≤ 2. The pairing

$$\text{(pointwise) } P : H_z \times H_{j(z)} \to \mathbb{C}, \quad \text{for all } z \in \mathbb{P}^1, \tag{6.5}$$

$$\text{(for sections) } P : \mathcal{O}(H) \times j^*\mathcal{O}(H) \to \mathcal{O}_{\mathbb{P}^1}, \quad \mathcal{O}_{\mathbb{P}^1}\text{-linear}$$

© Springer International Publishing AG 2017
M.A. Guest, C. Hertling, *Painlevé III: A Case Study in the Geometry of Meromorphic Connections*, Lecture Notes in Mathematics 2198,
DOI 10.1007/978-3-319-66526-9_6

is symmetric, nondegenerate, and flat on $H|_{\mathbb{C}^*}$. Symmetric means that

$$P(a(z), b(-z)) = P(b(-z), a(z))$$

for $a(z) \in H_z, b(-z) \in H_{-z}$, and flatness means

$$z\partial_z P(a(z), b(-z)) = P(\nabla_{z\partial_z} a(z), b(-z)) + P(a(z), \nabla_{z\partial_z} b(-z))$$

for $a \in \Gamma(U, \mathcal{O}(H)), b \in \Gamma(j(U), \mathcal{O}(H))$.

(b) A P_{3D6}-TEP bundle is a P_{3D6} bundle which is also a TEP bundle.

Remarks 6.2

(i) The definition of TEP bundles here differs from the definition of TEP structures in [He03] and [HS07]. The TEP structures there are restrictions to $\mathbb{C}$ of TEP bundles here.

(ii) The name TEP is intended to be self-referencing (cf. [He03, HS07]): T = twistor $\sim$ holomorphic vector bundle on $\mathbb{P}^1$, E = *extension* $\sim$ a meromorphic connection, P = pairing. By adding letters, one obtains richer structures (see Chaps. 7 and 16).

(iii) In formula (6.10) and later, we write the matrix of the pairing P for a pair (v_1, v_2) of vectors in H_z and a pair (v_3, v_4) of vectors in H_{-z} as

$$P((v_1 \; v_2)^t, (v_3 \; v_4)) = P(\begin{pmatrix} v_1 \\ v_2 \end{pmatrix}, (v_3 \; v_4)) = \begin{pmatrix} P(v_1, v_3) & P(v_1, v_4) \\ \mathbb{P}(v_2, v_3) & P(v_2, v_4) \end{pmatrix}.$$

Theorem 6.3

(a) *A P_{3D6} bundle can be enriched to a P_{3D6}-TEP bundle if and only if it is*

$$\text{irreducible or completely reducible, and} \tag{6.6}$$

$$\alpha_0^1 = \alpha_0^2 = \alpha_\infty^1 = \alpha_\infty^2 = 0. \tag{6.7}$$

(b) *All TEP structures on a given P_{3D6} bundle are isomorphic.*

(c) *Only two completely reducible P_{3D6} bundles satisfying (6.7) exist. In both cases, each flat rank 1 subbundle can be equipped with a TEP structure, which is unique up to rescaling. The orthogonal sums of these TEP structures are the only TEP structures on the P_{3D6} bundle. Therefore, such TEP structures are parametrized by $\mathbb{C}^* \times \mathbb{C}^*$.*

(d) *The TEP structures on an irreducible P_{3D6} bundle with (6.7) differ only by rescaling. Therefore they are parametrized by $\mathbb{C}^*$.*

(e) *Any P_{3D6}-TEP bundle has bases $\underline{e}_0^\pm = (e_0^{\pm 1}, e_0^{\pm 2})$ and $\underline{e}_\infty^\pm = (e_\infty^{\pm 1}, e_\infty^{\pm 2})$ such that*

$$e_0^{\pm j} \text{ and } e_\infty^{\pm j} \text{ are flat bases of } L_0^{\pm j} \text{ and } L_\infty^{\pm j}, \tag{6.8}$$

$$e_{0/\infty}^{+1}|_{\widehat{I}_{0/\infty}^a} = e_{0/\infty}^{-1}|_{\widehat{I}_{0/\infty}^a} \quad \text{and} \quad e_{0/\infty}^{+2}|_{\widehat{I}_{0/\infty}^b} = e_{0/\infty}^{-2}|_{\widehat{I}_{0/\infty}^b} \tag{6.9}$$

(this is (2.12) and its analogue at ∞ in the case of (6.7)) and

$$P((\underline{e}_0^\pm)'(z), \underline{e}_\infty^\mp(-z)) = P((\underline{e}_\infty^\pm)'(z), \underline{e}_\infty^\mp(-z)) = 1_2, \tag{6.10}$$

$$\det B(\beta) = 1 \text{ for } \beta \text{ and } B(\beta) \text{ as in (2.23), (2.24).} \tag{6.11}$$

If $\underline{e}_0^\pm, \underline{e}_\infty^\pm$ is a 4-tuple of such bases, then all others are obtained by modifying signs. Altogether there are eight 4-tuples of such bases, namely:

$$\varepsilon_0(e_0^{\pm 1}, \varepsilon_1 e_0^{\pm 2}), \ \varepsilon_0\varepsilon_2(e_\infty^{\pm 1}, \varepsilon_1 e_\infty^{\pm 2}) \text{ for } \varepsilon_0, \varepsilon_1, \varepsilon_2 \in \{\pm 1\}. \tag{6.12}$$

For any 4-tuple of such bases, the P_{3D6} numerical tuple satisfies

$$s_0^a = s_0^b = -s_\infty^a = -s_\infty^b =: s,$$

$$B(\beta) = \begin{pmatrix} b_1 & b_2 \\ -b_2 & b_1 + sb_2 \end{pmatrix} \quad \text{(with } \det B(\beta) = 1\text{)}. \tag{6.13}$$

(f) *The automorphism group of a P_{3D6}-TEP bundle is*

$$\mathrm{Aut}(H, \nabla, P) = \{\pm \mathrm{id}\} \quad \text{(irreducible case)}$$

$$\mathrm{Aut}(H, \nabla, P) = \{\pm \mathrm{id}, \pm G\} \quad \text{(completely reducible case)} \tag{6.14}$$

with

$$G : \underline{e}_0^\pm \mapsto (e_0^{\pm 1}, -e_0^{\pm 2}), \ \underline{e}_\infty^\pm \mapsto \varepsilon_2(e_\infty^{\pm 1}, -e_\infty^{\pm 2}) \text{ and} \tag{6.15}$$

$$\varepsilon_2 = 1 \text{ in the case } (C_{11}), (C_{22}) \text{ (i.e. } b_2 = b_3 = 0),$$

$$\varepsilon_2 = -1 \text{ in the case } (C_{12}), (C_{21}) \text{ (i.e. } b_1 = b_4 = 0).$$

(g) *Any P_{3D6} numerical tuple of an irreducible P_{3D6} bundle with (6.7) satisfies either $s_0^a s_0^b = s_\infty^a s_\infty^b \neq 0$ or $s_0^a = s_0^b = s_\infty^a = s_\infty^b = 0$.*

Proof We shall give the proofs in this order: (a)($\Rightarrow$), (g), (e), (a)($\Leftarrow$), (c), (d), (b), (f).

(a)($\Rightarrow$): Let $(H, \nabla, u_0^1, u_\infty^1, P)$ be a P_{3D6}-TEP bundle. First we discuss the pole at 0 (the pole at ∞ will be analogous). Let $\underline{e}_0^\pm = (e_0^{\pm 1}, e_0^{\pm 2})$ be bases of $L|_{\widehat{I}_0^\pm}$ consisting of flat generating sections of $L_0^{\pm 1}, L_0^{\pm 2}$ with (2.12). Choose one branch of $\log z$ on $\widehat{I}_0^+$ and extend it counterclockwise to $\widehat{I}_0^-$. We shall write $(ze^{\pi i})^{\alpha_0^j}$ instead of $(-z)^{\alpha_0^j}$ to indicate this extension.

Choose a basis $\underline{\varphi}$ of $\mathcal{O}(H)_0$ as in Remark 2.5 (ii), with $A_0^\pm \in GL(2, \mathcal{A}_{I_0^\pm}), \widehat{A}_0^+ = \widehat{A}_0^-$ and $\widehat{A}_0^\pm(0) = 1_2$. Then, for $z \in \widehat{I}_0^+$, the matrix $P((\varphi(z))', \varphi(-z)) \in GL(2, \mathbb{C}\{z\})$

is

$$(A_0^+(z))' \begin{pmatrix} \gamma^{11}(z) & \gamma^{12}(z) \\ \gamma^{21}(z) & \gamma^{22}(z) \end{pmatrix} A_0^-(-z) \tag{6.16}$$

with $\gamma^{ij}(z) = z^{\alpha_0^i + \alpha_0^j} e^{\pi i \alpha_0^j} e^{-(u_0^i - u_0^j)/z} P(e_0^{+i}(z), e_0^{-j}(-z))$.

Because $\widehat{I}_0^+$ contains the Stokes line $\mathbb{R}_{>0}\,\zeta_0$, both functions $e^{-(u_0^1 - u_0^2)/z}$ and $e^{-(u_0^2 - u_0^1)/z}$ are unbounded on $\widehat{I}_0^+$. Therefore

$$P(e_0^{+1}, e_0^{-2}) = 0 = P(e_0^{+2}, e_0^{-1}),$$
$$P(e_0^{+1}, e_0^{-1}) \neq 0, \; P(e_0^{+2}, e_0^{-2}) \neq 0. \tag{6.17}$$

Therefore the splittings $L|_{\widehat{I}_0^+} = L_0^{\pm 1} \oplus L_0^{\pm 2}$ are dual to one another with respect to the pairing. Thus L_0^{+1} and L_0^{-1} glue to a rank 1 subbundle of L if and only if L_0^{+2} and L_0^{-2} glue to a rank 1 subbundle of L. Since $z^{2\alpha_0^i}$ must be holomorphic and nonvanishing at 0, $\alpha_0^1 = \alpha_0^2 = 0$.

Everything so far holds analogously at ∞. This establishes (6.7). If (H, ∇) is reducible, then one condition (C_{ij}) from Lemma 3.1 (a) holds. But because of the duality of the subbundles, at 0 as well as at ∞, then also $(C_{3-i,3-j})$ holds, and (H, ∇) is completely reducible by Lemma 3.1 (b). (a)($\Rightarrow$) is proved.

(g) Let (H, ∇) be an irreducible P_{3D6} bundle satisfying (6.7). Choose bases $\underline{e}_0^{\pm}$ and $\underline{e}_\infty^{\pm}$ as in the construction of P_{3D6} numerical tuples from P_{3D6} monodromy tuples in Chap. 2. Because of (2.22), the monodromy matrices with respect to the bases $\underline{e}_0^+$ and $\underline{e}_\infty^-$ are

$$S_0^b(S_0^a)^{-1} = \begin{pmatrix} 1 & -s_0^a \\ s_0^b & 1 - s_0^a s_0^b \end{pmatrix}, \; (S_\infty^b)^{-1} S_\infty^a = \begin{pmatrix} 1 & s_\infty^a \\ -s_\infty^b & 1 - s_\infty^a s_\infty^b \end{pmatrix}. \tag{6.18}$$

Therefore

$$s_0^a s_0^b = 2 - \mathrm{tr}(\mathrm{Mon}) = s_\infty^a s_\infty^b. \tag{6.19}$$

If $s_0^a = s_0^b = 0$ (or $s_\infty^a = s_\infty^b = 0$) then $\mathrm{Mon} = \mathrm{id}$ and also $s_\infty^a = s_\infty^b = 0$ (or $s_0^a = s_0^b = 0$).

Suppose $s_0^a \neq 0$ and $s_0^b = 0$ (the other case $s_0^a = 0$ and $s_0^b \neq 0$ is analogous). Then the monodromy is unipotent with a 2×2 Jordan block. As $s_0^b = 0$, L_0^{+1} and L_0^{-1} glue to a flat rank 1 subbundle of L. This must be the bundle of eigenspaces of the monodromy.

Since $s_\infty^a s_\infty^b = s_0^a s_0^b = 0$, the same argument applies at ∞. For some $k \in \{1, 2\}$, L_∞^{+k} and L_∞^{-k} must glue to the bundle of eigenspaces of the monodromy. Thus (H, ∇) is reducible, a contradiction.

(e) First, let us ignore (6.11). Then (6.17) and the analogous condition at ∞ show the existence of bases $\underline{e}_0^{\pm}$ and $\underline{e}_\infty^{\pm}$ with (6.8), (6.9), and (6.10). It is also clear that each of

$$(e_0^{+1}, e_0^{-1}),\ (e_0^{+2}, e_0^{-2}),\ (e_\infty^{+1}, e_\infty^{-1}),\ (e_\infty^{+2}, e_\infty^{-2})$$

is unique up to a sign. The following calculation, which will also be useful for (a)($\Leftarrow$), will show that $\det B(\beta) = \pm 1$. Recall (2.26) and (2.21) and the meaning of $\underline{e}_0^+(ze^{\pi i})$ explained after (2.21). For $z \in \widehat{I}_0^+$ and $\widetilde{z} = z\, e^{\beta} \in \widehat{I}_\infty^-$

$$
\begin{aligned}
1_2 &= P((\underline{e}_\infty^+(-\widetilde{z}))^t, \underline{e}_\infty^-(\widetilde{z})) \\
&= P((\underline{e}_\infty^-(\widetilde{z}e^{\pi i})\,(S_\infty^a)^{-1})^t, \underline{e}_\infty^-(\widetilde{z})) \\
&= ((S_\infty^a)^{-1})^t\, P((\underline{e}_\infty^-(\widetilde{z}e^{\pi i}))^t, \underline{e}_\infty^-(\widetilde{z})) \\
&= ((S_\infty^a)^{-1})^t\, P((\underline{e}_0^+(ze^{\pi i})\, B(\beta))^t, \underline{e}_0^+(z)\, B(\beta)) \\
&= ((S_\infty^a)^{-1})^t\, (B(\beta))^t\, P((\underline{e}_0^+(ze^{\pi i}))^t, \underline{e}_0^-(-ze^{\pi i}))\,(S_0^b)^{-1}\, B(\beta) \\
&= ((S_\infty^a)^{-1})^t\, (B(\beta))^t\, 1_2\,(S_0^b)^{-1}\, B(\beta). \quad\quad (6.20)
\end{aligned}
$$

This shows that $\det B(\beta) = \pm 1$. If $\det B(\beta) = -1$, one can replace $(e_\infty^{+2}, e_\infty^{-2})$ by $(-1)(e_\infty^{+2}, e_\infty^{-2})$. The new 4-tuple of bases (or the old if $\det B(\beta)$ was already 1) satisfies (6.10) and (6.11). The possible sign changes in (6.12) are also clear now.

It remains to prove (6.13). If $z \in \widehat{I}_0^a$ then $-z \in \widehat{I}_0^b$, and (2.21) and (6.10) show

$$
\begin{aligned}
1_2 &= P((e_0^-)^t(z), e_0^+(-z)) = (S_0^a)^t\, P((e_0^+)^t(z), e_0^-(-z))\,(S_0^b)^{-1} \\
&= (S_0^a)^t\,(S_0^b)^{-1} = \begin{pmatrix} 1 & 0 \\ s_0^a & 1 \end{pmatrix}\begin{pmatrix} 1 & 0 \\ -s_0^b & 1 \end{pmatrix} = \begin{pmatrix} 1 & 0 \\ s_0^a - s_0^b & 1 \end{pmatrix},
\end{aligned}
$$

thus $s_0^a = s_0^b$. Analogously one obtains $s_\infty^a = s_\infty^b$. Define $s := s_0^a = s_0^b$ and

$$S := \begin{pmatrix} 1 & s \\ 0 & 1 \end{pmatrix} = S_0^a = (S_0^b)^t.$$

Because of (g), $s^2 = s_0^a s_0^b = s_\infty^a s_\infty^b$, thus $s_\infty^a = s$ or $s_\infty^a = -s$.
If $s_\infty^a = s \neq 0$ then (6.20) leads to

$$1_2 = (S^{-1})^t\, B(\beta)^t\, 1_2\, (S^t)^{-1}\, B(\beta),$$
$$B(\beta)^{-1}\, S^t = (S^{-1})^t\, B(\beta)^t,$$

and, with $B(\beta) = \left(\begin{smallmatrix} b_1 & b_2 \\ b_3 & b_4 \end{smallmatrix} \right)$,

$$\begin{pmatrix} b_4 - sb_2 & -b_2 \\ -b_3 + sb_1 & b_1 \end{pmatrix} = \begin{pmatrix} b_4 & -b_2 \\ -b_3 & b_1 \end{pmatrix} \begin{pmatrix} 1 & 0 \\ s & 1 \end{pmatrix}$$

$$= \begin{pmatrix} 1 & 0 \\ -s & 1 \end{pmatrix} \begin{pmatrix} b_1 & b_3 \\ b_2 & b_4 \end{pmatrix} = \begin{pmatrix} b_1 & b_3 \\ b_2 - sb_1 & b_4 - sb_3 \end{pmatrix},$$

so $b_3 = -b_2$, $sb_1 = -sb_1$, $b_1 = 0, -sb_2 = -sb_3$, $b_2 = b_3 = 0, b_4 = 0$, a contradiction. Thus $s_\infty^a = -s$. Now (6.20) gives

$$B(\beta)^{-1} S^t = S^t B(\beta),$$

which is one of the equivalent conditions in Lemma 5.2 (a). This establishes (6.13).

(a)($\Leftarrow$): The 1st step is the construction of bases $\underline{e}_0^\pm, \underline{e}_\infty^\pm$ which satisfy (6.8), (6.9), (6.11) and (6.13). The 2nd step is the definition of P on $L|_{\widehat{T_0^+}} \times L|_{\widehat{T_0^-}}$. The 3rd step is to show that P extends to H and defines a TEP structure and that the bases from the 1st step satisfy also (6.10).

1st step: Because of (g) there are three cases:

1st case: (H, ∇) is irreducible and $s_0^a s_0^b = s_\infty^a s_\infty^b \neq 0$.
2nd case: (H, ∇) is irreducible and $s_0^a = s_0^b = s_\infty^a = s_\infty^b = 0$.
3rd case: (H, ∇) is completely reducible (and $s_0^a = s_0^b = s_\infty^a = s_\infty^b = 0$).

1st case: Choose an $s \in \mathbb{C}^*$ with $s^2 = 2 - \text{tr Mon}$. It is unique up to a sign. The action of $(\mathbb{C}^*)^4$ in (2.29) on the P_{3D6} numerical tuples shows that we can choose flat generating sections $e_0^{\pm j}, e_\infty^{\pm j}$ of $L_0^{\pm j}, L_\infty^{\pm j}$ with (6.8) and (6.9) and with

$$s = s_0^a = s_0^b = -s_\infty^a = -s_\infty^b \tag{6.21}$$

$$\text{and} \quad \det B(\beta) = 1 \text{ for some (or any) } \beta. \tag{6.22}$$

Then

$$S_0^a = \begin{pmatrix} 1 & s \\ 0 & 1 \end{pmatrix} =: S, \quad S_0^b = S^t, \quad S_\infty^a = S^{-1}, \quad S_\infty^b = (S^t)^{-1}. \tag{6.23}$$

The monodromy matrices of e_0^+ and e_∞^- are both

$$S_0^b(S_0^a)^{-1} = S^t S^{-1} = \text{Mon}_0^{mat} := \begin{pmatrix} 1 & -s \\ s & 1 - s^2 \end{pmatrix} = (S_\infty^b)^{-1} S_\infty^a. \tag{6.24}$$

Condition (2.26) for $B(\beta)$ becomes

$$\text{Mon}_0^{mat} B(\beta) = B(\beta) \text{Mon}_0^{mat}.$$

Lemma 5.2 (a) gives the second half of (6.13),

$$B(\beta) = \begin{pmatrix} b_1 & b_2 \\ -b_2 & b_1 + sb_2 \end{pmatrix}.$$

2nd case: Here and in the 3rd case Mon = id, condition (2.26) is empty, and the subbundles L_0^{+j} and L_0^{-j} and L_∞^{+j} and L_∞^{-j} glue to rank 1 subbundles L_0^j and L_∞^j ($j = 1, 2$) of L.

Choose generating flat sections e_0^j and e_∞^j of L_0^j and L_∞^j. Then $\underline{e}_0 = (e_0^1, e_0^2)$ and $\underline{e}_\infty = (e_\infty^1, e_\infty^2)$ are flat global bases of L. In the irreducible case

$$\underline{e}_\infty = \underline{e}_0 B \text{ with } B = \begin{pmatrix} b_1 & b_2 \\ b_3 & b_4 \end{pmatrix} \text{ and all } b_i \neq 0 \tag{6.25}$$

(here B is independent of the choice of β). We seek $(\lambda_1, \lambda_2, \lambda_3, \lambda_4) \in (\mathbb{C}^*)^4$ such that

$$\widetilde{B} := \begin{pmatrix} \lambda_1^{-1} & 0 \\ 0 & \lambda_2^{-1} \end{pmatrix} B \begin{pmatrix} \lambda_3 & 0 \\ 0 & \lambda_4 \end{pmatrix} = \begin{pmatrix} \frac{\lambda_3}{\lambda_1} b_1 & \frac{\lambda_4}{\lambda_1} b_2 \\ \frac{\lambda_3}{\lambda_2} b_3 & \frac{\lambda_4}{\lambda_2} b_4 \end{pmatrix} = \begin{pmatrix} \widetilde{b}_1 & \widetilde{b}_2 \\ \widetilde{b}_3 & \widetilde{b}_4 \end{pmatrix}$$

satisfies $\widetilde{b}_4 = \widetilde{b}_1, \widetilde{b}_3 = -\widetilde{b}_2, \det \widetilde{B} = 1$, i.e.,

$$\frac{\lambda_4}{\lambda_2} b_4 = \frac{\lambda_3}{\lambda_1} b_1, \quad \frac{\lambda_3}{\lambda_2} b_3 = -\frac{\lambda_4}{\lambda_1} b_2, \quad 1 = \frac{\lambda_3 \lambda_4}{\lambda_1 \lambda_2} \det B. \tag{6.26}$$

This is equivalent to

$$\left(\frac{\lambda_2}{\lambda_1}\right)^2 = -\frac{b_4 b_3}{b_1 b_2}, \quad \frac{\lambda_3}{\lambda_4} = \frac{\lambda_1}{\lambda_2} \frac{b_4}{b_1}, \quad \lambda_3 \lambda_4 = \lambda_1 \lambda_2 \det B^{-1}.$$

After making an arbitrary choice of λ_1, we see that λ_2 exists and is unique up to a sign, and then λ_3 and λ_4 exist and are unique up to a common sign.

3rd case: Generating sections e_0^j and e_∞^j as in the 2nd case are chosen. In the completely reducible case

$$\underline{e}_\infty = \underline{e}_0 B \text{ with } B = \begin{pmatrix} b_1 & b_2 \\ b_3 & b_4 \end{pmatrix} \text{ and} \tag{6.27}$$

either $b_2 = b_3 = 0$ or $b_1 = b_4 = 0$.

In both cases (6.26) is solvable. In the case $b_2 = b_3 = 0$, we have $\widetilde{B} = \varepsilon \mathbf{1}_2$ where $\varepsilon \in \{\pm 1\}$ has to be chosen. Then one just needs $\lambda_4/\lambda_2 = \varepsilon b_4^{-1}, \lambda_3/\lambda_1 = \varepsilon b_1^{-1}$. In the case $b_1 = b_4 = 0$, we have $\widetilde{B} = \varepsilon \begin{pmatrix} 0 & 1 \\ -1 & 0 \end{pmatrix}$ where $\varepsilon \in \{\pm 1\}$ has to be chosen. Then one just needs $\lambda_4/\lambda_1 = \varepsilon b_2^{-1}, \lambda_3/\lambda_2 = -\varepsilon b_3^{-1}$.

2nd step: In the 1st step bases $\underline{e}_0^{\pm}, \underline{e}_{\infty}^{\pm}$ were constructed which satisfy (6.8), (6.9), (6.11) and (6.13). Define P on $L|_{\widehat{I_0^+}} \times L|_{\widehat{I_0^-}}$ by

$$P((\underline{e}_0^+)^t(z), \underline{e}_0^-(-z)) := \mathbf{1}_2.$$

3rd step: If $z \in \widehat{I_0^a}$ then $-z \in \widehat{I_0^b}$, and (2.21) gives

$$P((\underline{e}_0^-)^t(z), \underline{e}_0^+(-z)) = (S_0^a)^t \, P((\underline{e}_0^+)^t(z), e_0^-(-z)) \, (S_0^b)^{-1}$$
$$= S^t \, \mathbf{1}_2 \, (S^t)^{-1} = \mathbf{1}_2. \tag{6.28}$$

If $z \in \widehat{I_0^b}$ then $-z \in \widehat{I_0^a}$, and (2.21) gives

$$P((\underline{e}_0^-)^t(z), \underline{e}_0^+(-z)) = (S_0^b)^t \, P((\underline{e}_0^+)^t(z), e_0^-(-z)) \, (S_0^a)^{-1}$$
$$= S \, \mathbf{1}_2 \, S^{-1} = \mathbf{1}_2. \tag{6.29}$$

Therefore P is well defined on $H|_{\mathbb{C}^*}$, nondegenerate and symmetric, and satisfies the first half of (6.10). The calculation in (6.20) (without the first equality $\mathbf{1}_2 = \ldots$) gives

$$P((\underline{e}_{\infty}^+)^t(-\widetilde{z}), \underline{e}_{\infty}^-(\widetilde{z})) = S^t \, B(\beta)^t \, \mathbf{1}_2 \, (S^t)^{-1} \, B(\beta).$$

Because of (6.13) and Lemma 5.2 (a) this is $\mathbf{1}_2$. With the symmetry of P this establishes the second half of (6.10).

It remains to show that P extends in a nondegenerate way to 0 and ∞. Choose a basis φ of $\mathcal{O}(H)_0$ as in the proof of (a)($\Rightarrow$). The matrix $P((\varphi(z))^t, \varphi(-z))$ in (6.16) for $z \in \widehat{I_0^+}$ becomes

$$(A_0^+(z))^t \, \mathbf{1}_2 \, A_0^-(-z).$$

With $\widehat{A_0^+} = \widehat{A_0^-}$ and $\widehat{A_0^{\pm}}(0) = \mathbf{1}_2$ this proves the claim. The extension to ∞ is shown analogously. (a)($\Leftarrow$) is proved.

(c) In the notation of the 2nd and 3rd case in the proof of (a)($\Leftarrow$), the two completely reducible bundles are the bundles with

$$\text{either } \; b_2 = b_3 = 0, \; L_0^1 = L_{\infty}^1, \; L_0^2 = L_{\infty}^2, \; \text{i.e. } (C_{11}), (C_{22}),$$
$$\text{or } \; b_1 = b_4 = 0, \; L_0^1 = L_{\infty}^2, \; L_0^2 = L_{\infty}^1, \; \text{i.e. } (C_{12}), (C_{21}).$$

In both cases, (e) shows that any TEP structure splits into orthogonal TEP structures on the two flat rank 1 subbundles. They can be rescaled independently: One can choose $(\lambda_1, \lambda_2, \lambda_3, \lambda_4)$ in the 3rd case in the proof of (a)($\Leftarrow$) such that the matrix

$\widetilde{B}$ in (6.27) there stays as described, but the pairing of a given TEP structure takes up in the new basis the factors λ_1^2 and λ_2^2 on the first and second flat rank 1 bundle, respectively. Therefore all TEP structures are isomorphic.

(d) One has to consider separately the 1st case and the 2nd case in the proof of (a)($\Leftarrow$) and see in both cases that the possible choices of bases $\underline{e}_0^{\pm}$ and $\underline{e}_\infty^{\pm}$ differ only by the signs in (6.11) and a common scalar (which absorbs the sign ε_0). In both cases this is easy.

(b) This follows from (c) and (d).

(f) An automorphism of a P_{3D6}-TEP bundle maps a 4-tuple $\underline{e}_0^{\pm}, \underline{e}_\infty^{\pm}$ of bases in (e) to a 4-tuple $\widetilde{\underline{e}}_0^{\pm}, \widetilde{\underline{e}}_\infty^{\pm}$ which must be one of the eight 4-tuples in (6.12). Then

$$s = \widetilde{s} = \varepsilon_1 s, \quad b_1 = \widetilde{b}_1 = \varepsilon_2 b_1, \quad b_2 = \widetilde{b}_2 = \varepsilon_1 \varepsilon_2 b_2.$$

Conversely, any $(\varepsilon_0, \varepsilon_1, \varepsilon_2) \in (\{\pm 1\})^3$ with these properties induces an automorphism. In the irreducible case $\varepsilon_1 = \varepsilon_2 = 1$ and $\varepsilon_0 \in \{\pm 1\}$. In the completely reducible case $s = 0, b_2 = 0$: $\varepsilon_2 = 1$, $\varepsilon_0, \varepsilon_1 \in \{\pm 1\}$. In the completely reducible case $s = 0, b_1 = 0$: $\varepsilon_1 \varepsilon_2 = 1$, $\varepsilon_0, \varepsilon_1 = \varepsilon_2 \in \{\pm 1\}$. $\square$

Chapter 7
P_{3D6}-TEJPA Bundles and Moduli Spaces of Their Monodromy Tuples

We are concerned with P_{3D6} bundles which satisfy (6.6) and (6.7). They can be equipped with TEP structures, and the TEP structure is unique up to isomorphism. Therefore from now on we consider P_{3D6}-TEP bundles. As explained in the previous chapter, they possess eight distinguished 4-tuples of bases $\underline{e}_0^{\pm}, \underline{e}_{\infty}^{\pm}$ (Theorem 6.3 (e)).

This is good, but not good enough. We want to distinguish two of the 4-tuples, which differ only by a global sign. This may be regarded as a "marking" or "framing". It can be expressed very elegantly in the choice of two isomorphisms A and J, which are unique up to signs. The choice of the signs corresponds to the marking. The isomorphisms A, J express certain symmetries of P_{3D6}-TEP bundles (together with a reality condition, J is related to the R in TERP structures, whose relation to real solutions of Painlevé III (0,0,4,-4) will be developed in Chaps. 16 and 17).

We recall from (6.1) and (6.2) the involutions

$$j : \mathbb{P}^1 \to \mathbb{P}^1, \ z \mapsto -z,$$

$$\rho_c : \mathbb{P}^1 \to \mathbb{P}^1, \ z \mapsto \tfrac{1}{c}z$$

of $\mathbb{P}^1$.

Definition 7.1

(a) A TEPA bundle is a TEP bundle together with a $\mathbb{C}$-linear isomorphism

$$\text{(pointwise)} \ A : H_z \to H_{j(z)}, \quad \text{for all } z \in \mathbb{P}^1, \tag{7.1}$$

$$\text{(for sections)} \ A : \mathcal{O}(H) \to j^*\mathcal{O}(H), \quad \mathcal{O}_{\mathbb{P}^1}\text{-linear}$$

© Springer International Publishing AG 2017
M.A. Guest, C. Hertling, *Painlevé III: A Case Study in the Geometry of Meromorphic Connections*, Lecture Notes in Mathematics 2198, DOI 10.1007/978-3-319-66526-9_7

which is flat on $H|_{\mathbb{C}^*}$ and which satisfies

$$A^2 = -\mathrm{id} \quad \text{and} \tag{7.2}$$

$$P(A\,a, A\,b) = P(a, b) \text{ for all } a, b \in \mathcal{O}(H). \tag{7.3}$$

(b) A TEJPA bundle is a TEPA bundle together with some $c \in \mathbb{C}^*$ and a $\mathbb{C}$-linear isomorphism

$$\text{(pointwise) } J : H_z \to H_{\rho_c(z)}, \quad \text{for all } z \in \mathbb{P}^1, \tag{7.4}$$

$$\text{(for sections) } J : \mathcal{O}(H) \to \rho_c^* \mathcal{O}(H), \quad \mathcal{O}_{\mathbb{P}^1}\text{-linear}$$

which is flat on $H|_{\mathbb{C}^*}$ and which satisfies

$$J^2 = \mathrm{id}, \tag{7.5}$$

$$P(J\,a, J\,b) = P(a, b) \quad \text{for all } a, b \in \mathcal{O}(H), \text{ and} \tag{7.6}$$

$$A \circ J = -J \circ A. \tag{7.7}$$

(c) A P_{3D6}-TEPA bundle is a P_{3D6} bundle which is also a TEPA bundle. A P_{3D6}-TEJPA bundle is a P_{3D6} bundle which is, with $c = u_\infty^1/u_0^1$, also a TEJPA bundle.

Remarks 7.2

  (i) TEJP bundles and P_{3D6}-TEJP bundles are defined in the obvious way. One can also extract from $\widetilde{J} := A \circ J$ in a TEJPA bundle the properties of $\widetilde{J}$ and define TE$\widetilde{J}$P bundles and P_{3D6}-TE$\widetilde{J}$P bundles. We shall not need these, but we make a comment on them in Remark 7.4 (i) and (ii).
 (ii) If a P_{3D6} bundle is also a TEJPA bundle then automatically $c = \pm u_\infty^1/u_0^1$. The choice $c = u_\infty^1/u_0^1$ in Definition 7.1 (c) is good enough and avoids a pointless discussion of both cases.
(iii) The existence of A is probably specific to P_{3D6} bundles. On the other hand, the cousin of J, the R in TERP structures, exists in great generality.
 (iv) If (H, ∇, P) is a TEP bundle then also $j^*(H, \nabla, P)$ is a TEP bundle. In the case of a TEPA bundle, A is an isomorphism of TEP bundles.
  (v) Let (H, ∇, P, A, J) be a TEJPA bundle. Then the pole at ∞ has order ≤ 2, $\rho_c^*(H, \nabla, P)$ is a TEP bundle, and J is an isomorphism of TEP bundles.

Theorem 7.3

(a) A P_{3D6}-TEP bundle can be enriched to a P_{3D6}-TEPA bundle if and only if

$$u_0^1 + u_0^2 = 0 = u_\infty^1 + u_\infty^2. \tag{7.8}$$

The isomorphism A is then unique, up to a sign.
(b) A P_{3D6}-TEPA bundle can be enriched to a P_{3D6}-TEJPA bundle. The isomorphism J is unique up to a sign.

It follows that a P_{3D6}-TEP bundle can be enriched in four ways to a P_{3D6}-TEJPA bundle. Given $(H, \nabla, u_0^1, u_\infty^1, P, A, J)$, these are

$$(H, \nabla, u_0^1, u_\infty^1, P, \varepsilon_1 A, \varepsilon_2 J) \quad \text{with } \varepsilon_1, \varepsilon_2 \in \{\pm 1\}. \tag{7.9}$$

(c) Let $(H, \nabla, u_0^1, u_\infty^1, P, A, J)$ be a P_{3D6}-TEJPA bundle. It has a 4-tuple of bases $\underline{e}_0^\pm, \underline{e}_\infty^\pm$, unique up to a global sign, with (6.8), (6.9) and

$$P((\underline{e}_0^+)^t, \underline{e}_0^-) = \mathbf{1}_2, A(\underline{e}_0^+) = \underline{e}_0^- \begin{pmatrix} 0 & -1 \\ 1 & 0 \end{pmatrix}, J(\underline{e}_0^+) = \underline{e}_\infty^+ \begin{pmatrix} 1 & 0 \\ 0 & -1 \end{pmatrix}. \tag{7.10}$$

This is one of the eight 4-tuples in Theorem 6.3 (e). It satisfies (6.10), (6.11), (6.13) and

$$A(\underline{e}_0^\pm(z)) = \underline{e}_0^\mp(-z) \begin{pmatrix} 0 & -1 \\ 1 & 0 \end{pmatrix}, A(\underline{e}_\infty^\pm(z)) = \underline{e}_\infty^\mp(-z) \begin{pmatrix} 0 & -1 \\ 1 & 0 \end{pmatrix}, \tag{7.11}$$

$$J(\underline{e}_0^\pm(z)) = \underline{e}_\infty^\pm(\rho_c(z)) \begin{pmatrix} 1 & 0 \\ 0 & -1 \end{pmatrix}, J(\underline{e}_\infty^\pm(z)) = \underline{e}_0^\pm(\rho_c(z)) \begin{pmatrix} 1 & 0 \\ 0 & -1 \end{pmatrix}. \tag{7.12}$$

(d) If the data in (c) and a P_{3D6}-TEJPA bundle in (7.9) for some $\varepsilon_1, \varepsilon_2 \in \{\pm 1\}$ are given, the 4-tuple of bases as in (c) for this P_{3D6}-TEJPA bundle is

$$\varepsilon_0(\underline{e}_0^{\pm 1}, \varepsilon_1 \underline{e}_0^{\pm 2}), \varepsilon_0 \varepsilon_2(\underline{e}_\infty^{\pm 1}, \varepsilon_1 \underline{e}_\infty^{\pm 2}) \quad \text{with arbitrary } \varepsilon_0 \in \{\pm 1\}. \tag{7.13}$$

Thus the four possible enrichments to P_{3D6}-TEJPA bundles of a P_{3D6}-TEP bundle correspond to the four equivalence classes which are obtained from the eight 4-tuples in Theorem 6.3 (e) modulo a global sign.

(e) If the P_{3D6} bundle which underlies a P_{3D6}-TEJPA bundle is completely reducible, then the automorphism $G \in \text{Aut}(H, \nabla, P)$ from Theorem 6.3 (f) maps A and J to

$$\begin{aligned} -A \text{ and } J \text{ in the case } (C_{11}), (C_{22}), \\ -A \text{ and } -J \text{ in the case } (C_{12}), (C_{21}). \end{aligned} \tag{7.14}$$

Proof

(a) Let $(H, \nabla, u_0^1, u_\infty^1, P)$ be a P_{3D6}-TEP bundle. First we shall prove (a)($\Rightarrow$), then (a)($\Leftarrow$), then the uniqueness of A up to sign.

(a)($\Rightarrow$): The bundle $j^*(H, \nabla, P)$ is almost a P_{3D6}-TEP bundle. All that is missing is the choice of one of the eigenvalues $\{-u_0^1, -u_0^2\}$ at 0, and one of the eigenvalues $\{-u_\infty^1, -u_\infty^2\}$ at ∞. If an isomorphism $A : (H, \nabla, P) \to j^*(H, \nabla, P)$ exists then $\{u_0^1, u_0^2\} = \{-u_0^1, -u_0^2\}$ and $\{u_\infty^1, u_\infty^2\} = \{-u_\infty^1, -u_\infty^2\}$. Since $u_0^1 \neq u_0^2$ and $u_\infty^1 \neq u_\infty^2$, this implies $u_0^1 + u_0^2 = 0$ and $u_\infty^1 + u_\infty^2 = 0$, hence (7.8).

(a)($\Leftarrow$): Suppose that (7.8) holds. Then $(j^*H, j^*\nabla, u_0^1, u_\infty^1, j^*P)$ is a P_{3D6}-TEP bundle. Let us use a tilde to indicate the data of its P_{3D6} monodromy tuple:

$$\widetilde{I}_0^\pm = j^*I_0^\mp, \quad \widetilde{I}_\infty^\pm = j^*I_\infty^\mp, \quad \widetilde{I}_0^{a/b} = j^*I_0^{b/a}, \quad \widetilde{I}_\infty^{a/b} = j^*I_\infty^{b/a},$$
$$\widetilde{L} = j^*L, \quad \widetilde{L}_0^{\pm k} = j^*L_0^{\mp 3-k}, \quad \widetilde{L}_\infty^{\pm k} = j^*L_\infty^{\mp 3-k}. \tag{7.15}$$

Let $\underline{e}_0^\pm, \underline{e}_\infty^\pm$ be one of the eight 4-tuples of bases in Theorem 6.3 (e) for the P_{3D6}-TEP bundle $(H, \nabla, u_0^1, u_\infty^1, P)$.

Claim The 4-tuple of bases

$$\widetilde{\underline{e}}_0^\pm(z) := \underline{e}_0^\mp(-z) \begin{pmatrix} 0 & -1 \\ 1 & 0 \end{pmatrix}, \quad \widetilde{\underline{e}}_\infty^\pm(z) := \underline{e}_\infty^\mp(-z) \begin{pmatrix} 0 & -1 \\ 1 & 0 \end{pmatrix} \tag{7.16}$$

is one of the eight 4-tuples in Theorem 6.3 (e) for the P_{3D6}-TEP bundle $(j^*H, j^*\nabla, u_0^1, u_\infty^1, j^*P)$, and

$$\widetilde{s} = s, \quad \widetilde{B}(\beta) = B(\beta). \tag{7.17}$$

The claim implies that there exists an isomorphism $A : (H, \nabla, P) \to j^*(H, \nabla, P)$ with

$$A(\underline{e}_0^\pm(z)) = \underline{e}_0^\mp(-z) \begin{pmatrix} 0 & -1 \\ 1 & 0 \end{pmatrix}, \quad A(\underline{e}_\infty^\pm(z)) := \underline{e}_\infty^\mp(-z) \begin{pmatrix} 0 & -1 \\ 1 & 0 \end{pmatrix}. \tag{7.18}$$

Then (6.10) for the new and old 4-tuples shows that $P(A\,a, A\,b) = P(a, b)$, and $A^2 = -\mathrm{id}$ follows from (7.18). It remains to prove the claim.

Proof of the Claim: Equation (6.9) for $\widetilde{\underline{e}}_0^\pm$ follows from

$$\widetilde{\underline{e}}_0^{+1}|_{I_0^a} = j^*\underline{e}_0^{-2}|_{j^*I_0^b} = j^*\underline{e}_0^{+2}|_{j^*I_0^b} = \widetilde{\underline{e}}_0^{-1}|_{I_0^a},$$
$$\widetilde{\underline{e}}_0^{+2}|_{I_0^b} = -j^*\underline{e}_0^{-1}|_{j^*I_0^a} = -j^*\underline{e}_0^{+1}|_{j^*I_0^a} = \widetilde{\underline{e}}_0^{-2}|_{I_0^b}$$

and the analogue at ∞ is similar. Next, (6.10) for $\widetilde{\underline{e}}_0^\pm$ follows from

$$P((\widetilde{\underline{e}}_0^\pm)^t(z), \widetilde{\underline{e}}_0^\mp(-z)) = \begin{pmatrix} 0 & -1 \\ 1 & 0 \end{pmatrix}^t P((\underline{e}_0^\mp)^t(-z), \underline{e}_0^\pm(z)) \begin{pmatrix} 0 & -1 \\ 1 & 0 \end{pmatrix}$$
$$= \begin{pmatrix} 0 & -1 \\ 1 & 0 \end{pmatrix}^t 1_2 \begin{pmatrix} 0 & -1 \\ 1 & 0 \end{pmatrix} = 1_2$$

and the analogue at ∞ is similar. The following calculation shows $\widetilde{B}(\beta) = B(\beta)$ (and thus $\det \widetilde{B}(\beta) = 1$, which is (6.11)). Choose $z \in \widetilde{I}_0^a$ and β as in (2.23) and define $y := ze^\beta$. Then z and y satisfy

$$-z \in \widehat{I}_0^b, \ y \in \widehat{I}_\infty^a, \ -y = -ze^\beta \in \widehat{I}_\infty^b.$$

Recall (2.21):

$$(\widetilde{e}_0^+(-z) \text{ extended along } [\beta] \text{ to } (-y)) \, \widetilde{B}(\beta)$$

$$= \widetilde{e}_\infty^-(-y) = \underline{e}_\infty^+(y) \begin{pmatrix} 0 & -1 \\ 1 & 0 \end{pmatrix}$$

$$= \underline{e}_\infty^-(y) \, (S_\infty^a)^{-1} \begin{pmatrix} 0 & -1 \\ 1 & 0 \end{pmatrix}$$

$$= (\underline{e}_0^+(z) \text{ extended along } [\beta] \text{ to } y) \, B(\beta)(S_\infty^a)^{-1} \begin{pmatrix} 0 & -1 \\ 1 & 0 \end{pmatrix}$$

$$= (\underline{e}_0^-(z) \text{ extended along } [\beta] \text{ to } y) \, (S_0^a)^{-1} B(\beta)(S_\infty^a)^{-1} \begin{pmatrix} 0 & -1 \\ 1 & 0 \end{pmatrix}$$

$$= (\widetilde{e}_0^+(-z) \text{ extended along } [\beta] \text{ to } (-y))$$

$$\times \begin{pmatrix} 0 & -1 \\ 1 & 0 \end{pmatrix}^{-1} (S_0^a)^{-1} B(\beta)(S_\infty^a)^{-1} \begin{pmatrix} 0 & -1 \\ 1 & 0 \end{pmatrix}$$

$$= (\widetilde{e}_0^+(-z) \text{ extended along } [\beta] \text{ to } (-y)) \, B(\beta).$$

The last equality uses (6.13) and Lemma 5.2 (a). The facts proved up to now show that $\widetilde{e}_0^\pm, \widetilde{e}_\infty^\pm$ is one of the eight 4-tuples from Theorem 6.3 (e) for $(j^*H, j^*\nabla, u_0^1, u_\infty^1, j^*P)$. Therefore $\widetilde{s}_0^a = \widetilde{s}_0^b = -\widetilde{s}_\infty^a = -\widetilde{s}_\infty^b =: \widetilde{s}$ holds. Finally $\widetilde{s} = s$ follows with $z \in \widehat{I}_0^a$ and $-z \in \widehat{I}_0^b$ from the following calculation:

$$\widetilde{e}_0^-(-z) = \underline{e}_0^+(z) \begin{pmatrix} 0 & -1 \\ 1 & 0 \end{pmatrix} = \underline{e}_0^-(z)(S_0^a)^{-1} \begin{pmatrix} 0 & -1 \\ 1 & 0 \end{pmatrix}$$

$$= \widetilde{e}_0^+(-z) \begin{pmatrix} 0 & -1 \\ 1 & 0 \end{pmatrix}^{-1} (S_0^a)^{-1} \begin{pmatrix} 0 & -1 \\ 1 & 0 \end{pmatrix}$$

$$= \widetilde{e}_0^+(-z) \begin{pmatrix} 1 & 0 \\ s & 1 \end{pmatrix}.$$

The claim is proved. $\hfill (\square)$

The uniqueness of A up to a sign: Suppose $A^{(1)}$ is a second isomorphism which gives a TEPA bundle. Then $A^{(1)} \circ A \in \mathrm{Aut}(H, \nabla, P)$. If (H, ∇) is irreducible then $\mathrm{Aut}(H, \nabla, P) = \{\pm \mathrm{id}\}$ by Theorem 6.3 (f), and $A^{(1)} = \pm A^{-1} = \mp A$. If (H, ∇) is completely reducible then $\mathrm{Aut}(H, \nabla, P) = \{\pm \mathrm{id}, \pm G\}$ with G as in Theorem 6.3

(f). But then

$$A \circ G = -G \circ A, \quad \text{thus } (G \circ A^{-1})^2 = (G \circ A)^2 = \text{id} \neq -\text{id},$$

hence $A^{(1)} \circ A = \pm G$ is impossible.

(b) The proof is similar to the proof of (a). First the existence of J is shown, then the uniqueness up to a sign. Observe that

$$c = \frac{u_\infty^1}{u_0^1} = \frac{-u_\infty^1}{-u_0^1} = \frac{u_\infty^2}{u_0^2}, \quad \frac{u_0^j}{z} = u_\infty^j \, \rho_c(z).$$

Therefore $(\rho_c^* H, \rho_c^* \nabla, u_0^1, u_\infty^1, \rho_c^* P)$ is a P_{3D6}-TEP bundle. The data of its P_{3D6} monodromy tuple are denoted with a tilde.

$$I_0^\pm = \rho_c^* I_\infty^\pm, \; I_\infty^\pm = \rho_c^* I_0^\pm, \; I_0^{a/b} = \rho_c^* I_\infty^{a/b}, \; I_\infty^{a/b} = \rho_c^* I_0^{a/b},$$
$$\widetilde{L} = j^* L, \; \widetilde{L}_0^{\pm j} = \rho_c^* L_\infty^{\pm j}, \; \widetilde{L}_\infty^{\pm j} = \rho_c^* L_0^{\pm j}. \tag{7.19}$$

Let $e_0^\pm, e_\infty^\pm$ be one of the eight 4-tuples of bases in Theorem 6.3 (e) for the P_{3D6}-TEP bundle $(H, \nabla, u_0^1, u_\infty^1, P)$, which satisfies additionally (7.18).

Claim The 4-tuple of bases

$$\widetilde{e}_0^\pm(z) := e_\infty^\pm(\rho_c(z)) \begin{pmatrix} 1 & 0 \\ 0 & -1 \end{pmatrix}, \; \widetilde{e}_\infty^\pm(z) := e_0^\pm(\rho_c(z)) \begin{pmatrix} 1 & 0 \\ 0 & -1 \end{pmatrix} \tag{7.20}$$

is one of the eight 4-tuples in Theorem 6.3 (e) for the P_{3D6}-TEP bundle $(\rho_c^* H, \rho_c^* \nabla, u_0^1, u_\infty^1, \rho_c^* P)$, and

$$\widetilde{s} = s, \quad \widetilde{B}(\beta) = B(\beta). \tag{7.21}$$

Proof of the Claim: Equation (6.9) for $\widetilde{e}_0^\pm$ follows from

$$\widetilde{e}_0^{+1}|_{I_0^a} = \rho_c^* e_\infty^{+1}|_{\rho_c^* I_\infty^a} = \rho_c^* e_\infty^{-1}|_{\rho_c^* I_\infty^a} = \widetilde{e}_0^{-1}|_{I_0^a},$$
$$\widetilde{e}_0^{+2}|_{I_0^b} = -\rho_c^* e_\infty^{+2}|_{\rho_c^* I_\infty^b} = -\rho_c^* e_\infty^{-2}|_{\rho_c^* I_\infty^b} = \widetilde{e}_0^{-2}|_{I_0^b}$$

and the analogue at ∞ is similar. Next, (6.10) for $\widetilde{e}_0^\pm$ follows from

$$P((\widetilde{e}_0^\pm)^t(z), \widetilde{e}_0^\mp(-z)) = \begin{pmatrix} 1 & 0 \\ 0 & -1 \end{pmatrix}^t P((e_\infty^\pm)^t(\rho_c(z)), e_\infty^\mp(-\rho_c(z))) \begin{pmatrix} 1 & 0 \\ 0 & -1 \end{pmatrix}$$

$$= \begin{pmatrix} 1 & 0 \\ 0 & -1 \end{pmatrix}^t 1_2 \begin{pmatrix} 1 & 0 \\ 0 & -1 \end{pmatrix} = 1_2$$

and the analogue at ∞ is similar. The following calculation shows $\widetilde{B}(\beta) = B(\beta)$ (and thus $\det \widetilde{B}(\beta) = 1$, which is (6.11)). Choose $y \in \widehat{I}_0^a$ and β as in (2.23) and define $z := ye^\beta$. Then

$$z \in \widehat{I}_\infty^a, \quad \rho_c(y) = \rho_c(z)e^\beta, \quad \rho_c(z) \in \widehat{I}_0^a, \quad \rho_c(y) \in \widehat{I}_\infty^a.$$

Recall (2.21). The fifth equality in the following calculation uses the fact that ρ_c inverts the path $[\beta]$. The last equality uses (6.13) and Lemma 5.2 (a).

$$(\widetilde{\underline{e}}_0^+ (\rho_c(z)) \text{ extended along } [\beta] \text{ to } \rho_c(y)) \, \widetilde{B}(\beta)$$

$$= \widetilde{\underline{e}}_\infty^- (\rho_c(y)) = \underline{e}_0^- (y) \begin{pmatrix} 1 & 0 \\ 0 & -1 \end{pmatrix}$$

$$= \underline{e}_0^+ (y) S_0^a \begin{pmatrix} 1 & 0 \\ 0 & -1 \end{pmatrix}$$

$$= (\underline{e}_\infty^- (z) \text{ extended along } [-\beta] \text{ to } y) B(\beta)^{-1} S_0^a \begin{pmatrix} 1 & 0 \\ 0 & -1 \end{pmatrix}$$

$$= (\underline{e}_\infty^+ (z) \text{ extended along } [-\beta] \text{ to } y) S_\infty^a B(\beta)^{-1} S_0^a \begin{pmatrix} 1 & 0 \\ 0 & -1 \end{pmatrix}$$

$$= (\rho_c^* (\underline{e}_\infty^+ (z)) \text{ extended along } [\beta] \text{ to } \rho_c(y))$$
$$\times S_\infty^a B(\beta)^{-1} S_0^a \begin{pmatrix} 1 & 0 \\ 0 & -1 \end{pmatrix}$$

$$= (\widetilde{\underline{e}}_0^+ (\rho_c(z)) \text{ extended along } [\beta] \text{ to } \rho_c(y))$$
$$\times \begin{pmatrix} 1 & 0 \\ 0 & -1 \end{pmatrix} S_\infty^a B(\beta)^{-1} S_0^a \begin{pmatrix} 1 & 0 \\ 0 & -1 \end{pmatrix}$$

$$= (\widetilde{\underline{e}}_0^+ (\rho_c(z)) \text{ extended along } [\beta] \text{ to } \rho_c(y)) B(\beta).$$

The facts proved up to now show that $\widetilde{\underline{e}}_0^\pm, \widetilde{\underline{e}}_\infty^\pm$ is one of the eight 4-tuples from Theorem 6.3 (e) for $(\rho_c^* H, \rho_c^* \nabla, u_0^1, u_\infty^1, \rho_c^* P)$. Therefore $\widetilde{s}_0^a = \widetilde{s}_0^b = -\widetilde{s}_\infty^a = -\widetilde{s}_\infty^b =: \widetilde{s}$ holds. Finally $\widetilde{s} = s$ follows with $z \in \widehat{I}_0^a$ and $\rho_c(z) \in \widehat{I}_\infty^a$ from the calculation

$$\widetilde{\underline{e}}_\infty^- (\rho_c(z)) = \underline{e}_0^- (z) \begin{pmatrix} 1 & 0 \\ 0 & -1 \end{pmatrix} = \underline{e}_0^+ (z) S_0^a \begin{pmatrix} 1 & 0 \\ 0 & -1 \end{pmatrix}$$

$$= \widetilde{\underline{e}}_\infty^+ (\rho_c(z)) \begin{pmatrix} 1 & 0 \\ 0 & -1 \end{pmatrix}^{-1} S_0^a \begin{pmatrix} 1 & 0 \\ 0 & -1 \end{pmatrix}$$

$$= \widetilde{\underline{e}}_\infty^+ (\rho_c(z)) \begin{pmatrix} 1 & -s \\ 0 & 1 \end{pmatrix}.$$

The claim is proved. $\hspace{1cm} (\Box)$

The claim implies that an isomorphism $J : (H, \nabla, P) \to \rho_c^*(H, \nabla, P)$ exists with

$$J(\underline{e}_0^{\pm}(z)) = \underline{e}_{\infty}^{\pm}(\rho_c(z)) \begin{pmatrix} 1 & 0 \\ 0 & -1 \end{pmatrix}, \ J(\underline{e}_{\infty}^{\pm}(z)) = \underline{e}_0^{\pm}(\rho_c(z)) \begin{pmatrix} 1 & 0 \\ 0 & -1 \end{pmatrix}. \tag{7.22}$$

Equation (6.10) for the new and the old 4-tuple shows $P(J a, J b) = P(a, b)$. Equation (7.22) implies $J^2 = \mathrm{id}$, and $A \circ J = -J \circ A$ follows from (7.22) and (7.18).

The uniqueness of J up to a sign: Suppose $J^{(1)}$ is a second isomorphism which gives a TEJPA structure. Then $J^{(1)} \circ J \in \mathrm{Aut}(H, \nabla, P)$. If (H, ∇) is irreducible then $\mathrm{Aut}(H, \nabla, P) = \{\pm\mathrm{id}\}$ by Theorem 6.3 (f), and $J^{(1)} = \pm J^{-1} = \pm J$. If (H, ∇) is completely reducible then $\mathrm{Aut}(H, \nabla, P) = \{\pm\mathrm{id}, \pm G\}$ with G as in Theorem 6.3 (f). But then $A \circ G = -G \circ A$. Then $J^{(1)} = \pm G \circ J^{-1} = \pm G \circ J$ is impossible because

$$A \circ (G \circ J) = -G \circ A \circ J = (G \circ J) \circ A, \ \text{but } A \circ J^{(1)} = -J^{(1)} \circ A.$$

(c) The existence of a 4-tuple with all properties in (c) follows from the constructions in (a) and (b). $\varepsilon_0(e_1^{\pm}, \varepsilon_1 e_2^{\pm})$ is determined up to the signs $\varepsilon_0, \varepsilon_1 \in \{\pm 1\}$ by (6.9) and $P((e^+)^t, e^-) = \mathbf{1}_2$. Then $A(\underline{e}_0^+) = \underline{e}_0^- \begin{pmatrix} 0 & -1 \\ 1 & 0 \end{pmatrix}$ fixes the sign $\varepsilon_1 = 1$. $J(\underline{e}_0^+) = \underline{e}_{\infty}^+ \begin{pmatrix} 1 & 0 \\ 0 & -1 \end{pmatrix}$ determines $\underline{e}_{\infty}^+$. Equation (6.9) determines $\underline{e}_{\infty}^-$.

(d) This follows from the uniqueness in (c), from (7.9), and from (7.10).

(e) This follows from (d) and Theorem 6.3 (f). $\square$

Remarks 7.4

(i) Given a P_{3D6}-TEP bundle, a J which enriches it to a P_{3D6}-TEJP bundle is unique up to a sign in the irreducible cases and in the completely reducible case $(C_{12}), (C_{21})$, because $\mathrm{Aut}(H, \nabla, P) = \{\pm\mathrm{id}\}$ in the irreducible cases and $G \circ J = -J \circ G$ in the case $(C_{12}), (C_{21})$, so then $(G \circ J)^2 = -\mathrm{id}$ and $G \circ J$ is not a candidate for another J. But in the completely reducible case $(C_{11}), (C_{21})$, we have $G \circ J = J \circ G$ by (7.13), and $\pm J, \pm G \circ J$ give four possible enrichments to a P_{3D6}-TEJP bundle.

(ii) This apparent asymmetry between the two completely reducible cases is resolved by considering also $\widetilde{J}$ as in Remark 7.2 (i). But the asymmetry between the reducible and the irreducible cases is not resolvable. It leads to the two A_1-singularities in the moduli space M_{3T}^{mon} in Theorem 7.6

Now we turn to moduli spaces of P_{3D6}-TEJPA bundles and P_{3D6}-TEP bundles. Using the distinguished 4-tuples of bases in Theorem 7.3 (c) and their P_{3D6} numerical tuples, we shall see that these moduli spaces are quotients of algebraic manifolds and come equipped with foliations, in a way which makes the isomonodromic families transparent.

We introduce now the fundamental moduli spaces

$$M_{3TJ} = \{\text{isomorphism classes of } P_{3D6}\text{-TEJPA bundles}\},$$

$$M_{3T} = \{\text{isomorphism classes of } P_{3D6}\text{-TEP bundles}\},$$

and we denote by pr^u the natural projections

$$pr^u : M_{3TJ} \to \mathbb{C}^* \times \mathbb{C}^*, (H, \nabla, u_0^1, u_\infty^1, P, A, J) \mapsto (u_0^1, u_\infty^1),$$
$$pr^u : M_{3T} \to \mathbb{C}^* \times \mathbb{C}^*, (H, \nabla, u_0^1, u_\infty^1, P) \mapsto (u_0^1, u_\infty^1),$$

(7.23)

with fibres $M_{3TJ}(u_0^1, u_\infty^1)$, $M_{3T}(u_0^1, u_\infty^1)$, respectively.

Consider the covering

$$c^{path} : \mathbb{C} \times \mathbb{C}^* \to \mathbb{C}^* \times \mathbb{C}^*, (\beta, u_0^1) \mapsto (u_0^1, \tfrac{1}{4}e^{-\beta}/u_0^1).$$

(7.24)

It is the covering in which the triples $(\beta, u_0^1, u_\infty^1)$ with $e^{-\beta} = 4u_0^1 u_\infty^1$ live, which is (2.23) in the case $u_0^2 = -u_0^1, u_\infty^2 = -u_\infty^1$. Recall for $s \in \mathbb{C}$ the definition in (5.1) of the matrices

$$S(s) = \begin{pmatrix} 1 & s \\ 0 & 1 \end{pmatrix}, \quad \text{Mon}_0^{mat}(s) = S^t S^{-1} = \begin{pmatrix} 1 & -s \\ s & 1-s^2 \end{pmatrix}.$$

Define the affine algebraic surface

$$V^{mon} := \{(s, b_1, b_2) \in \mathbb{C}^3 \mid b_1^2 + b_2^2 + s b_1 b_2 = 1\}$$

(7.25)

and the canonically isomorphic space

$$V^{mat} = \{(s, B) \in \mathbb{C} \times SL(2, \mathbb{C}) \mid B = \begin{pmatrix} b_1 & b_2 \\ -b_2 & b_1 + s b_2 \end{pmatrix}\}.$$

(7.26)

Theorem 7.5

(a) V^{mon} and V^{mat} are smooth.

(b) The map

$$\Phi^{path} : (c^{path})^* M_{3TJ} \to \mathbb{C} \times \mathbb{C}^* \times V^{mat}$$

(7.27)

$$((H, \nabla, u_0^1, u_\infty^1, P, A, J), \beta) \mapsto (\beta, u_0^1, s, B(\beta))$$

is a bijection. Here s and $B(\beta)$ are the data from (6.13) which are associated to the distinguished 4-tuple of bases $e_0^\pm, e_\infty^\pm$ in Theorem 7.3 (c) for the P_{3D6}-TEJPA bundle $(H, \nabla, u_0^1, u_\infty^1, P, A, J)$.

Φ^{path} equips the set $(c^{path})^* M_{3TJ}$ with the structure of an affine algebraic manifold, which is denoted by $((c^{path})^* M_{3TJ})^{mon}$.

(c) Φ^{path} maps the isomorphism classes of P_{3D6}-TEJPA bundles in $(c^{path})^* M_{3TJ}$ to the orbits of the action of the group $\mathbb{Z}$ on $\mathbb{C} \times \mathbb{C}^* \times V^{mat}$, whose generator $[1]$ acts by the algebraic automorphism

$$m_{[1]} : (\beta, u_0^1, s, B) \mapsto (\beta + 2\pi i, u_0^1, s, (\text{Mon}_0^{mat})^{-1} B).$$

(7.28)

The quotient $\mathbb{C} \times \mathbb{C}^* \times V^{mat}/\langle m_{[1]} \rangle$ *is an analytic manifold.* Φ^{path} *induces a bijection*

$$\Phi_{3TJ} : M_{3TJ} \rightarrow \mathbb{C} \times \mathbb{C}^* \times V^{mat}/\langle m_{[1]} \rangle. \tag{7.29}$$

This equips M_{3TJ} *with the structure of an analytic manifold, which is denoted by* M_{3TJ}^{mon}. *The projection* $pr^u : M_{3TJ}^{mon} \rightarrow \mathbb{C}^* \times \mathbb{C}^*$ *is an analytic morphism, the fibres are algebraic manifolds* $M_{3TJ}^{mon}(u_0^1, u_\infty^1)$, *isomorphic to* V^{mat}. *Any choice of* β *(with (2.23)) induces an isomorphism* $M_{3TJ}^{mon}(u_0^1, u_\infty^1) \rightarrow V^{mat}$.

(d) *The trivial foliation on* $\mathbb{C} \times \mathbb{C}^* \times V^{mat}$ *with leaves* $\mathbb{C} \times \mathbb{C}^* \times \{(s, B)\}$ *induces foliations on* $\mathbb{C} \times \mathbb{C}^* \times V^{mat}/\langle m_{[1]} \rangle$ *and on* M_{3TJ}^{mon}.

   A family of P_{3D6}-*TEJPA bundles over a (complex analytic) base manifold* T *is isomonodromic if and only if the induced holomorphic map* $T \rightarrow M_{3TJ}^{mon}$ *takes values in one leaf.*

(e) *A leaf in* M_{3TJ}^{mon} *contains only finitely many branches over* $\mathbb{C}^* \times \mathbb{C}^*$ *if and only if* Mon_0^{mat} *has finite order.*

Proof

(a) The three partial derivatives of $g := b_1^2 + b_2^2 + sb_1b_2$,

$$\frac{\partial g}{\partial b_1} = 2b_1 + sb_2, \quad \frac{\partial g}{\partial b_2} = 2b_2 + sb_1, \quad \frac{\partial g}{\partial s} = -b_1b_2,$$

   do not vanish simultaneously on V^{mon}.

(b) The map Φ^{path} is injective: the data $(\beta, u_0^1, s, B(\beta))$ contain a P_{3D6} numerical tuple of the P_{3D6} bundle and determine P, A and J by (7.10), and they contain β.

   The map Φ^{path} is surjective: setting $\alpha_0^j = \alpha_\infty^j = 0$, the data (β, u_0^1, s, B) satisfy condition (2.26) of a P_{3D6} numerical tuple by Lemma 5.2 (a), so they define a P_{3D6}-bundle. In order to see that (7.10) defines a P_{3D6}-TEJPA bundle one has to go through the constructions of P, A and J in the proofs of Theorem 6.3 (a) and Theorem 7.3 (a) and (b).

(c) The first statement including (7.28) follows from (2.28). The quotient $\mathbb{C} \times \mathbb{C}^* \times V^{mat}$ is not an algebraic manifold, because $\langle m_{[1]} \rangle \cong \mathbb{Z}$ does not have finite order. But it is an analytic manifold. The rest is clear.

(d) A family of P_{3D6}-TEJPA bundles is isomonodromic if the underlying family of P_{3D6}-TEP bundles is isomonodromic and if P, A and J are flat (in the parameters). Then the (up to a global sign) distinguished 4-tuple of bases $\underline{e}_0^\pm, \underline{e}_\infty^\pm$ varies flatly. As this 4-tuple defines s and $B(\beta)$ and the underlying family of P_{3D6} bundles is isomonodromic, s and $B(\beta)$ are constant.

   Conversely, a family over one leaf is an isomonodromic family of P_{3D6} bundles by the discussion in Chap. 4, and the bases $\underline{e}_0^\pm, \underline{e}_\infty^\pm$ used for its construction vary flatly. By (7.10) P, A and J are flat.

(e) This is obvious. $\qquad\qquad\qquad\qquad\qquad\qquad\qquad\qquad\qquad\qquad\qquad\qquad\qquad\qquad$ $\square$

Define algebraic automorphisms R_1, R_2 and R_3 of $\mathbb{C} \times \mathbb{C}^* \times V^{mon}$ by

$$R_1 : (\beta, u_0^1, s, b_1, b_2) \mapsto (\beta, u_0^1, -s, b_1, -b_2)$$

$$R_2 : (\beta, u_0^1, s, b_1, b_2) \mapsto (\beta, u_0^1, s, -b_1, -b_2) \tag{7.30}$$

$$R_3 = R_1 \circ R_2 : (\beta, u_0^1, s, b_1, b_2) \mapsto (\beta, u_0^1, -s, -b_1, b_2).$$

Then $G^{mon} := \{id, R_1, R_2, R_3\}$ is a group isomorphic to $\mathbb{Z}_2 \times \mathbb{Z}_2$. The action on $\mathbb{C} \times \mathbb{C}^* \times V^{mat}$ commutes with $m_{[1]}$ and respects the fibres of the projection to $\mathbb{C} \times \mathbb{C}^*$. Denote the induced automorphisms on $\mathbb{C} \times \mathbb{C}^* \times V^{mat}/\langle m_{[1]}\rangle$, V^{mat}, $((c^{path})^* M_{3TJ})^{mon}$, M_{3TJ}^{mon} and on the fibres $M_{3TJ}^{mon}(u_0^1, u_\infty^1)$ of pr^u also by R_1, R_2 and R_3.

Theorem 7.6

(a) *Consider the action of G^{mon} on V^{mon}. The orbits of the four points $(0, \pm 1, 0)$ and $(0, 0, \pm 1)$ have length 2, all other orbits have length 4. The quotient is an affine algebraic variety with two A_1-singularities at the orbits of $(0, \pm 1, 0)$ and $(0, 0, \pm 1)$ and smooth elsewhere. With $(y_1, y_2, y_3) = (s^2, b_1^2, b_2^2)$ it is*

$$V^{mon}/G^{mon} \cong \{(y_1, y_2, y_3) \in \mathbb{C}^3 \mid y_1 y_2 y_3 - (y_2 + y_3 - 1)^2 = 0\}. \tag{7.31}$$

(b) *The set of isomorphism classes of P_{3D6}-TEP bundles M_{3T} is M_{3TJ}/G^{mon}. The natural bijections*

$$\Phi_{3T}^{path} : (c^{path})^* M_{3T} \to \mathbb{C} \times \mathbb{C}^* \times V^{mon}/G^{mon},$$

$$\Phi_{3T} : M_{3T} \to \mathbb{C} \times \mathbb{C}^* \times V^{mon}/G^{mon} \times \langle m_{[1]}\rangle$$

induces on $(c^{path})^* M_{3T}$ the structure of an algebraic variety, denoted by $((c^{path})^* M_{3T})^{mon}$, and on M_{3T} the structure of an analytic variety, denoted by M_{3T}^{mon}, with

$$M_{3T}^{mon} = M_{3TJ}^{mon}/G^{mon}. \tag{7.32}$$

The two completely reducible P_{3D6}-TEP bundles in $M_{3T}(u_0^1, u_\infty^1)$ lie at the two A_1-singularities of $M_{3T}^{mon}(u_0^1, u_\infty^1) = M_{3TJ}^{mon}(u_0^1, u_\infty^1)/G^{mon}$.

Proof

(a) As G^{mon} is finite and V^{mon} is an affine algebraic variety, the quotient is a geometric quotient and an affine algebraic variety. The statement on the lengths of the orbits is easily checked. It implies immediately that the quotient of the smooth V^{mon} has two A_1-singularities and is smooth elsewhere.

It remains to calculate equation(s) for the quotient. With $(y_1, y_2, y_3, y_4) = (s^2, b_1^2, b_2^2, sb_1b_2)$ one obtains

$$\mathrm{Spec}(V^{mon}) = \mathbb{C}[s, b_1, b_2]/(b_1^2 + b_2^2 + sb_1b_2 - 1),$$

$$\mathrm{Spec}(V^{mon}/G^{mon}) = (\mathrm{Spec}(V^{mon}))^{G^{mon}}$$

$$= (\text{image of } \mathbb{C}[y_1, y_2, y_3, y_4] \text{ in } \mathrm{Spec}(V^{mon}))$$

$$= \mathbb{C}[y_1, y_2, y_3, y_4]/(y_2 + y_3 + y_4 - 1, y_1y_2y_3 - y_4^2)$$

$$= \mathbb{C}[y_1, y_2, y_3]/(y_1y_2y_3 - (y_2 + y_3 - 1)^2).$$

(b) Consider a P_{3D6}-TEJPA bundle $(H, \nabla, u_0^1, u_\infty^1, P, A, J)$ and, for some β, its image $(\beta, u_0^1, s, b_1, b_2)$ in $\mathbb{C} \times \mathbb{C}^* \times V^{mon}$ under Φ^{path}. By Theorem 7.3 (b) there are altogether four enrichments of the underlying P_{3D6}-TEP bundle to P_{3D6}-TEJPA bundles, they differ only by signs $\varepsilon_1, \varepsilon_2 \in \{\pm 1\}$ and are listed in (7.9). Their distinguished 4-tuples of bases from Theorem 7.3 (c) are listed in (7.13). From this one reads off that their images under Φ^{path} in $\mathbb{C} \times \mathbb{C}^* \times V^{mon}$ are $(\beta, u_0^1, \varepsilon_1 s, \varepsilon_2 b_1, \varepsilon_1 \varepsilon_2 b_2)$. This shows that $M_{3T} = M_{3TJ}/G^{mon}$.

The four enrichments of a P_{3D6}-TEP bundle with irreducible P_{3D6}-bundle are non-isomorphic because $\mathrm{Aut}(H, \nabla, P) = \{\pm \mathrm{id}\} = \mathrm{Aut}(H, \nabla, P, A, J)$.

In the completely reducible cases the four enrichments form pairs of isomorphic enrichments by Theorem 7.3 (e). Therefore in each space $M_{3T}^{mon}(u_0^1, u_\infty^1)$ the completely reducible P_{3D6}-TEP bundles lie at the two A_1-singularities. □

Chapter 8
Normal Forms of P_{3D6}-TEJPA Bundles and Their Moduli Spaces

Any P_{3D6}-TEJPA bundle is pure or a $(1, -1)$-twistor (see Remark 4.1 (iv) for these notions). This follows from Theorems 4.2 and 6.3 (a). In this chapter we shall give an elementary independent proof. But our main purpose is to give normal forms. They allow us to classify P_{3D6}-TEJPA bundles in a new way. This induces a new structure as algebraic manifold M_{3TJ}^{ini} on the set M_{3TJ} of isomorphism classes of P_{3D6}-TEJPA bundles and a stratification into an open submanifold of pure bundles and two codimension 1 submanifolds of $(1, -1)$-twistors, though the two codimension 1 submanifolds intersect each subspace $M_{3TJ}^{ini}(u_0^1, u_\infty^1)$ in four codimension 1 submanifolds. $M_{3TJ}^{ini}(u_0^1, u_\infty^1)$ has four natural charts isomorphic to $\mathbb{C}^2$, each containing all pure bundles and one of the four families of $(1, -1)$-twistors.

Theorem 8.2 will consider $M_{3TJ}(u_0^1, u_\infty^1)$ for a fixed pair (u_0^1, u_∞^1). Theorem 8.4 will consider M_{3TJ}. The proofs use repeatedly some observations of a general nature. They are collected in the following remarks.

Remarks 8.1

(i) The proof of the existence and uniqueness of the normal forms in Theorem 8.2 is not so difficult. They are largely determined by the properties of the values at 0 and ∞ of the sections with which they are defined. And these properties can easily be rewritten using the correspondence in (2.19), in terms of properties of flat generating sections of the bundles $L_0^\pm, L_\infty^\pm$. This is made precise in (ii).

(ii) Consider a P_{3D6}-TEJPA bundle. Let $\underline{e}_0^\pm$ and $\underline{e}_\infty^\pm$ be the (up to a global sign, unique) 4-tuple of bases from Theorem 7.3 (c). By the correspondence in (2.19) they correspond to bases $\underline{v}_0$ of H_0 and $\underline{v}_\infty$ of H_∞. These bases satisfy

$$[z\nabla_{z\partial_z}]\underline{v}_0 = \underline{v}_0\, u_0^1 \begin{pmatrix} 1 & 0 \\ 0 & -1 \end{pmatrix}, \; [-\nabla_{\partial_z}]\underline{v}_\infty = \underline{v}_\infty\, u_\infty^1 \begin{pmatrix} 1 & 0 \\ 0 & -1 \end{pmatrix}, \tag{8.1}$$

$$P(\underline{v}_0^t, \underline{v}_0) = P(\underline{v}_\infty^t, \underline{v}_\infty) = \mathbf{1}_2, \tag{8.2}$$

© Springer International Publishing AG 2017
M.A. Guest, C. Hertling, *Painlevé III: A Case Study in the Geometry of Meromorphic Connections*, Lecture Notes in Mathematics 2198,
DOI 10.1007/978-3-319-66526-9_8

$$A(\underline{v_0}) = \underline{v_0} \begin{pmatrix} 0 & -1 \\ 1 & 0 \end{pmatrix}, \quad A(\underline{v_\infty}) = \underline{v_\infty} \begin{pmatrix} 0 & -1 \\ 1 & 0 \end{pmatrix}, \tag{8.3}$$

$$J(\underline{v_0}) = \underline{v_\infty} \begin{pmatrix} 1 & 0 \\ 0 & -1 \end{pmatrix}, \quad J(\underline{v_\infty}) = \underline{v_0} \begin{pmatrix} 1 & 0 \\ 0 & -1 \end{pmatrix}. \tag{8.4}$$

If any basis $\widetilde{\underline{v}}_0$ of H_0 satisfies the same equations (8.1) to (8.3) as $\underline{v_0}$, then it coincides up to a global sign with $\underline{v_0}$. An analogous statement holds for a basis $\widetilde{\underline{v}}_\infty$ of H_∞.

(iii) Consider the matrix

$$C := \begin{pmatrix} 1 & 1 \\ -i & i \end{pmatrix}, \quad \text{hence} \quad C^{-1} = \frac{1}{2}\begin{pmatrix} 1 & i \\ 1 & -i \end{pmatrix}. \tag{8.5}$$

It satisfies

$$C^{-1}\begin{pmatrix} 1 & 0 \\ 0 & -1 \end{pmatrix} C = \begin{pmatrix} 0 & 1 \\ 1 & 0 \end{pmatrix},$$

$$C^t \mathbf{1}_2 C = \begin{pmatrix} 0 & 2 \\ 2 & 0 \end{pmatrix}, \tag{8.6}$$

$$C^{-1}\begin{pmatrix} 0 & -1 \\ 1 & 0 \end{pmatrix} C = \begin{pmatrix} i & 0 \\ 0 & -i \end{pmatrix}.$$

Therefore the bases $\underline{w_0} := \underline{v_0}C$ and $\underline{w_\infty} := \underline{v_\infty}C$ satisfy

$$[z\nabla_{z\partial_z}]\underline{w_0} = \underline{w_0}\, u_0^1 \begin{pmatrix} 0 & 1 \\ 1 & 0 \end{pmatrix}, \, [-\nabla_{\partial_z}]\underline{w_\infty} = \underline{w_\infty}\, u_\infty^1 \begin{pmatrix} 0 & 1 \\ 1 & 0 \end{pmatrix}, \tag{8.7}$$

$$P(\underline{w_0^t}, \underline{w_0}) = P(\underline{w_\infty^t}, \underline{w_\infty}) = \begin{pmatrix} 0 & 2 \\ 2 & 0 \end{pmatrix}, \tag{8.8}$$

$$A(\underline{w_0}) = \underline{w_0} \begin{pmatrix} i & 0 \\ 0 & -i \end{pmatrix}, \quad A(\underline{w_\infty}) = \underline{w_\infty} \begin{pmatrix} i & 0 \\ 0 & -i \end{pmatrix}, \tag{8.9}$$

$$J(\underline{w_0}) = \underline{w_\infty} \begin{pmatrix} 0 & 1 \\ 1 & 0 \end{pmatrix}, \quad J(\underline{w_\infty}) = \underline{w_0} \begin{pmatrix} 0 & 1 \\ 1 & 0 \end{pmatrix}. \tag{8.10}$$

If any basis $\widetilde{\underline{w}}_0$ of H_0 satisfies the same equations (8.7) to (8.9) as $\underline{w_0}$, then it coincides up to a global sign with $\underline{w_0}$. An analogous statement holds for a basis $\widetilde{\underline{w}}_\infty$ of H_∞.

(iv) Let $(H, \nabla, u_0^1, u_\infty^1, P, A, J)$ be a P_{3D6}-TEJPA bundle. A and J act on the space $\Gamma(\mathbb{P}^1, \mathcal{O}(H))$ of global sections of H. Since $A^2 = -\,\mathrm{id}$ and $J^2 = \mathrm{id}$, they act semisimply with eigenvalues in $\{\pm i\}$ and $\{\pm 1\}$. Since $AJ = -JA$, they exchange eigenspaces. Therefore the four eigenspaces all have the same dimension, which is half the dimension of $\Gamma(\mathbb{P}^1, \mathcal{O}(H)$. Thus this dimension is even, and H is a $(k, -k)$-twistor with $k = 0$ or $k > 0$ and odd ($\deg H = 0$ by (3.1) and (6.7)). Theorem 8.2 (a) will show $k \in \{0, 1\}$.

If $\sigma \in \mathcal{O}(H)|_{\mathbb{C}^*}$ and $A(\sigma(z)) = \varepsilon_1 i \sigma(-z)$ with some $\varepsilon_1 \in \{\pm 1\}$, then for $k \in \mathbb{Z}$

$$A(z^k \sigma(z)) = (-1)^k \varepsilon_1 i (-z)^k \sigma(-z). \tag{8.11}$$

(v) Denote by $\mathcal{O}_{\mathbb{P}^1}(a, b)$ for $a, b \in \mathbb{Z}$ the sheaf of holomorphic functions on $\mathbb{P}^1$ with poles of order $\leq a$ at 0 and poles of order $\leq b$ at ∞.

Suppose that a P_{3D6}-TEJPA bundle $(H, \nabla, u_0^1, u_\infty^1, P, A, J)$ is a $(k, -k)$-twistor for some $k > 0$. One can choose sections $\sigma_1 \in \Gamma(\mathbb{P}^1, \mathcal{O}(H))$ and $\sigma_2 \in \Gamma(\mathbb{C}, \mathcal{O}(H))$ such that

$$\mathcal{O}(H) = \mathcal{O}_{\mathbb{P}^1}(0, k)\,\sigma_1 \oplus \mathcal{O}_{\mathbb{P}^1}(0, -k)\,\sigma_2.$$

Then $(\sigma_1(0), \sigma_2(0))$ is a basis of H_0, and for $a, b \in \mathbb{Z}$

$$\Gamma(\mathbb{P}^1, \mathcal{O}_{\mathbb{P}^1}(a, b)\,\mathcal{O}(H)) = \left(\mathbb{C}z^{-a}\sigma_1 \oplus \mathbb{C}z^{-a+1}\sigma_1 \oplus \cdots \oplus \mathbb{C}z^{k+b}\sigma_1\right)$$
$$\oplus \left(\mathbb{C}z^{-a}\sigma_2 \oplus \mathbb{C}z^{-a+1}\sigma_2 \cdots \oplus \mathbb{C}z^{-k+b}\sigma_2\right)$$

(the second sum is 0 if $-a > -k + b$). The section σ_1 is unique up to a scalar. Therefore it is an eigenvector of A, and

$$A(\sigma_1(z)) = \varepsilon_1 i \sigma_1(-z) \quad \text{for some unique } \varepsilon_1 \in \{\pm 1\}.$$

The section σ_2 is not unique. But it can also be chosen as an eigenvector of A, with

$$A(\sigma_2(z)) = -\varepsilon_1 i \sigma_2(-z).$$

The reason for this is that $\Gamma(\mathbb{P}^1, \mathcal{O}_{\mathbb{P}^1}(0, k)\,\mathcal{O}(H))$ is the sum of the two $k + 1$-dimensional eigenspaces of A with eigenvalues $\pm \varepsilon_1 i$,

$$\mathbb{C}\sigma_1 \oplus \mathbb{C}z^2\sigma_1 \oplus \cdots \oplus \mathbb{C}z^{2k}\sigma_1$$
$$\text{and } \mathbb{C}z\sigma_1 \oplus \mathbb{C}z^3\sigma_1 \oplus \cdots \oplus \mathbb{C}z^{2k-1}\sigma_1 \oplus \widetilde{\sigma}_2$$

for some section $\widetilde{\sigma}_2$. One can choose $\sigma_2 = \widetilde{\sigma}_2$.

(vi) Suppose that $(H, \nabla, u_0^1, u_\infty^1, P, A, J)$ is a P_{3D6}-TEJPA bundle, and $\underline{\sigma}$ is basis of $\mathcal{O}(H)_0$ with

$$A(\underline{\sigma}(z)) = \underline{\sigma}(-z) \begin{pmatrix} i & 0 \\ 0 & -i \end{pmatrix}.$$

Then (8.11) shows that the flatness $\nabla_{z\partial_z} \circ A = A \circ \nabla_{z\partial_z}$ of A is equivalent to

$$\nabla_{z\partial_z}\underline{\sigma}(z) = \underline{\sigma}(z)\frac{1}{z}\begin{pmatrix} a(z) & b(z) \\ c(z) & d(z) \end{pmatrix} \tag{8.12}$$

with $a(z), d(z) \in z\mathbb{C}\{z^2\}, b(z), c(z) \in \mathbb{C}\{z^2\}.$

Because of $P(A\,a, A\,b) = P(a, b)$ and $A^2 = -\,\mathrm{id}$, the eigenspaces of A are isotropic with respect to P. This isotropy, the symmetry of P, $P(A\,a, A\,b) = P(a, b)$ and $A(\underline{\sigma}(z)) = \underline{\sigma}(-z)\begin{pmatrix} i & 0 \\ 0 & -i \end{pmatrix}$ show that

$$P(\underline{\sigma}(z)^t, \underline{\sigma}(-z)) = e(z)\begin{pmatrix} 0 & 1 \\ 1 & 0 \end{pmatrix} \text{ with } e(z) \in \mathbb{C}\{z^2\}. \tag{8.13}$$

The flatness of P is equivalent to

$$\begin{aligned} z\partial_z e(z)\begin{pmatrix} 0 & 1 \\ 1 & 0 \end{pmatrix} &= P(\nabla_{z\partial_z}\underline{\sigma}(z)^t, \underline{\sigma}(-z)) + P(\underline{\sigma}(z)^t, \nabla_{z\partial_z}\underline{\sigma}(-z)) \\ &= \frac{1}{z}\begin{pmatrix} a(z) & c(z) \\ b(z) & d(z) \end{pmatrix}e(z)\begin{pmatrix} 0 & 1 \\ 1 & 0 \end{pmatrix} \\ &\quad + \frac{1}{-z}e(z)\begin{pmatrix} 0 & 1 \\ 1 & 0 \end{pmatrix}\begin{pmatrix} a(-z) & b(-z) \\ c(-z) & d(-z) \end{pmatrix} \\ &= \frac{1}{z}e(z)(a(z) + d(z))\begin{pmatrix} 0 & 1 \\ 1 & 0 \end{pmatrix}. \end{aligned} \tag{8.14}$$

Therefore $a(z) + d(z) = 0$ if and only if $z\partial_z e(z) = 0$.

Theorem 8.2

(a) Any P_{3D6}-TEJPA bundle is pure or a $(1, -1)$-twistor.

(b) Any pure P_{3D6}-TEJPA bundle has four normal forms (any of which can be chosen), and these are listed in (8.15) to (8.18). They are indexed by $(\varepsilon_1, \varepsilon_2) \in \{\pm 1\}^2$ or by $k \in \{0, 1, 2, 3\}$, the correspondence between $(\varepsilon_1, \varepsilon_2)$ and k being given by

$(\varepsilon_1, \varepsilon_2)$	$(1, 1)$	$(-1, 1)$	$(1, -1)$	$(-1, -1)$
k	0	1	2	3

*In each case, the basis $\underline{\sigma}_k$ of $\Gamma(\mathbb{P}^1, \mathcal{O}(H))$ is unique, and $(f_k, g_k) \in \mathbb{C}^* \times \mathbb{C}$
below is also unique.*

$$\nabla_{z\partial_z}\underline{\sigma}_k = \underline{\sigma}_k \left[\frac{u_0^1}{z} \begin{pmatrix} 0 & 1 \\ 1 & 0 \end{pmatrix} - g_k \begin{pmatrix} 1 & 0 \\ 0 & -1 \end{pmatrix} - z u_\infty^1 \begin{pmatrix} 0 & f_k^{-2} \\ f_k^2 & 0 \end{pmatrix} \right], \tag{8.15}$$

$$P((\underline{\sigma}_k)^t(z), \underline{\sigma}_k(-z)) = \begin{pmatrix} 0 & 2 \\ 2 & 0 \end{pmatrix}, \tag{8.16}$$

$$A(\underline{\sigma}_k(z)) = \underline{\sigma}_k(-z)\,\varepsilon_1 \begin{pmatrix} i & 0 \\ 0 & -i \end{pmatrix}, \tag{8.17}$$

$$J(\underline{\sigma}_k(z)) = \underline{\sigma}_k(\rho_c(z))\,\varepsilon_2 \begin{pmatrix} 0 & f_k^{-1} \\ f_k & 0 \end{pmatrix}. \tag{8.18}$$

The bases are related by

$$\underline{\sigma}_0 = \underline{\sigma}_2, \quad \underline{\sigma}_1 = \underline{\sigma}_3 = \underline{\sigma}_0 \begin{pmatrix} 0 & 1 \\ 1 & 0 \end{pmatrix}. \tag{8.19}$$

(c) *Formulae (8.15) to (8.18) define for any $k \in \{0, 1, 2, 3\}$ and for any $(f_k, g_k) \in$
$\mathbb{C}^* \times \mathbb{C}$ a pure P_{3D6}-TEJPA bundle. Therefore the set of pure P_{3D6}-TEJPA
bundles has four natural charts with coordinates $(u_0^1, u_\infty^1, f_k, g_k)$, and each chart
is isomorphic to $\mathbb{C}^* \times \mathbb{C}^* \times \mathbb{C}^* \times \mathbb{C}$. With this structure as an affine algebraic
manifold the set is called M_{3TJ}^{reg}. Obviously the restriction of pr^u to M_{3TJ}^{reg},*

$$pr^u : M_{3TJ}^{reg} \to \mathbb{C}^* \times \mathbb{C}^*, P_{3D6}\text{-TEJPA bundle} \mapsto (u_0^1, u_\infty^1),$$

*is an algebraic morphism. The fibres are called $M_{3TJ}^{reg}(u_0^1, u_\infty^1)$. The coordinates
of the four charts are related by*

$$(f_1, g_1) = (f_0^{-1}, -g_0),$$
$$(f_2, g_2) = (-f_0, g_0), \tag{8.20}$$
$$(f_3, g_3) = (-f_0^{-1}, -g_0).$$

(d) *Fix a pair $(u_0^1, u_\infty^1) \in \mathbb{C}^* \times \mathbb{C}^*$ and a square root $\sqrt{c}$ of $c = u_\infty^1/u_0^1$. Each
P_{3D6}-TEJPA bundle which is a $(1, -1)$-twistor has a unique normal form given
by (8.21) to (8.24). Here, $k \in \{0, 1, 2, 3\}$, $(\varepsilon_1, \varepsilon_2) \in \{\pm 1\}^2$ (k and $(\varepsilon_1, \varepsilon_2)$ are
related as in (b)), $\widetilde{g}_k \in \mathbb{C}$ and the basis ψ are unique. ψ is a basis of $H|_{\mathbb{C}}$ such
that $(z\psi_1, z^{-1}\psi_2)$ is a basis of $H|_{\mathbb{P}^1 - \{0\}}$. The pair $[\varepsilon_1, \varepsilon_2\sqrt{c}]$ is called the type
of the P_{3D6}-TEJPA bundle. The proof will give a different characterization of*

this type.

$$\nabla_{z\partial_z}\underline{\psi} = \underline{\psi} \left[\frac{u_0^1}{z} \begin{pmatrix} 0 & 1 \\ 1 & 0 \end{pmatrix} - \frac{1}{2} \begin{pmatrix} 1 & 0 \\ 0 & -1 \end{pmatrix} \right. \tag{8.21}$$

$$\left. -\widetilde{g}_k \sqrt{c} \, z \begin{pmatrix} 0 & 1 \\ 0 & 0 \end{pmatrix} + u_\infty^1 c z^3 \begin{pmatrix} 0 & 1 \\ 0 & 0 \end{pmatrix} \right]$$

$$P((\underline{\psi})^t(z), \underline{\psi}(-z)) = \begin{pmatrix} 0 & 2 \\ 2 & 0 \end{pmatrix}, \tag{8.22}$$

$$A(\underline{\psi}(z)) = \underline{\psi}(-z)\,\varepsilon_1 \begin{pmatrix} i & 0 \\ 0 & -i \end{pmatrix}, \tag{8.23}$$

$$J(\underline{\psi}(z)) = \underline{\psi}(\rho_c(z))\,\varepsilon_2 \begin{pmatrix} -\frac{1}{\sqrt{c}}z^{-1} & 0 \\ 0 & \sqrt{c}z \end{pmatrix}. \tag{8.24}$$

(e) *Fix a pair $(u_0^1, u_\infty^1) \in \mathbb{C}^* \times \mathbb{C}^*$ and a square root $\sqrt{c}$ of $c = u_\infty^1/u_0^1$. Formulae (8.21) to (8.24) define for any $k \in \{0, 1, 2, 3\}$ and the corresponding $(\varepsilon_1, \varepsilon_2) \in \{\pm 1\}^2$ and for any $\widetilde{g}_k \in \mathbb{C}$ a P_{3D6}-TEJPA bundle which is a $(1, -1)$-twistor.*

    Therefore the set of P_{3D6}-TEJPA bundles in $M_{3TJ}(u_0^1, u_\infty^1)$ which are $(1, -1)$-twistors has four components, one for each type, and each component has a coordinate $\widetilde{g}_k$ and is, with this coordinate, isomorphic to $\mathbb{C}$. With this structure as an affine algebraic manifold the set is called $M_{3TJ}^{[\varepsilon_1, \varepsilon_2 \sqrt{c}]}(u_0^1, u_\infty^1)$. The union of the four sets is called $M_{3TJ}^{sing}(u_0^1, u_\infty^1)$.

(f) *Fix a pair $(u_0^1, u_\infty^1) \in \mathbb{C}^* \times \mathbb{C}^*$, a square root $\sqrt{c}$ of $c = u_\infty^1/u_0^1$, a $k \in \{0, 1, 2, 3\}$ and the corresponding pair $(\varepsilon_1, \varepsilon_2) \in \{\pm 1\}^2$. Each P_{3D6}-TEJPA bundle in $M_{3TJ}^{reg}(u_0^1, u_\infty^1) \cup M_{3TJ}^{[\varepsilon_1, \varepsilon_2 \sqrt{c}]}(u_0^1, u_\infty^1)$ has a unique normal form listed in (8.25) to (8.28).*

    Here, $(f_k, \widetilde{g}_k) \in \mathbb{C} \times \mathbb{C}$ and the basis $\underline{\varphi}_k$ are unique. $\underline{\varphi}_k$ is a basis of $H|_{\mathbb{C}}$ such that $(z\varphi_{k,1} - \frac{f_k}{\sqrt{c}}\varphi_{k,2}, z^{-1}\varphi_{k,2})$ is a basis of $H|_{\mathbb{P}^1 - \{0\}}$.

$$\nabla_{z\partial_z}\underline{\varphi}_k = \underline{\varphi}_k \left[\frac{u_0^1}{\cdot z} \begin{pmatrix} 0 & 1 \\ 1 & 0 \end{pmatrix} - \frac{1}{2}(1 + \widetilde{g}_k f_k) \begin{pmatrix} 1 & 0 \\ 0 & -1 \end{pmatrix} \right.$$

$$\left. -\widetilde{g}_k \sqrt{c} \, z \begin{pmatrix} 0 & 1 \\ 0 & 0 \end{pmatrix} - u_\infty^1 f_k^2 z \begin{pmatrix} 0 & 0 \\ 1 & 0 \end{pmatrix} \right. \tag{8.25}$$

$$\left. + u_\infty^1 \sqrt{c} f_k z^2 \begin{pmatrix} 1 & 0 \\ 0 & -1 \end{pmatrix} + u_\infty^1 c z^3 \begin{pmatrix} 0 & 1 \\ 0 & 0 \end{pmatrix} \right]$$

$$P((\underline{\varphi}_k)^t(z), \underline{\varphi}_k(-z)) = \begin{pmatrix} 0 & 2 \\ 2 & 0 \end{pmatrix}, \tag{8.26}$$

$$A(\underline{\varphi}_k(z)) = \underline{\varphi}_k(-z) \, \varepsilon_1 \begin{pmatrix} i & 0 \\ 0 & -i \end{pmatrix}, \tag{8.27}$$

$$J(\underline{\varphi}_k(z)) = \underline{\varphi}_k(\rho_c(z)) \, \varepsilon_2 \begin{pmatrix} -\frac{1}{\sqrt{c}} z^{-1} & 0 \\ f_k & \sqrt{c} z \end{pmatrix}. \tag{8.28}$$

(g) *In the situation of (f), $\mathbb{C}^2$ with coordinates $(f_k, \widetilde{g}_k)$ is a chart for $M_{3TJ}^{reg}(u_0^1, u_\infty^1) \cup$ $M_{3TJ}^{[\varepsilon_1, \varepsilon_2 \sqrt{c}]}(u_0^1, u_\infty^1)$. Here the points $(0, \widetilde{g}_k)$ correspond to the points in $M_{3TJ}^{[\varepsilon_1, \varepsilon_2 \sqrt{c}]}(u_0^1, u_\infty^1)$ with coordinate $\widetilde{g}_k$. The points $(f_k, \widetilde{g}_k) \in \mathbb{C}^* \times \mathbb{C}$ correspond to the points*

$$(f_k, g_k) = (f_k, -u_0^1 \frac{\sqrt{c}}{f_k} + \frac{1}{2} + \frac{f_k}{2} \widetilde{g}_k) \tag{8.29}$$

in $M_{3TJ}^{reg}(u_0^1, u_\infty^1)$ with coordinates (f_k, g_k) from (b) and (c). For $(0, \widetilde{g}_k)$, the bases in (f) and (d) are related by $\psi = \underline{\varphi}_k$. This means that the normal form in (d) is the restriction to $f_k = 0$ of the normal form in (f). For $(f_k, \widetilde{g}_k)$ the bases in (f) and (b) are related by

$$\underline{\varphi}_k = \underline{\sigma}_k \begin{pmatrix} 1 & \frac{\sqrt{c}}{f_k} z \\ 0 & 1 \end{pmatrix} \tag{8.30}$$

Therefore $M_{3TJ}(u_0^1, u_\infty^1)$ is covered by four affine charts isomorphic to $\mathbb{C}^2$ and with coordinates $(f_k, \widetilde{g}_k)$. Each chart contains $M_{3TJ}^{reg}(u_0^1, u_\infty^1)$ and one of the four components $M_{3TJ}^{[\varepsilon_1, \varepsilon_2 \sqrt{c}]}(u_0^1, u_\infty^1)$. With the induced structure as an algebraic manifold it is called $M_{3TJ}^{ini}(u_0^1, u_\infty^1)$. The coordinate changes between the charts can be read off from (8.29) and (8.20).

Proof

(a) Let $(H, \nabla, u_0^1, u_\infty^1, P, A, J)$ be a P_{3D6}-TEJPA bundle which is a $(k, -k)$-twistor for some $k > 0$. We shall show $k = 1$.

1st proof: If (H, ∇) is irreducible, this follows immediately from Theorem 4.2. If (H, ∇) is completely reducible, then the last statement in Lemma 3.1 (a) shows that H is a $(0, 0)$-twistor, so this case is impossible as $k > 0$.

2nd proof: We give a 2nd proof. First, because the following proof is elementary and independent of Theorem 4.2. Second, because the proof is a preparation of the proof of (d).

Choose sections σ_1 and σ_2 as in Remark 8.1 (v), that means with

$$\mathcal{O}(H) = \mathcal{O}_{\mathbb{P}^1}(0,k)\,\sigma_1 \oplus \mathcal{O}_{\mathbb{P}^1}(0,-k)\,\sigma_2, \tag{8.31}$$

$$A(\underline{\sigma}(z)) = \underline{\sigma}(-z)\,\varepsilon_1 \begin{pmatrix} i & 0 \\ 0 & -i \end{pmatrix} \quad \text{for some } \varepsilon_1 \in \{\pm 1\}. \tag{8.32}$$

The section σ_1 is unique up to rescaling. The section σ_2 is unique up to adding an element of $\oplus_{l=0}^{k-1}\mathbb{C}z^{2l+1}\sigma_1$ and rescaling. The sections σ_1 and σ_2 can be chosen such that

$$\underline{\sigma}(0) = \underline{w}_0 \quad \text{if } \varepsilon_1 = 1,$$

$$\underline{\sigma}(0) = \underline{w}_0 \begin{pmatrix} 0 & 1 \\ 1 & 0 \end{pmatrix} \quad \text{if } \varepsilon_1 = -1 \tag{8.33}$$

($\underline{w}_0$ is defined in Remark 8.1 (iii)). This determines σ_1 uniquely, but σ_2 only up to adding an element as above. $\underline{\sigma}$ is a basis of $H|_{\mathbb{C}}$, and (8.7) and (8.12) give

$$\nabla_{z\partial_z}\underline{\sigma}(z) = \underline{\sigma}\left[\frac{u_0^1}{z}\begin{pmatrix} 0 & 1 \\ 1 & 0 \end{pmatrix} + \begin{pmatrix} a & b \\ \widetilde{c} & d \end{pmatrix}\right] \tag{8.34}$$

for some $a, b \in \mathbb{C}\{z^2\}, \widetilde{c}, d \in z\,\mathbb{C}\{z^2\}$.

$(z^k\sigma_1, z^{-k}\sigma_2)$ is a basis of $H|_{\mathbb{P}^1-\{0\}}$ with

$$\nabla_{z\partial_z}(z^k\sigma_1, z^{-k}\sigma_2) = (z^k\sigma_1, z^{-k}\sigma_2)\left[\frac{u_0^1}{z}\begin{pmatrix} 0 & z^{-2k} \\ z^{2k} & 0 \end{pmatrix}\right.$$
$$\left. + \begin{pmatrix} a & bz^{-2k} \\ \widetilde{c}z^{2k} & d \end{pmatrix} + \begin{pmatrix} k & 0 \\ 0 & -k \end{pmatrix}\right]. \tag{8.35}$$

But (H, ∇) has a pole of order 2 at ∞. Therefore

$$u_0^1 z^{2k-1} + \widetilde{c}(z)z^{2k} \in \mathbb{C} \oplus \mathbb{C}z.$$

This is only possible if $k = 1$ (and then $\widetilde{c}(z) = 0$). This finishes the 2nd proof of (a).

(b) It is obvious that the normal forms for $k = 1, 2, 3$ are obtained from the normal form for $k = 0$ by the base change in (8.19) and the change of the scalars in (8.20). Therefore it is sufficient to prove existence and uniqueness of the normal form for $k = 0$.

First we assume existence and prove uniqueness. Formulae (8.15) to (8.17) and the last statement of Remark 8.1 (iii) show $\underline{\sigma}(0) = \pm\underline{w}_0$. As H is pure, the basis $\underline{\sigma}$ is uniquely determined by $\underline{\sigma}(0) = \pm\underline{w}_0$. Then also (f_0, g_0) is unique.

Now we show existence. Choose that basis $\underline{\sigma}$ of $\Gamma(\mathbb{P}^1, \mathcal{O}(H))$ which satisfies $\underline{\sigma}(0) = \underline{w}_0$. As P is holomorphic on $\mathcal{O}(H) \times j^* \mathcal{O}(H)$, it takes constant values on sections of $\Gamma(\mathbb{P}^1, \mathcal{O}(H))$, and A maps global sections to global sections. Therefore (8.8) and (8.9) imply (8.16) and (8.17) with $\varepsilon_1 = 1$.

By Remark 8.1 (iv), J exchanges the eigenspaces of A in $\Gamma(\mathbb{P}^1, \mathcal{O}(H))$. Therefore, and because of $J^2 = \mathrm{id}$, a scalar $f_0 \in \mathbb{C}^*$ exists such that (8.18) holds with $\varepsilon_2 = 1$. Comparison with (8.10) shows

$$\underline{\sigma}(\infty) = \underline{w}_\infty \begin{pmatrix} f_0 & 0 \\ 0 & f_0^{-1} \end{pmatrix}.$$

Comparison with (8.7) shows

$$[-\nabla_{\partial_z}]\underline{\sigma}(\infty) = \underline{\sigma}(\infty) \, u_\infty^1 \begin{pmatrix} 0 & f_0^{-2} \\ f_0^2 & 0 \end{pmatrix}.$$

This equation together with (8.7) for $\underline{\sigma}(0) = \underline{w}_0$ and (8.12) shows

$$\nabla_{z\partial_z}\underline{\sigma}_0 = \underline{\sigma}_0 \left[\frac{u_0^1}{z} \begin{pmatrix} 0 & 1 \\ 1 & 0 \end{pmatrix} - \begin{pmatrix} a & 0 \\ 0 & d \end{pmatrix} - z\, u_\infty^1 \begin{pmatrix} 0 & f_0^{-2} \\ f_0^2 & 0 \end{pmatrix} \right]$$

for some $a, d \in \mathbb{C}$. Now (8.16) and (8.14) show $a + d = 0$. One sets $g_0 := a$. Existence of the normal form for $k = 0$ is proved.

(c) Everything is clear as soon as the first statement is proved. As above, it is sufficient to prove the first statement for the normal form with $k = 0$. So we have to show that (8.15) to (8.18) define in the case $k = 0$ for any $(f_0, g_0) \in \mathbb{C}^* \times \mathbb{C}$ a P_{3D6}-TEJPA bundle.

The conditions

$$A^2 = -\,\mathrm{id}, \quad J^2 = \mathrm{id}, \quad AJ = -JA,$$

$$P(A\,a, A\,b) = P(a, b), \quad P(J\,a, J\,b) = P(a, b)$$

are obviously satisfied, and P is symmetric and nondegenerate. It remains to show that P, A and J are flat.

A is flat because (8.12) holds with $a(z) = -d(z) = -g_0 z$, $b(z) = u_0^1 - z^2\, u_\infty^1 f_0^{-2}$, $c(z) = u_0^1 - z^2\, u_\infty^1 f_0^2$. And P is flat because (8.14) holds with $e(z) = 2$ and $a(z) + d(z) = 0$. Finally, the flatness of J is equivalent to

$$\nabla_{z\partial_z} J(\underline{\sigma}(z)) = J(\nabla_{z\partial_z}\underline{\sigma}(z)),$$

which holds because of the following calculations. The first one uses $z\partial_z = -\rho_c(z)\partial_{\rho_c(z)}$.

$$\nabla_{z\partial_z} J(\underline{\sigma}(z))$$

$$= \nabla_{z\partial_z}\sigma(\rho_c(z))\begin{pmatrix} 0 & f_0^{-1} \\ f_0 & 0 \end{pmatrix} = -\nabla_{\rho_c(z)\partial_{\rho_c(z)}}\sigma(\rho_c(z))\begin{pmatrix} 0 & f_0^{-1} \\ f_0 & 0 \end{pmatrix}$$

$$= -\underline{\sigma}(\rho_c(z))\left[\frac{u_0^1}{\rho_c(z)}\begin{pmatrix} 0 & 1 \\ 1 & 0 \end{pmatrix} - g_0\begin{pmatrix} 1 & 0 \\ 0 & -1 \end{pmatrix}\right.$$

$$\left. - \rho_c(z)\, u_\infty^1\begin{pmatrix} 0 & f_0^{-2} \\ f_0^2 & 0 \end{pmatrix}\right]\begin{pmatrix} 0 & f_0^{-1} \\ f_0 & 0 \end{pmatrix}$$

$$= \underline{\sigma}(\rho_c(z))\left[-z\,u_\infty^1\begin{pmatrix} f_0 & 0 \\ 0 & f_0^{-1} \end{pmatrix} - g_0\begin{pmatrix} 0 & -f_0^{-1} \\ f_0 & 0 \end{pmatrix} + \frac{u_0^1}{z}\begin{pmatrix} f_0^{-1} & 0 \\ 0 & f_0 \end{pmatrix}\right]$$

and

$$J(\nabla_{z\partial_z}\underline{\sigma}(z))$$

$$= \sigma(\rho_c(z))\begin{pmatrix} 0 & f_0^{-1} \\ f_0 & 0 \end{pmatrix}\left[\frac{u_0^1}{z}\begin{pmatrix} 0 & 1 \\ 1 & 0 \end{pmatrix} - g_0\begin{pmatrix} 1 & 0 \\ 0 & -1 \end{pmatrix} - z\,u_\infty^1\begin{pmatrix} 0 & f_0^{-2} \\ f_0^2 & 0 \end{pmatrix}\right]$$

$$= \underline{\sigma}(\rho_c(z))\left[\frac{u_0^1}{z}\begin{pmatrix} f_0^{-1} & 0 \\ 0 & f_0 \end{pmatrix} - g_0\begin{pmatrix} 0 & -f_0^{-1} \\ f_0 & 0 \end{pmatrix} - z\,u_\infty^1\begin{pmatrix} f_0 & 0 \\ 0 & f_0^{-1} \end{pmatrix}\right].$$

(d) First, existence of the normal form will be shown, then uniqueness.

Choose σ_1 and σ_2 as in the 2nd proof of part (a). Then (8.31) to (8.35) hold with $k = 1$. First we shall refine the choice of σ_2 and then call the new basis $\underline{\psi}$.

J acts on

$$\Gamma(\mathbb{P}^1, \mathcal{O}(H)) = \mathbb{C}\sigma_1 \oplus \mathbb{C}z\sigma_1,$$

it maps

$$\Gamma(\mathbb{P}^1, \mathcal{O}_{\mathbb{P}^1}(0,1)\,\mathcal{O}(H)) = \mathbb{C}\sigma_1 \oplus \mathbb{C}z\sigma_1 \oplus \mathbb{C}z^2\sigma_1 \oplus \mathbb{C}\sigma_2$$

to

$$\Gamma(\mathbb{P}^1, \mathcal{O}_{\mathbb{P}^1}(1,0)\,\mathcal{O}(H)) = \mathbb{C}z^{-1}\sigma_1 \oplus \mathbb{C}\sigma_1 \oplus \mathbb{C}z\sigma_1 \oplus \mathbb{C}z^{-1}\sigma_2,$$

and it exchanges in any case the eigenspaces of A. Therefore

$$J(\underline{\sigma}(z)) = \underline{\sigma}(\rho_c(z))\begin{pmatrix} \gamma_1 z^{-1} & \gamma_2 \\ 0 & \gamma_3 z \end{pmatrix}$$

$$\text{with}\quad \gamma_1, \gamma_3 \in \mathbb{C}^*, \gamma_2 \in \mathbb{C}, \gamma_1\gamma_3 = -1.$$

$J^2 = \mathrm{id}$ shows $\gamma_1^2 c = 1 = \gamma_3^2/c$. We had chosen a fixed root $\sqrt{c}$. Therefore there is a unique $\varepsilon_2 \in \{\pm 1\}$ with

$$(\gamma_1, \gamma_3) = \varepsilon_2 \left(-\frac{1}{\sqrt{c}}, \sqrt{c} \right).$$

Now we choose the new basis of $H|_C$

$$\underline{\psi} := \underline{\sigma} \begin{pmatrix} 1 & \frac{1}{2}\varepsilon_2 \sqrt{c}\gamma_2 z \\ 0 & 1 \end{pmatrix}.$$

It satisfies (8.24).

As $(z\sigma_1)(0) = 0$ it also satisfies $\underline{\psi}(0) = \underline{w}_0$. Comparison with (8.10) shows

$$(z\psi_1, z^{-1}\psi_2)(\infty)\varepsilon_2 \begin{pmatrix} -\sqrt{c} & 0 \\ 0 & \frac{1}{\sqrt{c}} \end{pmatrix} = J(\underline{\psi}(0)) = J(\underline{w}_0) = \underline{w}_\infty \begin{pmatrix} 0 & 1 \\ 1 & 0 \end{pmatrix},$$

so $\quad (z\psi_1, z^{-1}\psi_2)(\infty) = \underline{w}_\infty \, \varepsilon_2 \begin{pmatrix} 0 & \sqrt{c} \\ -\frac{1}{\sqrt{c}} & 0 \end{pmatrix}.$

Comparison with (8.7) shows

$$[-\nabla_{\partial_z}](z\psi_1, z^{-1}\psi_2)(\infty) = (z\psi_1, z^{-1}\psi_2)(\infty) \begin{pmatrix} 0 & -u_\infty^1 c \\ -u_0^1 & 0 \end{pmatrix}. \qquad (8.36)$$

The equations (8.31) to (8.35) hold with $k = 1$ also for the new basis $\underline{\psi}$ instead of the basis $\underline{\sigma}$. As (H, ∇) has a pole of order 2 at ∞, (8.35) shows

$$\widetilde{c} = 0, \ a, d \in \mathbb{C}, \ b = b_1 z + b_3 z^3 \text{ with } b_1, b_3 \in \mathbb{C}.$$

Comparison of (8.35) and (8.36) gives $b_3 = u_\infty^1 c$. This gives for the moment the following approximation of (8.21),

$$\nabla_{z\partial_z}\underline{\psi} = \underline{\psi} \left[\frac{u_0^1}{z}\begin{pmatrix} 0 & 1 \\ 1 & 0 \end{pmatrix} + \begin{pmatrix} a & 0 \\ 0 & d \end{pmatrix} \right. \qquad (8.37)$$

$$\left. + b_1 z \begin{pmatrix} 0 & 1 \\ 0 & 0 \end{pmatrix} + u_\infty^1 c z^3 \begin{pmatrix} 0 & 1 \\ 0 & 0 \end{pmatrix} \right].$$

Because of (8.23) for $\underline{\psi}$ and because of Remark 8.1 (vi), ψ_1 and ψ_2 are isotropic with respect to $\overline{P}$. Because of

$$P(\psi_1(z), \psi_2(-z)) = -P(z\psi_1(z), (-z)^{-1}\psi_2(-z))$$

the function $P(\psi_1(z), \psi_2(-z))$ is also holomorphic at ∞, thus constant. Together
with $\underline{\psi}(0) = \underline{w}_0$ and (8.8) this establishes (8.22). Now (8.14) shows $a + d = 0$.

It remains to show $a = -\frac{1}{2}$. We use the flatness of J for this. The matrices
M_1 and M_2 in

$$J(\underline{\psi}(z)) M_1 = \nabla_{z\partial_z} J(\underline{\psi}(z)) = J(\nabla_{z\partial_z} \underline{\psi}(z)) = J(\underline{\psi}(z)) M_2$$

coincide. The constant part of M_1 can be read off from (8.37) and (8.24) and is,
using $\nabla_{z\partial_z} = -\nabla_{\rho_c(z)} \partial_{\rho_c(z)}$, $\left(\begin{smallmatrix} -a-1 & 0 \\ 0 & -d+1 \end{smallmatrix} \right)$. The constant part of M_2 can be read
off (8.37) and (8.24) and is $\left(\begin{smallmatrix} a & 0 \\ 0 & d \end{smallmatrix} \right)$. This shows $a = -\frac{1}{2}$. Existence of the normal
form is proved.

It remains to prove the uniqueness of the normal form. Equations (8.31)
and (8.32) show that ε_1 is unique. Equations (8.21) to (8.23) and Remark 8.1
(iii) show

$$\underline{\psi}(0) = \underline{w}_0 \quad \text{if } \varepsilon_1 = 1,$$

$$\underline{\psi}(0) = \underline{w}_0 \begin{pmatrix} 0 & 1 \\ 1 & 0 \end{pmatrix} \quad \text{if } \varepsilon_1 = -1.$$

This and (8.23) and the condition that $(z\psi_1, z^{-1}\psi_2)$ is a basis of $H|_{\mathbb{P}^1-\{0\}}$
determines ψ_1 uniquely and ψ_2 up to addition of an element of $\mathbb{C}z\psi_1$.
Now (8.24) and the argument in the proof of the existence determine ε_2 and
show that also ψ_2 is unique.

(e) As in the proof of (c), everything is clear as soon as the first statement is
shown, i.e., that (8.21) to (8.24) define a P_{3D6}-TEJPA bundle for any k, $(\varepsilon_1, \varepsilon_2)$
(compatible as in (b)) and any $\widetilde{g}_k \in \mathbb{C}$.

One can check this in a similar way to the proof of (c). Only the flatness of J
requires a lengthy calculation. Alternatively one can refer to part (f). There the
formulae in (d) turn up as the special case $f_k = 0$.

(f) For the pure P_{3D6}-TEJPA bundles in (b), one can calculate in a straightforward
way their data ∇, P, A and J with respect to the new basis $\underline{\varphi}_k$ (on $H|_\mathbb{C}$) and the
new scalars $(f_k, \widetilde{g}_k)$, starting from the normal forms in (b) with respect to $\underline{\sigma}_k$
and (f_k, g_k), and using the formulae (8.30) and (8.29). We omit the calculation.
It leads to (8.25)–(8.28).

This calculation and the uniqueness of the normal forms in (b) show the
existence and uniqueness of the normal forms in (f) for pure P_{3D6}-TEJPA
bundles. Part (d) and the fact that (d) and (f) for $f_k = 0$ coincide, show the
existence and uniqueness of the normal form in (f) for non-pure P_{3D6}-TEJPA
bundles.

Then (e) becomes obvious. As the formulae (8.25) to (8.28) give P_{3D6}-TEJPA
bundles for all $(f_k, \widetilde{g}_k) \in \mathbb{C}^* \times \mathbb{C}$, they give P_{3D6}-TEJPA bundles also for all
values $(0, \widetilde{g}_k) \in \{0\} \times \mathbb{C}$.

(g) This is also clear now. $\square$

Remark 8.3 The group $G^{mon} = \{id, R_1, R_2, R_3\}$ was introduced before Theorem 7.6. It acts on M_{3TJ}^{mon}, and the orbits are the sets of (four or two) P_{3D6}-TEJPA bundles whose underlying P_{3D6}-TEP bundles are isomorphic. Theorem 8.2 (b)+(c) shows that it acts also on the algebraic manifold M_{3TJ}^{reg}, via

$$R_1 : (u_0^1, u_\infty^1, f_0, g_0) \mapsto (u_0^1, u_\infty^1, f_0^{-1}, -g_0) \tag{8.38}$$

$$R_2 : (u_0^1, u_\infty^1, f_0, g_0) \mapsto (u_0^1, u_\infty^1, -f_0, g_0)$$

$$R_3 : (u_0^1, u_\infty^1, f_0, g_0) \mapsto (u_0^1, u_\infty^1, -f_0^{-1}, -g_0).$$

In other words, for any P_{3D6}-TEJPA bundle $T \in M_{3TJ}^{reg}$,

$$(f_k, g_k)(T) = (f_0, g_0)(R_k(T)), \quad (f_k, g_k)(R_k(T)) = (f_0, g_0)(T). \tag{8.39}$$

The action of G^{mon} extends to an action on $M_{3TJ}^{ini}(u_0^1, u_\infty^1)$. For $T \in M_{3TJ}^{reg}(u_0^1, u_\infty^1) \cup M^{[+,+\sqrt{c}]}(u_0^1, u_\infty^1)$ the image $R_k(T)$ is in $M_{3TJ}^{reg}(u_0^1, u_\infty^1) \cup M^{[\varepsilon_1, \varepsilon_2 \sqrt{c}]}(u_0^1, u_\infty^1)$ with the $(\varepsilon_1, \varepsilon_2)$ which corresponds to k as in Theorem 8.2 (b), and

$$(f_k, \widetilde{g}_k)(R_k(T)) = (f_0, \widetilde{g}_0)(T). \tag{8.40}$$

In Theorem 8.2 (d)–(g), only P_{3D6}-TEJPA bundles with fixed (u_0^1, u_∞^1) were considered, because of the appearance of $\sqrt{c}$. Now we shall consider the full space M_{3TJ}, and the pull-back by a 2:1-covering, which will resolve the ambiguity of the square root $\sqrt{c}$. Consider the covering

$$c^{2:1} : \mathbb{C}^* \times \mathbb{C}^* \to \mathbb{C}^* \times \mathbb{C}^* \tag{8.41}$$

$$(x, y) \mapsto (x/y, xy) = (u_0^1, u_\infty^1).$$

The covering c^{path} from (7.24) factorizes through it via

$$c^{path} : \mathbb{C} \times \mathbb{C}^* \to \mathbb{C}^* \times \mathbb{C}^* \xrightarrow{c^{2:1}} \mathbb{C}^* \times \mathbb{C}^*, \tag{8.42}$$

$$(\beta, u_0^1) \mapsto (\tfrac{1}{2}e^{-\beta/2}, \tfrac{1}{2}e^{-\beta/2}/u_0^1), (x, y) \mapsto (x/y, xy).$$

The following theorem is a direct consequence of Theorem 8.2.

Theorem 8.4

(a) *The set $(c^{2:1})^* M_{3TJ}$ carries a natural structure as an algebraic manifold, with which it is called $((c^{2:1})^* M_{3TJ})^{ini}$. It is obtained by gluing the four affine charts in Theorem 8.2 (g) for the spaces $M_{3TJ}^{ini}(u_0^1, u_\infty^1)$ to four affine charts for $(c^{2:1})^* M_{3TJ}$. Here $y = \sqrt{c}$ with y as in (8.41). Each chart is isomorphic to $\mathbb{C}^* \times \mathbb{C}^* \times \mathbb{C} \times \mathbb{C}$ with coordinates $(x, y, f_k, \widetilde{g}_k)$. The set of non-pure P_{3D6}-TEJPA*

*bundles consists of the four hyperplanes, one in each chart, with $f_k = 0$,
$k = 0, 1, 2, 3$. Here they can be denoted by $((c^{2:1})^* M_{3TJ})^{[k]}$.*

(b) *The set M_{3TJ} inherits from $((c^{2:1})^* M_{3TJ})^{ini}$ the structure of an algebraic
manifold, with which it is called M_{3TJ}^{ini}. Then $((c^{2:1})^* M_{3TJ})^{ini} = (c^{2:1})^* M_{3TJ}^{ini}$.
But a description of M_{3TJ}^{ini} in charts is not obvious.*

The set of non-pure P_{3D6}-TEJPA bundles consists only of two hypersurfaces. They
can be denoted $M_{3TJ}^{[\varepsilon_1]}$. Each of them intersects each fibre $M_{3TJ}^{ini}(u_0^1, u_\infty^1)$ in the two
hypersurfaces $M_{3TJ}^{[\varepsilon_1, \varepsilon_2 \sqrt{c}]}(u_0^1, u_\infty^1)$.

Theorem 4.2 makes a stronger statement than Theorem 8.2 (a). It gives a solution
of the inverse monodromy problem for trace free P_{3D6} bundles. We state the special
case of P_{3D6}-TEJPA bundles (or P_{3D6}-TEP bundles) in the following lemma again,
and we provide an elementary proof, independent of the proofs in [Heu09] and
[Ni09].

Lemma 8.5 *No isomonodromic family in M_{3TJ} is contained in $M_{3TJ}^{sing} =$
$\cup_{\varepsilon_1 = \pm 1} M_{3TJ}^{[\varepsilon_1]}$. The intersection of any isomonodromic family with M_{3TJ}^{sing} is empty
or a hypersurface in the isomonodromic family.*

Proof The first statement implies the second, because M_{3TJ}^{mon} and M_{3TJ}^{ini} give the same
complex analytic manifold, and the isomonodromic families as well as M_{3TJ}^{sing} are
analytic subvarieties of this complex manifold.

We assume that an isomonodromic family is contained in $M_{3TJ}^{[k]} \subset (c^{2:1})^* M_{3TJ}^{ini}$
for some $k \in \{0, 1, 2, 3\}$. Then the normal form in Theorem 8.2 (d) for $u_0^1 = u_\infty^1 =:$
$x \in U$ with $U \subset \mathbb{C}^*$ simply connected must give for some holomorphic function
$\widetilde{g}_k \in \mathcal{O}_U$ an isomonodromic family with basis of sections $\underline{\psi}(z, x)$, $x \in U$. Because
of (8.21) to (8.23), the bases $\underline{e}_0^\pm(z, x)$ which correspond by (2.19) to $\underline{\psi}(0) C^{-1}$, are
flat. The basis $\underline{\sigma} := \underline{\psi} C^{-1}$ can be written as in (2.15),

$$\underline{\psi}(z, x) = \underline{e}_0^\pm(z, x) \begin{pmatrix} e^{-x/z} & 0 \\ 0 & e^{x/z} \end{pmatrix} A_0^\pm(z, x) C$$

with $\widehat{A}_0^+ = \widehat{A}_0^-$ and $\widehat{A}_0^\pm(0, x) = \mathbf{1}_2$ for all x.

This formula and the flatness of $\underline{e}_0^\pm$ give

$$\nabla_{x\partial_x} \underline{\sigma}(z, x) = \underline{\sigma}(z, x) \left[-\frac{x}{z} \begin{pmatrix} 0 & 1 \\ 1 & 0 \end{pmatrix} + \begin{pmatrix} a & b \\ \widetilde{c} & d \end{pmatrix} \right]$$

with $a, b, \widetilde{c}, d \in \mathcal{O}_U[z]$. Now one proceeds as in the proof of Theorem 8.2 (d). One
uses (8.21) to (8.23) and arguments as in Remark 8.1 (vi) and derives that $\widetilde{c} = 0$,

$a = -d \in \mathcal{O}_U, b = b_1 z - xz^3$ with $b_1 \in \mathcal{O}_U$ and

$$\nabla_{x\partial_x}\underline{\sigma}(z,x) = \underline{\sigma}(z,x)\left[-\frac{x}{z}\begin{pmatrix} 0 & 1 \\ 1 & 0 \end{pmatrix} + a\begin{pmatrix} 1 & 0 \\ 0 & -1 \end{pmatrix} + \begin{pmatrix} 0 & b_1 z - xz^3 \\ 0 & 0 \end{pmatrix}\right].$$

Now one can calculate

$$(\nabla_{x\partial_x}\nabla_{z\partial_z} - \nabla_{z\partial_z}\nabla_{x\partial_x})\underline{\sigma}(z,x).$$

The calculation shows that it cannot be 0, although it must be 0. Because of this contradiction, the assumption above is wrong.   □

Chapter 9
Generalities on the Painlevé Equations

The Painlevé equations are six families P_I, P_{II}, P_{III}, P_{IV}, P_V, P_{VI} of second order differential equations in

$$U = \mathbb{C}, \quad \mathbb{C}^* = \mathbb{C} - \{0\}, \text{ or } \mathbb{C} - \{0, 1\}$$

of the form

$$f_{xx} = R(x, f, f_x)$$

where R is holomorphic for $x \in U$ and rational in f and f_x. The following table lists the number of essential parameters in R and the subset U.

	P_{VI}	P_V	P_{IV}	P_{III}	P_{II}	P_I
parameters	4	3	2	2	1	0
U	$\mathbb{C}-\{0,1\}$	$\mathbb{C}^*$	$\mathbb{C}$	$\mathbb{C}^*$	$\mathbb{C}$	$\mathbb{C}$

The equations are distinguished by the following *Painlevé property*: Any local holomorphic solution extends to a global multi-valued meromorphic solution in U (single-valued if $U = \mathbb{C}$). This means that solutions branch only at 0, 1 for P_{VI}, 0 for P_V and P_{III}, and on U the only singularities are poles. The positions of the poles in U depend on the solution, so they are *movable*. In other words, the Painlevé equations are distinguished by the property that the only movable singularities of solutions are poles. For proofs of the Painlevé property which do not use isomonodromic families of connections, see [GLSh02] and references there. In Theorem 10.3 we shall rewrite in our language the proof in [FN80] for $P_{III}(0, 0, 4, -4)$ which uses isomonodromic families.

© Springer International Publishing AG 2017
M.A. Guest, C. Hertling, *Painlevé III: A Case Study in the Geometry of Meromorphic Connections*, Lecture Notes in Mathematics 2198, DOI 10.1007/978-3-319-66526-9_9

Between 1895 and 1910, Painlevé and Gambier discovered 50 such families of differential equations, which included P_I to P_{VI}. But the general solutions of the other 44 equations could be reduced to those of P_I to P_{VI}, to rational or elliptic functions, or to solutions of linear or first order differential equations.

In this monograph we are interested in a special case of the Painlevé III equations. A priori, there are four parameters $(\alpha, \beta, \gamma, \delta) \in \mathbb{C}^4$: the equation $P_{III}(\alpha, \beta, \gamma, \delta)$ is

$$f_{xx} = \frac{f_x^2}{f} - \frac{1}{x}f_x + \frac{1}{x}(\alpha f^2 + \beta) + \gamma f^3 + \delta\frac{1}{f}. \tag{9.1}$$

Let us choose locally a logarithm $\varphi = 2\log f$ of f, i.e. $f = e^{\varphi/2}$. Then φ branches at the poles and zeros of f. Equation (9.1) becomes

$$(x\partial_x)^2\varphi = 2x(\alpha e^{\varphi/2} + \beta e^{-\varphi/2}) + 2x^2(\gamma e^\varphi + \delta e^{-\varphi}) \tag{9.2}$$

which is more symmetric than (9.1).

One sees that

$$
\begin{aligned}
&f \text{ and } \varphi + 4\pi ik \text{ are solutions for } (\alpha, \beta, \gamma, \delta) \iff \\
&-f \text{ and } \varphi + 2\pi i + 4\pi ik \text{ are solutions for } (-\alpha, -\beta, \gamma, \delta) \iff \\
&f^{-1} \text{ and } -\varphi + 4\pi ik \text{ are solutions for } (-\beta, -\alpha, -\delta, -\gamma) \iff \\
&-f^{-1} \text{ and } -\varphi + 2\pi i + 4\pi ik \text{ are solutions for } (\beta, \alpha, -\delta, -\gamma).
\end{aligned} \tag{9.3}
$$

Rescaling x by $r \in \mathbb{C}^*$ and f by $s \in \mathbb{C}^*$ shows

$$f \text{ and } \varphi \text{ are solutions for } (\alpha, \beta, \gamma, \delta) \iff \tag{9.4}$$

$$sf(rx) \text{ and } \varphi(rx) + 2\log s \text{ are solutions for } (\tfrac{r}{s}\alpha, rs\beta, \tfrac{r^2}{s^2}\gamma, r^2 s^2\delta).$$

In [OKSK06] the equations (9.1) are split into four cases:

$$
\begin{aligned}
P_{III}(D_6) &\qquad \gamma\delta \neq 0 \\
P_{III}(D_7) &\quad \gamma = 0, \alpha\delta \neq 0 \text{ or } \delta = 0, \beta\gamma \neq 0 \\
P_{III}(D_8) &\qquad \gamma = 0, \delta = 0, \alpha\beta \neq 0 \\
P_{III}(Q) &\qquad \alpha = 0, \gamma = 0 \text{ or } \beta = 0, \delta = 0
\end{aligned}
$$

Using the symmetries (9.3) and the rescalings (9.4), these four cases can be reduced to the following:

$$
\begin{aligned}
P_{III}(D_6) &\qquad\qquad (\alpha, \beta, 4, -4) \\
P_{III}(D_7) &\qquad\qquad (2, \beta, 0, -4) \\
P_{III}(D_8) &\qquad\qquad (4, -4, 0, 0) \\
P_{III}(Q) &\quad (0, 1, 0, \delta), (0, 0, 0, 1), (0, 0, 0, 0)
\end{aligned}
$$

All cases for $P_{III}(Q)$ are solvable by quadratures, so they are usually ignored. The cases $P_{III}(D_7)$ and $P_{III}(D_8)$ are not treated in [Ok79, Ok86, FN80, JM81], but in [OKSK06, OO06, PS09]. Within the cases $P_{III}(D_6)(\alpha, \beta, 4, -4)$, the case $P_{III}(0, 0, 4, -4)$ lies at the center of the parameter space, as it is the fixed point of the group of the four symmetries in (9.3). This group acts on the space of its (global multi-valued meromorphic) solutions. In this monograph we are interested in this case. Then (9.2) is the radial sinh-Gordon equation

$$(x\partial_x)^2\varphi = 16x^2 \sinh\varphi. \tag{9.5}$$

Chapter 10 will connect two objects associated to $P_{III}(0, 0, 4, -4)$: the *spaces of initial conditions* of Okamoto [Ok79] and certain *isomonodromic families of meromorphic connections*. The interrelations between these objects have been studied thoroughly in the case of P_{VI} by M.-H. Saito and others. But for P_I to P_V only first steps have been done in [PS09]. We shall treat the case $P_{III}(0, 0, 4, -4)$ in Chap. 10. It will also be the basis for the later chapters.

The rational function $R(x, f, f_x)$ in f and f_x in the Painlevé equations contains only the inverses indicated in the following table:

	P_{VI}	P_V	P_{IV}	P_{III}	P_{II}	P_I
	$f^{-1}, (f-1)^{-1}, (f-x)^{-1}$	$f^{-1}, (f-1)^{-1}$	f^{-1}	f^{-1}		
U'	$\mathbb{C} - \{0, 1, x_0\}$	$\mathbb{C} - \{0, 1\}$	$\mathbb{C}^*$	$\mathbb{C}^*$	$\mathbb{C}$	$\mathbb{C}$

Therefore at any point $x_0 \in U$, any pair $(f(x_0), f_x(x_0)) \in U' \times \mathbb{C}$ with U' as shown, determines a unique local holomorphic solution f, and, by the Painlevé property, a unique global multi-valued meromorphic solution. Thus $U' \times \mathbb{C}$ is a naive space of initial conditions at x_0 for the solutions of the Painlevé equation. But it misses the initial conditions for the solutions which are singular at x_0. Okamoto [Ok79] gave complete spaces of initial conditions for all $x_0 \in U$ for most of the Painlevé equations. He missed some cases, including $P_{III}(D_7)$ and $P_{III}(D_8)$. These two cases were treated in [OKSK06].

In Chap. 10 we shall describe the spaces of initial conditions for the case $P_{III}(0, 0, 4, -4)$ by four charts such that each extends the naive space $U' \times \mathbb{C} = \mathbb{C}^* \times \mathbb{C}$ by initial data for singular solutions. This is not what Okamoto did. He started with the compactification $\mathbb{P}^2$ of $U' \times \mathbb{C}$, blew it up several times in appropriate ways and threw out some superfluous hypersurfaces [Ok79]. Descriptions by charts are provided in [ShT97, MMT99, Mat97, NTY02, Te07] but they all build on [Ok79] and do not interpret the charts as extensions of $U' \times \mathbb{C}$ by initial data for singular solutions.

The relations between the Painlevé equations and isomonodromic families of meromorphic connections have a long history. Soon after the discovery of the Painlevé equations, Fuchs and Garnier found second order linear differential equations with rational coefficients whose isomonodromic families are governed by equations of type P_{VI} (Fuchs) or P_I to P_V (Garnier).

Their work was continued in [Ok86] and (almost) completed in [OO06]. Equation [Ok86] gives six types of second order linear differential equations, for generic members of the six Painlevé equations. Equation [OO06] adds four types. The ten types together cover all Painlevé equations which cannot be solved by quadratures. They can be classified by the orders of the poles of second order linear differential equations. The following table is essentially taken from [PS09] and is equivalent to data in [OO06].

	orders of poles	parameters
P_{VI}	1 1 1 1	4
P_V	1 1 2	3
$P_{V,deg}$	1 1 $\frac{3}{2}$	2
$P_{III}(D_6)$	2 2	2
$P_{III}(D_7)$	$\frac{3}{2}$ 2	1
$P_{III}(D_8)$	$\frac{3}{2}$ $\frac{3}{2}$	0
P_{IV}	1 3	2
$P_{II}(\sim P_{34})$	1 $\frac{5}{2}$	1
P_{II}	4	1
P_I	$\frac{7}{2}$	0

Remarkably, there are two types for P_{II}, namely type $P_{II}[1, \frac{5}{2}]$ and type $P_{II}[4]$. In [Ok86] the six types $P_{VI}, P_V, P_{III}(D_6), P_{IV}, P_{II}[4], P_I$ of the above ten are given.

Second order linear differential equations can be rewritten as first order linear systems in 2×2 matrices, and these are equivalent to trivial holomorphic vector bundles of rank 2 on $\mathbb{P}^1$ with meromorphic connections. Such data have been associated to the Painlevé equations in [FN80, JM81, IN86, FIKN06, PS09, PT14]. The vector bundle point of view is used in [PS09] and [PT14].

Reference [JM81] gives six types of such data. We expect that they are equivalent to the six types in [Ok86]. Reference [FIKN06, ch. 5] takes up the six types in [JM81]. Reference [FN80] considers only two types, a type equivalent to the type $P_{II}[1, \frac{5}{2}]$ and a new type for $P_{III}(0, 0, 4, -4)$. The type in [FN80] equivalent to the type $P_{II}[1, \frac{5}{2}]$ is the two-fold branched cover (branched at 0 and ∞ in $\mathbb{P}^1$) with poles of order 1 (at ∞) and 4 (at 0), such that the formal decomposition of Hukuhara and Turrittin exists at the pole of order 4. References [IN86] and [FIKN06, ch. 7–16] take up both types in [FN80] and connect Stokes data and the central connection matrix with the asymptotic behaviour near 0 and ∞ in U of solutions of P_{II} and $P_{III}(0, 0, 4, -4)$.

Remark 9.1 The relation between the type for $P_{III}(0, 0, 4, -4)$ in [FN80] and the type for $P_{III}(D_6)$ in [JM81] is quite remarkable and seems to have been unnoticed.

We claim that the following points (i) and (ii) are true. We intend to prove this elsewhere.

(i) The isomonodromic families for $P_{III}(0, 0, 4, -4)$ in [FN80] coincide essentially with the isomonodromic families for $P_{III}(0, 4, 4, -4)$ in [JM81], but the solutions of the Painlevé equations are built into the isomonodromic families in different ways.

(ii) One isomonodromic family gives rise to four solutions of $P_{III}(0, 0, 4, -4)$ (from the symmetries in (9.3)), and to one solution of $P_{III}(0, 4, 4, -4)$. These solutions are connected by the 4:1 folding transformation in [TOS05] and [Wi04] which is called $\psi^{[4]}_{III(D_6^{(1)})}$ in [TOS05] and which exists only for the solutions of $P_{III}(0, 0, 4, -4)$ and $P_{III}(0, 4, 4, -4)$.

Remarks 9.2

(i) References [IN86], [FIKN06, ch. 7–12], [Ni09] and this monograph work with the isomonodromic families for $P_{III}(0, 0, 4, -4)$ in [FN80]. Theorem 10.3 makes the relation between these isomonodromic families and the solutions of $P_{III}(0, 0, 4, -4)$ precise.

(ii) This relation for $P_{III}(0, 0, 4, -4)$ is not indicated in the table above of the above ten types, and it is not considered in [OO06, PS09]. Therefore we regard it as a new type. It is in this sense that [OO06] "almost" completes the work of Fuchs and Garnier.

(iii) If the isomonodromic family for $P_{III}(D_6)$ in [JM81] and the isomonodromic family for $P_{III}(0, 0, 4, -4)$ in [FN80] had been compared earlier, the 4:1 folding transformation $\psi^{[4]}_{III(D_6^{(1)})}$ in [TOS05, Wi04] might have been found earlier. Anyway, this comparison provides the isomonodromic interpretation of $\psi^{[4]}_{III(D_6^{(1)})}$ asked for in [FIKN06, ch 6 0.] (bottom of page 221).

(iv) In [TOS05] two other folding transformations $\psi^{[3]}_{IV}$ and $\psi^{[2]}_{II}$ are given, and according to [FIKN06, ch. 6] $\psi^{[3]}_{IV}$ is due to Okamoto (1986), and $\psi^{[2]}_{II}$ is due to Gambier (1910). They map $P_{IV}(0, -\frac{2}{9})$ to $P_{IV}(1, 0)$ and $P_{II}(0)$ to $P_{II}(-\frac{1}{2})$. It seems interesting to study the isomonodromic families for $P_{IV}(0, -\frac{2}{9})$ and $P_{II}(0)$ which one obtains via pull-back with $\psi^{[3]}_{IV}$ and $\psi^{[2]}_{II}$ from the isomonodromic families in [JM81] for $P_{IV}(1, 0)$ and $P_{II}(-\frac{1}{2})$. We believe that the relations of these isomonodromic families with $P_{IV}(0, -\frac{2}{9})$ and $P_{II}(0)$ have not yet been considered. They are not considered in [OO06] and [PS09].

Remark 9.3 Reference [OO06] gives more than the 10 types above. It discusses different ways to classify the Painlevé equations, which amount to different numbers: 5, 6, 8, 10, 14.

6 is the original number of families of Painlevé equations.

5 is obtained by unifying P_{II} and P_I in one family of equations.

By the scaling transformations which rescale x and f, these are separated into 14 types of equations. For example P_{III} is separated into $P_{III}(D_6)$, $P_{III}(D_7)$, $P_{III}(D_8)$, $P_{III}(Q)$.

4 of these 14 types can be solved by quadrature. The remaining 10 cases are given in the table above and are related to 10 types of second order linear differential equations.

But P_{II} and P_{34} are equivalent, and $P_{III}(D_6)$ and $P_{V,deg}$ are equivalent. That leads to 8 cases.

Chapter 10
Solutions of the Painlevé Equation
$P_{III}(0, 0, 4, -4)$

Now we come to the relation between P_{3D6}-TEJPA bundles and the Painlevé equation $P_{III}(0, 0, 4, -4)$. First, recall the covering $c^{2:1}$ from (8.41) and the covering c^{path} from (7.24) which factorizes through $c^{2:1}$:

$$c^{path} : \mathbb{C} \times \mathbb{C}^* \to \mathbb{C}^* \times \mathbb{C}^* \xrightarrow{c^{2:1}} \mathbb{C}^* \times \mathbb{C}^*, \qquad (10.1)$$

$$(\beta, u_0^1) \mapsto (\tfrac{1}{2}e^{-\beta/2}, \tfrac{e^{-\beta/2}}{2u_0^1}), (x, y) \xmapsto{c^{2:1}} (x/y, xy) = (u_0^1, \tfrac{e^{-\beta}}{4u_0^1}) = (u_0^1, u_\infty^1).$$

Recall also from (4.3) that within $(c^{2:1})^* M_{3TJ}^{mon}$, the families of P_{3D6}-TEJPA bundles where only y varies (but not (x, s, B)) are *inessential isomonodromic families* with trivial transversal monodromy. The next lemma combines this observation with the data in Theorems 7.5 and 8.4.

Lemma 10.1

(a) Define

$$M_{3FN} := ((c^{2:1})^* M_{3TJ})|_{y=1},$$

$$M_{3FN}(x) := ((c^{2:1})^* M_{3TJ})(x, 1) = M_{3TJ}(x, x),$$

$$M_{3FN}^{mon} := ((c^{2:1})^* M_{3TJ}^{mon})|_{y=1},$$

$$M_{3FN}^{mon}(x) := ((c^{2:1})^* M_{3TJ}^{mon})(x, 1) = M_{3TJ}^{mon}(x, x), \qquad (10.2)$$

$$M_{3FN}^{ini} := ((c^{2:1})^* M_{3TJ}^{ini})|_{y=1},$$

$$M_{3FN}^{ini}(x) := ((c^{2:1})^* M_{3TJ}^{ini})(x, 1) = M_{3TJ}^{ini}(x, x).$$

(Here, 3FN stands for the Flaschka-Newell version of P_{III}, which is exactly $P_{III}(0, 0, 4, -4)$.) Then M_{3FN}^{mon} and M_{3FN}^{ini} give M_{3FN} the same structure as an

© Springer International Publishing AG 2017

M.A. Guest, C. Hertling, *Painlevé III: A Case Study in the Geometry of Meromorphic Connections*, Lecture Notes in Mathematics 2198, DOI 10.1007/978-3-319-66526-9_10

analytic manifold, M_{3FN}^{ini} is an algebraic manifold, and the fibres $M_{3FN}(x)$ of the projection $pr_{3FN} : M_{3FN} \to \mathbb{C}^*$ carry two algebraic structures $M_{3FN}^{mon}(x)$ and $M_{3FN}^{ini}(x)$. There are canonical isomorphisms

$$(c^{2:1})^* M_{3TJ}^{mon} \cong M_{3FN}^{mon} \times \mathbb{C}^*, \quad M_{3TJ}^{mon}(x/y, xy) \cong M_{3FN}^{mon}(x) \ \forall \ y \in \mathbb{C}^*,$$
$$(c^{2:1})^* M_{3TJ}^{ini} \cong M_{3FN}^{ini} \times \mathbb{C}^*, \quad M_{3TJ}^{ini}(x/y, xy) \cong M_{3FN}^{ini}(x) \ \forall \ y \in \mathbb{C}^*, \tag{10.3}$$

where $\mathbb{C}^*$ is equipped with the coordinate y. M_{3FN} can be seen as the set of isomorphism classes of inessential isomonodromic families (over $\mathbb{C}^*$ with coordinate y and with trivial transversal monodromy) of P_{3D6}-TEJPA bundles.

(b) M_{3FN}^{mon} is the quotient of the algebraic manifold $\mathbb{C} \times V^{mat}$ by the action of $\langle m_{[1]}^2 \rangle$. The automorphism $m_{[1]}^2$ is algebraic, but the group $\langle m_{[1]}^2 \rangle$ is isomorphic to $\mathbb{Z}$, therefore the quotient is an analytic manifold, and M_{3FN}^{mon} is described by the isomorphism

$$\Phi^{mon} : M_{3FN}^{mon} \to \mathbb{C} \times V^{mat} / \langle m_{[1]}^2 \rangle, \tag{10.4}$$

$$(H, \nabla, x, x, P, A, J) \mapsto [(\beta, s, B(\beta))] \text{ for } \beta \text{ with } \tfrac{1}{2} e^{-\beta/2} = x.$$

Here s and $B(\beta)$ are associated to $(H, \nabla, x, x, P, A, J)$ as in Theorem 7.5 (b). And $m_{[1]}^2$ acts on

$$\mathbb{C} \times V^{mat} = \{(\beta, s, B) \in \mathbb{C} \times \mathbb{C} \times SL(2, \mathbb{C}) \mid B = \begin{pmatrix} b_1 & b_2 \\ -b_2 & b_1 + s b_2 \end{pmatrix}\} \tag{10.5}$$

by the algebraic automorphism

$$m_{[1]}^2 : (\beta, s, B) \mapsto (\beta + 4\pi i, s, (Mon_0^{mat})^{-2} B). \tag{10.6}$$

Any choice of β with $\tfrac{1}{2} e^{-\beta/2} = x$ induces an isomorphism $M_{3FN}^{mon}(x) \to V^{mat}$.

The trivial foliation on $\mathbb{C} \times V^{mat}$ with leaves $\mathbb{C} \times \{(s, B)\}$ induces foliations on $\mathbb{C} \times V^{mat} / \langle m_{[1]}^2 \rangle$ and on M_{3FN}^{mon}. It lifts by the first isomorphism in (10.3) to the foliation on $(c^{2:1})^* M_{3TJ}^{mon}$ from Theorem 7.5 (c). The leaves are the maximal isomonodromic families within M_{3FN}.

(c) M_{3FN}^{ini} is given by four natural charts. The charts have coordinates $(x, f_k, \widetilde{g}_k)$ for $k = 0, 1, 2, 3$ and are isomorphic to $\mathbb{C}^* \times \mathbb{C} \times \mathbb{C}$. Each chart intersects each other chart in $\mathbb{C}^* \times \mathbb{C}^* \times \mathbb{C}$. The coordinate changes are given by (8.20) and

$$(x, f_k, g_k) = (x, f_k, -\frac{x}{f_k} + \frac{1}{2} + \frac{f_k}{2} \widetilde{g}_k). \tag{10.7}$$

The intersection of all charts is (with the induced algebraic structure) M_{3FN}^{reg} and is in each chart $\mathbb{C}^* \times \mathbb{C}^* \times \mathbb{C}$. M_{3FN}^{reg} is the set of pure twistors in M_{3FN}. The *k-th chart ($k = 0, 1, 2, 3$) consists of M_{3FN}^{reg} and a smooth hypersurface $M_{3FN}^{[k]}$ isomorphic to $\mathbb{C}^* \times \{0\} \times \mathbb{C}$, which is the set of $(1, -1)$-twistors in this chart.*

Proof

(a) This follows immediately from Theorems 7.5 and 8.4 and the observation above on the inessential isomonodromic families of P_{3D6}-TEJPA bundles.

(b) This follows from Theorem 7.5 and (10.1) and $y = 1$, which give

$$u_\infty^1 = u_0^1 = \tfrac{1}{2}e^{-\beta/2} = x. \tag{10.8}$$

(c) It remains to see that the charts and coordinates in Theorem 8.4 are compatible with the isomorphisms in the second line of (10.3), in other words a P_{3D6}-TEJPA bundle $T \in M_{3TJ}^{ini}(x, x)$, and its pull-back $m_y^* T \in M_{3TJ}^{ini}(\tfrac{x}{y}, xy)$ with $m_y : \mathbb{C} \to \mathbb{C}, z \mapsto yz$, have the same coordinates, which are (f_0, g_0) if $T \in M_{3TJ}^{reg}(x, x)$ and which are $\widetilde{g}_k$ if $T \in M_{3TJ}^{[k]}(x, x)$. But this follows from inspection of (8.15)–(8.18) and (8.21)–(8.24): Writing $u_0^1 = x/y, u_\infty^1 = xy, \sqrt{c} = y$, in all these formulae z turns up only as part of the product yz. $\square$

Remarks 10.2 For any leaf in M_{3FN}^{mon}, the restrictions of $f_0, g_0, \widetilde{g}_0$ to the leaf will turn out (see Theorem 10.3 (a) and (b)) to be multi-valued meromorphic functions on the underlying punctured plane $\mathbb{C}^*$ with coordinate x. In order to deal with these multivalued functions properly, and in order to be more concrete, we fix some elementary facts and notation.

(i) A leaf in M_{3FN}^{mon} with a distinguished branch means a leaf where one branch over $\mathbb{C} - \mathbb{R}_{\leq 0}$ is distinguished. Leaves with distinguished branches are naturally parameterized by V^{mat}: There is the natural isomorphism $M_{3FN}^{mon}(1) \cong V^{mat}$ from Lemma 10.1 with $\beta = 0, x = 1$. Any leaf is mapped to the intersection point of its distinguished branch with $M_{3FN}^{mon}(1)$.

(ii) A (meromorphic) function h on a leaf is a multi-valued (meromorphic) function in $x \in \mathbb{C}^*$. Distinguishing a branch of the leaf distinguishes a branch of the multi-valued function. A multi-valued function on $\mathbb{C}^*$ with a distinguished branch over $\mathbb{C} - \mathbb{R}_{\leq 0}$ corresponds to a single-valued holomorphic function on $\mathbb{C}$ which gives for $\xi \in \mathbb{C}$ with $\Im(\xi) \in (-\pi, \pi)$ the value at $e^\xi = x$ of the distinguished branch of the multi-valued function.

(iii) In this way, the function f_0 on M_{3FN}^{ini} induces a family parameterized by V^{mat} of multi-valued functions $f_{mult}(., s, B)$ in $x \in \mathbb{C}^*$ with distinguished branches and the corresponding single-valued functions $f_{univ}(., s, B)$ in $\xi \in \mathbb{C}$ with $e^\xi = x$. More concretely, f_{univ} is the meromorphic function

$$f_{univ} : \mathbb{C} \times V^{mat} \to \mathbb{C}, \tag{10.9}$$

$$f_{univ} = f_0 \circ (\Phi_{mon}^{-1} \circ pr_{m_{[1]}}),$$

with $pr_{m_{[1]}} : \mathbb{C} \times V^{mat} \to \mathbb{C} \times V^{mat} / \langle m_{[1]}^2 \rangle.$

Similarly, g_0 and $\widetilde{g}_0$ give g_{mult}, g_{univ} and $\widetilde{g}_{mult}, \widetilde{g}_{univ}$.

(iv) The coordinates β and ξ are related by

$$\tfrac{1}{2}e^{-\beta/2} = x = e^{\xi}, \quad -\beta/2 = \xi + \log 2. \tag{10.10}$$

With respect to the coordinates (ξ, s, B) on $\mathbb{C} \times V^{mat}$, the action of $m_{[1]}$ on $\mathbb{C} \times V^{mat}$ takes the form

$$m_{[1]} : (\xi, s, B) \to (\xi - i\pi, s, \mathrm{Mon}_0^{mat}(s)^{-1} B) \tag{10.11}$$

Because of (7.28), f_{univ}, g_{univ} and $\widetilde{g}_{univ}$ are invariant under this action on $\mathbb{C} \times V^{mat}$. If one leaf with a distinguished branch has the parameter (s, B), the same leaf where the next branch after one anticlockwise turn around 0 is distinguished, has the parameter $(s, \mathrm{Mon}_0^{mat}(s)^{-2} B)$.

(v) The following properties are equivalent because of (iv):

(α) A multi-valued function $f_{mult}(., s, B)$ has only finitely many branches.

(β) The leaf in M_{3FN}^{mon} with distinguished branch and with parameter (s, B) has finitely many branches.

(γ) The matrix $\mathrm{Mon}_0^{mat}(s)$ has finite order. Equivalently: its eigenvalue $\lambda_+(s)$ is a root of unity and $\lambda_+(s) \neq -1$. Equivalently: the value $\alpha_+(s)$ is in $(-\tfrac{1}{2}, \tfrac{1}{2}) \cap \mathbb{Q}$.

In particular, (α) $-$ (γ) depend only on s, not on B. If they hold, then $s \in (-2, 2)$ and the number of branches of the leaf and of the function $f_{mult}(., s, B)$ is $\min(l \in \mathbb{Z}_{>0} \,|\, \lambda_+(s)^{2l} = 1\}$.

The two algebraic manifolds $\mathbb{C} \times V^{mat}$ (with quotient $\mathbb{C} \times V^{mat}/\langle m_{[1]}^2 \rangle \cong M_{3FN}^{mon}$) and M_{3FN}^{ini} are related by the coordinates (x, f_0, g_0) considered as holomorphic functions in $(\beta, s, B) \in \mathbb{C} \times V^{mat}$, i.e., by $x = \tfrac{1}{2}e^{-\beta/2}$ and f_{univ} and g_{univ}.

The following theorem shows that the dependence of f_{mult} and g_{mult} on x is governed by $P_{III}(0,0,4,-4)$. It contains the Painlevé property for $P_{III}(0,0,4,-4)$. It controls also the initial conditions of solutions at singular points. The theorem is essentially due to [FN80] and is also used in [IN86, FIKN06, Ni09], though the manifolds are treated here more carefully than in these references.

Theorem 10.3

(a) f_{mult}, g_{mult} and $\widetilde{g}_{mult}$ restrict for any $(s, B) \in V^{mat}$ to multi-valued meromorphic functions on $\mathbb{C}^*$ with coordinate x. The function $f_{mult}(., s, B)$ satisfies $P_{III}(0,0,4,-4)$, and together with $g_{mult}(., s, B)$ it satisfies

$$2f_{mult}g_{mult} = x\partial_x f_{mult}. \tag{10.12}$$

(b) Let us call a local solution f of $P_{III}(0,0,4,-4)$ singular at $x_0 \in \mathbb{C}^*$ if f has at x_0 a zero or a pole, and regular otherwise, i.e. if $f(x_0) \in \mathbb{C}^*$.

For any leaf in M_{3FN}^{mon} with distinguished branch, a branch (not necessarily the distinguished one) of the corresponding solution $f_{mult}(., s, B)$ is regular at

$x_0 \in \mathbb{C}^*$ *if the corresponding intersection point of the leaf with* $M_{3FN}^{mon}(x_0)$ *is in* $M_{3FN}^{reg}(x_0)$, *and singular otherwise. More precisely, the following holds if the intersection point of the leaf with* $M_{3FN}^{mon}(x_0)$ *is in* $M_{3FN}^{[k]}(x_0)$ *(k and* $(\varepsilon_1, \varepsilon_2)$ *are related as in Theorem 8.2 (b)):*

$$f_{mult}^{\varepsilon_1}(x_0, s, B) = 0,$$

$$(\varepsilon_2 \partial_x f_{mult}^{\varepsilon_1})(x_0, s, B) = -2,$$

$$(\varepsilon_2 \partial_x^2 f_{mult}^{\varepsilon_1})(x_0, s, B) = \frac{-2}{x_0}, \tag{10.13}$$

$$(\varepsilon_2 \partial_x^3 f_{mult}^{\varepsilon_1})(x_0, s, B) = \frac{2}{x_0^2} + \frac{8}{x_0} \widetilde{g}_{k,mult}(x_0, s, B).$$

So, $f_{mult}(., s, B)$ *has a simple zero at* x_0 *if* $k = 0, 2$ *(* $\iff \varepsilon_1 = 1$ *), a simple pole at* x_0 *if* $k = 1, 3$ *(* $\iff \varepsilon_1 = -1$ *), and the two types of simple zeros (respectively, simple poles) are distinguished by the sign of* $\partial_x f_{mult}(x_0, s, B) = -2\varepsilon_2$ *(respectively,* $(\partial_x f_{mult}^{-1})(x_0, s, B) = -2\varepsilon_2$ *).*

(c) *The leaves in* M_{3FN}^{mon} *(without/with distinguished branches) parameterize the global multi-valued solutions of* $P_{III}(0, 0, 4, -4)$ *(without/with distinguished branches).*

(d) *(The Painlevé property) Any local solution of* $P_{III}(0, 0, 4, -4)$ *extends to a global multi-valued meromorphic solution on* $\mathbb{C}^*$ *with only simple zeros and simple poles.*

(e) *The space* $M_{3FN}^{reg}(x_0) \cong \mathbb{C}^* \times \mathbb{C}$ *with the coordinates* (f_0, g_0) *is the space of initial conditions at* x_0

$$(f(x_0), \partial_x f(x_0)) = (f_0, \frac{2f_0}{x_0} g_0) \tag{10.14}$$

for at x_0 *regular local solutions* f. *For* $k = 0, 1, 2, 3$, *the space* $M_{3FN}^{[k]}(x_0) \cong \mathbb{C}$ *with the coordinate* $\widetilde{g}_k$ *is the space of the initial condition*

$$\varepsilon_2 \partial_x^3 f^{\varepsilon_1}(x_0) = \frac{2}{x_0^2} + \frac{8}{x_0} \widetilde{g}_k \tag{10.15}$$

for the at x_0 *singular local solutions* f *with* $f^{\varepsilon_1}(x_0) = 0, \varepsilon_2 \partial_x f^{\varepsilon_1}(x_0) = -2$. *This initial condition determines such a singular local solution.*

(f) *The algebraic manifolds* M_{3FN}^{ini} *and* M_{3FN}^{reg} *and the foliation on* M_{3FN}^{mon} *had been considered in a different way in [Ok79], in fact for all* $P_{III}(D_6)$ *equations. The spaces* $M_{3FN}^{ini}(x_0)$, $x_0 \in \mathbb{C}^*$, *are Okamoto's spaces of initial conditions. Descriptions of* M_{3FN}^{ini} *by four charts are given in [MMT99, NTY02, Te07], but part (b) and the four types of singular initial conditions are not made explicit there.*

Proof

(a) By Lemma 8.5, any leaf intersects $M^{sing}_{3FN} := \cup^3_{k=0} M^{[k]}_{3FN}$ only in a discrete set of points. On the intersection of the leaf with M^{reg}_{3FN}, the restrictions of f_0, g_0 and $\widetilde{g}_0$ to the leaf are multi-valued holomorphic functions in x, and f_0 takes there values only in $\mathbb{C}^*$. Part (b) will show that the restrictions $f_{mult}(., s, B)$, $g_{mult}(., s, B)$ and $\widetilde{g}_{mult}(., s, B)$ of f_0, g_0 and $\widetilde{g}_0$ to the leaf are meromorphic at the intersection points with M^{sing}_{3FN}, with only simple zeros and simple poles. It remains to establish the differential equations for $f_{mult}(., s, B)$.

Let $U \subset$ (a leaf) $\cap M^{reg}_{3FN}$ be a small open subset in a leaf, and let $U' := pr_x(U) \subset \mathbb{C}^*$ be the (isomorphic to U) open subset in $\mathbb{C}^*$. Theorem 8.2 (b) provides a basis $\underline{\sigma}_0$ of $\Gamma(\mathbb{P}^1, \mathcal{O}(H))$ for any P_{3D6}-TEJPA bundle in U, with (8.15)–(8.18) with $u^1_0 = u^1_\infty = x$. The proof of Theorem 8.2 (b), Remark 8.1 (ii)+(iii), Theorem 7.3 (c) and the correspondence in (2.19) provide a 4-tuple of bases $\underline{e}^\pm_0, \underline{e}^\pm_\infty$ with (6.8)–(6.11), (7.10)–(7.12), and matrices $A^\pm_0 \in GL(2, \mathcal{A}_{|I^\pm_0})$, $A^\pm_\infty \in GL(2, (\rho^*_1 \mathcal{A})_{|I^\pm_\infty})$ with $\widehat{A}^+_{0/\infty} = \widehat{A}^-_{0/\infty}, \widehat{A}^\pm_0(0) = \widehat{A}^\pm_\infty(\infty) = 1_2$ and

$$\underline{\sigma}_{0|\widehat{I}^\pm_0} = \underline{e}^\pm_0 \begin{pmatrix} e^{-x/z} & 0 \\ 0 & e^{x/z} \end{pmatrix} A^\pm_0 C \text{ near } 0, \tag{10.16}$$

$$\underline{\sigma}_{0|\widehat{I}^\pm_\infty} = \underline{e}^\pm_\infty \begin{pmatrix} e^{-xz} & 0 \\ 0 & e^{xz} \end{pmatrix} A^\pm_\infty C \begin{pmatrix} f_0 & 0 \\ 0 & f_0^{-1} \end{pmatrix} \text{ near } \infty. \tag{10.17}$$

With respect to the isomonodromic extension of the connection ∇ of the single P_{3D6}-TEJPA bundles in U', the $\underline{e}^\pm_0, \underline{e}^\pm_\infty$ are flat families (in x) of flat sections (in z). The basis $\underline{\sigma}_0$ and the matrices $A^\pm_0, A^\pm_\infty$ depend holomorphically on x and z.

By (8.15), the pole parts at 0 and ∞ of $\nabla_{z\partial_z}\underline{\sigma}_0$ are $\underline{\sigma}_0 \frac{x}{z} \begin{pmatrix} 0 & 1 \\ 1 & 0 \end{pmatrix}$ and $\underline{\sigma}_0 (-xz) \begin{pmatrix} 0 & f_0^{-2} \\ f_0^2 & 0 \end{pmatrix}$. Because of (10.16) and (10.17) and

$$x\partial_x e^{-x/z} = -\frac{x}{z} = -z\partial_z e^{-x/z}, \quad x\partial_x e^{-xz} = -xz e^{-xz} = z\partial_z e^{-xz},$$

the pole parts at 0 and ∞ of $\nabla_{x\partial_x}\underline{\sigma}_0$ are $\underline{\sigma}_0 \frac{-x}{z} \begin{pmatrix} 0 & 1 \\ 1 & 0 \end{pmatrix}$ and $\underline{\sigma}_0 (-xz) \begin{pmatrix} 0 & f_0^{-2} \\ f_0^2 & 0 \end{pmatrix}$. By calculations with $x\partial_x$ analogously to those with $z\partial_z$ in (8.12) and (8.14), one obtains that the part between the pole parts has the form $\underline{\sigma}_0\, a(x,z)\begin{pmatrix}1 & 0 \\ 0 & -1\end{pmatrix}$ for some holomorphic function $a(x,z)$. Thus

$$\nabla_{x\partial_x}\underline{\sigma}_0 = \underline{\sigma}_0\left[\frac{-x}{z}\begin{pmatrix}0 & 1 \\ 1 & 0\end{pmatrix} + a\begin{pmatrix}1 & 0 \\ 0 & -1\end{pmatrix} - xz\begin{pmatrix}0 & f_0^{-2} \\ f_0^{2} & 0\end{pmatrix}\right]. \qquad (10.18)$$

The flatness of ∇ gives together with (8.15) and (10.18)

$$0 = (\nabla_{z\partial_z}\nabla_{x\partial_x} - \nabla_{x\partial_x}\nabla_{z\partial_z})\underline{\sigma}_0$$

$$= \underline{\sigma}_0\left[2\frac{x}{z}(a-g_0)\begin{pmatrix}0 & -1 \\ 1 & 0\end{pmatrix} + (-2x^2(f_0^2 - f_0^{-2}) + x\partial_x g_0)\begin{pmatrix}1 & 0 \\ 0 & -1\end{pmatrix}\right.$$

$$\left. + 2(g_0 + a)xz\begin{pmatrix}0 & f_0^{-2} \\ -f_0^2 & 0\end{pmatrix} + xz\,x\partial_x\begin{pmatrix}0 & f_0^{-2} \\ f_0^2 & 0\end{pmatrix}\right]. \qquad (10.19)$$

Thus with $f_0 = e^{\varphi/2}$ for a chosen branch $\varphi = 2\log(f_0)$

$$a = g_0, \qquad (10.20)$$

$$x\partial_x g_0 = 2x^2(f_0^2 - f_0^{-2}) = 4x^2\sinh\varphi, \qquad (10.21)$$

$$x\partial_x f_0^2 = 4g_0 f_0^2, \quad x\partial_x f_0^{-2} = -4g_0 f_0^{-2}, \quad x\partial_x\varphi = 4g_0, \quad (10.22)$$

$$(x\partial_x)^2\varphi = x\partial_x(4g_0) = 16x^2\sinh\varphi,$$

which is the radial sinh-Gordon equation (9.5) and which is equivalent to $P_{III}(0,0,4,-4)$ for $f_0 = e^{\varphi/2}$.

(b) Consider a leaf in M_{3FN}^{mon} and the corresponding solution f_0 of $P_{III}(0,0,4,-4)$. By Theorem 8.2 (g) and (8.20) $f_0(x_0) \in \mathbb{C}^*$ holds (for one branch of the multi-valued function f_0) if and only if the corresponding intersection point of the leaf with $M_{3FN}^{ini}(x_0)$ is in $M_{3FN}^{reg}(x_0)$.

Consider an intersection point of a leaf in M_{3FN}^{mon} with $M_{3FN}^{[k]}$, a neighbourhood U in the leaf of this point and the functions $(f_0, g_0)|_U$. Then $(\varepsilon_2 f_0^{\varepsilon_1}, g_k)|_U = (f_k, g_k)|_U$ by (8.20), and this pair is mapped by R_k (in Remark 8.3) to $(f_0, g_0)|_{R_k(U)}$, and the intersection point is mapped to an intersection point of the leaf R_k (old leaf) with $M_{3FN}^{[0]}$. Therefore it is sufficient to prove (10.13) for $k = 0$.

Suppose now $k = 0$. Then the corresponding branch of f_0 satisfies $f_0(x_0) = 0$ by Theorem 8.2 (g). Equation (10.7) extends in the form

$$f_0 g_0 = -x + \frac{1}{2}f_0 + \frac{1}{2}f_0^2\widetilde{g}_0$$

from M_{3FN}^{reg} to $M_{3FN}^{[0]}$. Together with (10.12) this shows

$$\partial_x f_0 = -2 + \frac{f_0}{x} + \frac{f_0^2}{x}\widetilde{g}_0,$$

$$\partial_x f_0(x_0) = -2,$$

$$\partial_x^2 f_0 = -\frac{f_0}{x^2} + \frac{\partial_x f_0}{x} - \frac{f_0^2}{x^2}\widetilde{g}_0 + \frac{2f_0\partial_x f_0}{x}\widetilde{g}_0 + \frac{f_0^2}{x}\partial_x \widetilde{g}_0,$$

$$\partial_x^2 f_0(x_0) = \frac{-2}{x_0},$$

$$\partial_x^3 f_0(x_0) = \frac{2}{x_0^2} + \frac{8}{x_0}\widetilde{g}_0(x_0).$$

(c)+(d)+(e) The initial conditions $(f(x_0), \partial_x f(x_0))$ for a local regular solution f give rise to a leaf through the point $(f_0, g_0) = (f(x_0), \frac{x_0}{2f(x_0)}\partial_x f(x_0))$. Then the restriction of f_0 to this leaf is the extension of the local solution f to a global multi-valued solution in $\mathbb{C}^*$. It has only simple zeros and simple poles by (b). This establishes the Painlevé property. The other statements follow directly from (a) and (b) as well.

(f) On the one hand, the equality of M_{3FN}^{ini}, M_{3FN}^{reg} and the foliation on M_{3FN}^{mon} with data of Okamoto follows from the uniqueness of Okamoto's data, which is rather obvious. On the other hand, one can directly and easily compare the four charts for M_{3FN}^{ini} with those in [MMT99] or [Te07], who derive their charts from Okamoto's description of his data.   □

Remarks 10.4

(i) The construction of M_{3FN}^{ini}, M_{3FN}^{reg}, M_{3FN}^{mon} and the foliation on M_{3FN}^{mon} here is independent of [Ok79], in contrast to [MMT99, NTY02, Te07]. But [Ok79] provides for any space $M_{3FN}^{ini}(x_0)$ of initial conditions a natural compactification S and a divisor $Y \subset S$ of type $\widetilde{D}_6$ such that $M_{3FN}^{ini}(x_0) \cong S - Y$. It would be interesting to recover S and Y using generalizations of P_{3D6}-TEJPA bundles, and possibly derive new results on the solutions of $P_{III}(0,0,4,-4)$.

(ii) For any $P_{III}(D_6)$-equation, all solutions f satisfy a generalization of (10.13). This means that they have only simple zeros and simple poles, and there are two types of simple zeros and two types of simple poles, distinguished by the sign of $\partial_x f(x_0)$ or $\partial_x f^{-1}(x_0)$, and the initial condition at a simple zero (respectively, pole) is $\partial_x^3 f(x_0)$ (respectively, $\partial_x^3 f^{-1}(x_0)$). This can be proved by a power series ansatz.

(iii) By Theorem 10.3, the algebraic manifold M_{3FN}^{ini} has four algebraic hypersurfaces, one for each type of simple zeros or simple poles of the solutions, and these are precisely the hypersurfaces for $(1,-1)$-twistors (cf. Remark 4.4 (ii)).

The union of all spaces of initial conditions of all $P_{III}(D_6)$-equations is an algebraic [MMT99, Te07] manifold and has by (ii) four analytic (in fact, algebraic) hypersurfaces, one for each type of simple zeros or simple poles of the solutions. But in the linear system of [JM81] for the $P_{III}(D_6)$-equations, only one of the four hypersurfaces corresponds to $(1, -1)$-twistors. For the linear system for $P_{III}(0, 4, 4, -4)$ this follows from the 4:1 folding transformation which is mentioned in Remark 9.1 (ii) and which connects the P_{3D6}-TEJPA bundles for $P_{III}(0, 0, 4, -4)$ with the linear systems in [JM81] for $P_{III}(0, 4, 4, -4)$. The general case will be proved elsewhere, together with the claims in Remark 9.1.

(iv) In this monograph we do not make use of the Hamiltonian description of the $P_{III}(D_6)$-equations. For $P_{III}(0, 0, 4, -4)$ it is implicit in the coordinates $\widetilde{g}_k$ and the equations (10.7) and (10.12).

(v) The matrices $A_0^{\pm}(x, z)$ and $A_{\infty}^{\pm}(x, z)$ in (10.16) and (10.17) deserve more careful consideration. The properties of the basis σ_0 and the bases $\underline{e}_0^{\pm}, \underline{e}_{\infty}^{\pm}$ with respect to A, P and J imply the following. A gives for $A_0^{\pm}$

$$A_0^{\mp}(x, -z) = \begin{pmatrix} 0 & -1 \\ 1 & 0 \end{pmatrix} A_0^{\pm}(x, z) \begin{pmatrix} 0 & 1 \\ -1 & 0 \end{pmatrix}. \tag{10.23}$$

P gives for $A_0^{\pm}$

$$A_0^{\mp}(x, -z) = (A_0^{\pm}(x, z)^t)^{-1}. \tag{10.24}$$

Equation (10.23) and (10.24) are equivalent to (10.23) and $\det A_0^{\pm}(x, z) = 1$. J gives for $A_0^{\pm}$ and $A_{\infty}^{\pm}$

$$A_{\infty}^{\pm}(x, z) = \begin{pmatrix} 1 & 0 \\ 0 & -1 \end{pmatrix} A_0^{\pm}(x, \frac{1}{z}) \begin{pmatrix} 1 & 0 \\ 0 & -1 \end{pmatrix} \tag{10.25}$$

(here $c = 1$ because of $u_0^1 = u_{\infty}^1 = x$). Defining the matrices $B_{0/\infty}^{\pm}(x, z)$ by

$$C B_{0/\infty}^{\pm}(x, z) = A_{0/\infty}^{\pm}(x, z) C, \tag{10.26}$$

(10.23)–(10.25) are equivalent to

$$B_0^{\mp}(x, -z) = \begin{pmatrix} 1 & 0 \\ 0 & -1 \end{pmatrix} B_0^{\pm}(x, z) \begin{pmatrix} 1 & 0 \\ 0 & -1 \end{pmatrix}, \tag{10.27}$$

$$\det B_0^{\pm}(x, z) = 1, \tag{10.28}$$

$$B_{\infty}^{\pm}(x, z) = \begin{pmatrix} 0 & 1 \\ 1 & 0 \end{pmatrix} B_0^{\pm}(x, \frac{1}{z}) \begin{pmatrix} 0 & 1 \\ 1 & 0 \end{pmatrix}. \tag{10.29}$$

In particular

$$\widehat{B}_0^{\pm}(x,z) = \mathbf{1}_2 + z \begin{pmatrix} 0 & \beta_1 \\ \beta_2 & 0 \end{pmatrix} + z^2 \begin{pmatrix} \beta_3 & 0 \\ 0 & \beta_4 \end{pmatrix} + \dots \qquad (10.30)$$

where $\beta_1, \beta_2, \beta_3, \beta_4$ are holomorphic functions in x with $\beta_3 + \beta_4 - \beta_1\beta_2 = 0$, and

$$(\widehat{B}_0^{\pm}(x,z))^{-1} = \mathbf{1}_2 - z \begin{pmatrix} 0 & \beta_1 \\ \beta_2 & 0 \end{pmatrix} + z^2 \begin{pmatrix} \beta_4 & 0 \\ 0 & \beta_3 \end{pmatrix} + \dots \qquad (10.31)$$

Taking the derivatives of (10.16) and (10.17) by $z\partial_z$ and $x\partial_x$ gives (8.15) and (10.18) with

$$x(\beta_1 - \beta_2) = g_0 = a = -x(\beta_1 - \beta_2) + f_0^{-1}x\partial_x f_0, \qquad (10.32)$$

$$-2\beta_3 + \beta_2^2 - \frac{\beta_2}{x} = f_0^2 = 2\beta_3 - \beta_2^2 - \partial_x\beta_2, \qquad (10.33)$$

$$-2\beta_4 + \beta_1^2 - \frac{\beta_1}{x} = f_0^{-2} = 2\beta_4 - \beta_1^2 - \partial_x\beta_1. \qquad (10.34)$$

One recovers $2f_0g_0 = x\partial_x f_0$ from (10.32) and $x\partial_x g_0 = 2x^2(f_0^2 - f_0^{-2})$ from (10.32)–(10.34). This is an alternative proof of most of Theorem 10.3 (a).

(vi) Any control of the dependence of the (multi-valued) solutions $f_{mult}(x, s, B)$ in x of $P_{III}(0,0,4,-4)$ on the parameters s and B of the monodromy tuple would be very welcome. For example, are there differential equations governing the dependence on s or B? We do not see any and wonder whether there might be reasons (such as Umemura's results on irreducibility) which prevent their existence.

Remarks 10.5 For each u_0^1, u_∞^1 there are two completely reducible P_{3D6}-TEP bundles (Theorem 6.3 (c)), and above each of them are two P_{3D6}-TEJPA bundles (Theorem 7.6 (b)). The two P_{3D6}-TEJPA bundles over the P_{3D6}-TEP bundle with $(C_{11}), (C_{22})$ have monodromy data $(s, B) = (0, \pm\mathbf{1}_2)$, and they are fixed points of the action of R_1 on $M_{3TJ}^{mon}(u_0^1, u_\infty^1)$. The two P_{3D6}-TEJPA bundles over the P_{3D6}-TEP bundle with $(C_{12}), (C_{21})$ have monodromy data $(s, B) = (0, \pm\begin{pmatrix} 0 & 1 \\ -1 & 0 \end{pmatrix})$, and they are fixed points of the action of R_3 on $M_{3TJ}^{mon}(u_0^1, u_\infty^1)$.

As R_1, R_2 and R_3 map the coordinate f_0 on M_{3TJ}^{reg} to $f_0^{-1}, -f_0$ and $-f_0^{-1}$, respectively (remark (8.3)), we find

$$f_{mult}(x, 0, \pm\mathbf{1}_2) = \varepsilon_1 (\pm 1), \qquad (10.35)$$

$$f_{mult}(x, 0, \pm\begin{pmatrix} 0 & 1 \\ -1 & 0 \end{pmatrix}) = \varepsilon_2 (\pm i). \qquad (10.36)$$

for any x, with some yet to be determined signs $\varepsilon_1, \varepsilon_2 \in \{\pm 1\}$.

Consider the case $(s, B) = (0, 1_2)$. Then the four flat bases $e_0^\pm, e_\infty^\pm$ are restrictions of one global flat basis $\underline{e}$. It satisfies

$$J(\underline{e}(z)) = \underline{e}(\rho_c(z)) \begin{pmatrix} 1 & 0 \\ 0 & -1 \end{pmatrix}.$$

The basis $\underline{\sigma}_0$ from Theorem 8.2 (b) is

$$\underline{\sigma}_0(z) = \underline{e}(z) \begin{pmatrix} e^{-u_0^1/z - u_\infty^1 z} & 0 \\ 0 & e^{u_0^1/z + u_\infty^1 z} \end{pmatrix} C. \tag{10.37}$$

With

$$u_0^1/z = u_\infty^1 \, \rho_c(z) \text{ and } u_\infty^1 \, z = u_0^1/\rho_c(z)$$

one calculates

$$\underline{\sigma}_0(\rho_c(z)) \begin{pmatrix} 0 & f_0^{-1} \\ f_0 & 0 \end{pmatrix} = J(\underline{\sigma}_0(z))$$

$$= \underline{e}(\rho_c(z)) \begin{pmatrix} 1 & 0 \\ 0 & -1 \end{pmatrix} \begin{pmatrix} e^{-u_0^1/z - u_\infty^1 z} & 0 \\ 0 & e^{u_0^1/z + u_\infty^1 z} \end{pmatrix} C$$

$$= \underline{\sigma}_0(\rho_c(z)) \begin{pmatrix} 0 & 1 \\ 1 & 0 \end{pmatrix}. \tag{10.38}$$

This shows $\varepsilon_1 = 1$.

Consider the case $(s, B) = (0, \pm \begin{pmatrix} 0 & 1 \\ -1 & 0 \end{pmatrix})$. Then the two bases $e_0^\pm$ are restrictions of a global flat basis $\underline{e}_0$, the two bases $e_\infty^\pm$ are restrictions of a global flat basis $\underline{e}_\infty$, and

$$\underline{e}_\infty = \underline{e}_0 \begin{pmatrix} 0 & 1 \\ -1 & 0 \end{pmatrix}.$$

Again

$$J(\underline{e}_0(z)) = \underline{e}_\infty(\rho_c(z)) \begin{pmatrix} 1 & 0 \\ 0 & -1 \end{pmatrix}.$$

The basis $\underline{\sigma}_0$ from Theorem 8.2 (b) is

$$\underline{\sigma}_0(z) = \underline{e}_0(z) \begin{pmatrix} e^{-u_0^1/z + u_\infty^1 z} & 0 \\ 0 & e^{u_0^1/z - u_\infty^1 z} \end{pmatrix} C. \tag{10.39}$$

Similarly as above, one calculates

$$\underline{\sigma}_0(\rho_c(z)) \begin{pmatrix} 0 & f_0^{-1} \\ f_0 & 0 \end{pmatrix} = J(\underline{\sigma}_0(z)) = \underline{\sigma}_0(\rho_c(z)) \begin{pmatrix} 0 & -i \\ i & 0 \end{pmatrix}. \qquad (10.40)$$

This shows $\varepsilon_2 = 1$.

Chapter 11
Comparison with the Setting of Its, Novokshenov, and Niles

Theorem 10.3 fixes the relation between the solutions of $P_{III}(0, 0, 4, -4)$ and the isomonodromic families of P_{3D6}-TEJPA bundles. As said above, this relation is due to [FN80], though without the vector bundle language and the structural data P, A, J which incorporate the symmetries.

Good use of this relation has been made in [IN86, FIKN06, Ni09]. In particular, [IN86] connects for many, but not all, solutions of $P_{III}(0, 0, 4, -4)$ their asymptotic behaviour as $x \to 0$ and $x \to \infty$ with the Stokes data and the central connection matrix of the corresponding isomonodromic families of P_{3D6}-TEJPA bundles. Reference [FIKN06] rewrites and extends [IN86].

Reference [Ni09] builds on [IN86, FIKN06] and provides asymptotic formulae as $x \to 0$ for all solutions of $P_{III}(0, 0, 4, -4)$. These formulae and a proof of Theorem 4.2 for the case of $P_{III}(0, 0, 4, -4)$ are the two main results of [Ni09]. In Chap. 12 we shall rewrite Niles' asymptotic formulae for $x \to 0$ and simplify and slightly extend them.

As preparation for that, and in order to facilitate access to the rich results in [IN86, FIKN06], we shall now connect our setting with that in [Ni09] (and hence [IN86, FIKN06]).

As we have seen, equation $P_{III}(0, 0, 4, -4)$ (9.1) for f can be rewritten, using $f = e^{\varphi/2}$, as the radial sinh-Gordon equation $(x\partial_x)^2 \varphi = 16x^2 \sinh \varphi$ (9.5). In [Ni09, (1.1)] the sine-Gordon equation

$$(\partial^2_{x^{NI}} + \frac{1}{x^{NI}} \partial_{x^{NI}}) u^{NI}(x^{NI}) = -\sin u^{NI}(x^{NI}) \tag{11.1}$$

is used. Recall that $(x\partial_x)^2 = x^2(\partial_x^2 + \frac{1}{x}\partial_x)$. We have:

© Springer International Publishing AG 2017
M.A. Guest, C. Hertling, *Painlevé III: A Case Study in the Geometry of Meromorphic Connections*, Lecture Notes in Mathematics 2198, DOI 10.1007/978-3-319-66526-9_11

Lemma 11.1 *Equations (9.5) and (11.1) are related by*

$$x^{NI} = 4x, \tag{11.2}$$

$$u^{NI}(x^{NI}) = u^{NI}(4x) = u(x) = i\varphi(x) + \pi, \tag{11.3}$$

$$\varphi(x) = -iu(x) + i\pi.$$

Here $u(x)$ satisfies the sine-Gordon equation

$$(\partial_x^2 + \tfrac{1}{x}\partial_x)u = -16\sin u, \quad \text{i.e. } (x\partial_x)^2 u = -16x^2 \sin u. \tag{11.4}$$

and $f = e^{\varphi/2}$ satisfies

$$f(x) = e^{(-iu(x)+i\pi)/2} = ie^{-iu(x)/2}. \tag{11.5}$$

Remarks 11.2

(i) NI stands for Niles-Novokshenov-Its.
(ii) $u(x)$ is introduced in Lemma 11.1 so that we can rewrite below the formulae in
    [Ni09] with x and $u(x)$ instead of x^{NI} and $u^{NI}(x^{NI})$.

Reference [Ni09] considers in formulae (1.25)–(1.28) the first order matrix
differential equation

$$z\partial_z \Psi(z) = zA(z)\Psi(z) \quad \text{with} \tag{11.6}$$

$$zA(z) = \frac{i}{z} \begin{pmatrix} \cos u & i\sin u \\ -i\sin u & -\cos u \end{pmatrix} + g_0 \begin{pmatrix} 0 & 1 \\ 1 & 0 \end{pmatrix} + ix^2 z \begin{pmatrix} 1 & 0 \\ 0 & -1 \end{pmatrix}.$$

Here we have called x, $u(x)$, λ, w in [Ni09] x^{NI}, $u^{NI}(x^{NI})$, z, g_0 and replaced x^{NI} and
$u^{NI}(x^{NI})$ by $4x$ and $u(x)$.

We associate with (11.6) the trivial holomorphic vector bundle H of rank 2 on $\mathbb{P}^1$
with global basis $\underline{\rho} = (\rho_1, \rho_2)$ and meromorphic connection ∇ given by

$$\nabla_{z\partial_z}\underline{\rho} = \underline{\rho}(-zA(z)). \tag{11.7}$$

Then Ψ is a solution of (11.6) in a sector of $\mathbb{C}^*$ if and only if $\underline{\rho}\,\Psi$ is a flat basis of
(H, ∇) in this sector.

Lemma 11.3

(a) *The pair (H, ∇) takes the normal form (8.15)*

$$\nabla_{z\partial_z}\underline{\sigma}_0 = \underline{\sigma}_0 \left[\frac{u_0^1}{z} \begin{pmatrix} 0 & 1 \\ 1 & 0 \end{pmatrix} - g_0 \begin{pmatrix} 1 & 0 \\ 0 & -1 \end{pmatrix} - u_\infty^1 z \begin{pmatrix} 0 & f_0^{-2} \\ f_0^2 & 0 \end{pmatrix} \right] \tag{11.8}$$

of a pure P$_{3D6}$-TEJPA bundle with

$$f_0 = e^{\varphi/2} = ie^{-iu/2}, \quad \varphi = -iu + i\pi, \tag{11.9}$$

$$u_0^1 = -u_0^2 = -i, \ u_\infty^1 = -u_\infty^2 = ix^2, \tag{11.10}$$

$$\underline{\sigma}_0 = \underline{\rho}\, P_0 \begin{pmatrix} 1 & 0 \\ 0 & i \end{pmatrix} \quad C = \underline{\rho}\,(-i) \begin{pmatrix} e^{\varphi/2} & -e^{-\varphi/2} \\ e^{\varphi/2} & e^{-\varphi/2} \end{pmatrix}, \tag{11.11}$$

$$P_0 = \begin{pmatrix} \cos\frac{u}{2} & -i\sin\frac{u}{2} \\ -i\sin\frac{u}{2} & \cos\frac{u}{2} \end{pmatrix} = (-i) \begin{pmatrix} \sinh\frac{\varphi}{2} & \cosh\frac{\varphi}{2} \\ \cosh\frac{\varphi}{2} & \sinh\frac{\varphi}{2} \end{pmatrix} \tag{11.12}$$

and C as in (8.5).

(b) *(Definition) We enrich the pair (H, ∇) to the P$_{3D6}$-TEJPA bundle in (8.15)–(8.18) with $u_0^1, u_\infty^1, f_0, g_0$ as above.*

Furthermore, the two points in $(c^{2:1})^* M_{3TJ}^{reg}$ above this P$_{3D6}$-TEJPA bundle have the coordinates $(x, y, f_0, g_0) = (x, ix, f_0, g_0)$ and $(-x, -ix, f_0, g_0)$. We distinguish the point (x, ix, f_0, g_0).

(Lemma) Then f_0, φ and u restricted to the leaf through this point are solutions of $P_{III}(0, 0, 4, -4)$, (9.5) and (11.4), respectively.

Proof

(a) This is an elementary calculation, which we omit.

(b) Only the last statement is not a definition. It follows from Theorem 10.3. □

Lemma 11.3 connects the normal form (11.6) above from [Ni09, (1.25)–(1.28)] with the normal form in Theorem 8.2 (b) respectively with a point in $(c^{2:1})^* M_{3TJ}^{reg}$. Now we shall connect the Stokes data and the central connection matrix in [Ni09] with those in Theorems 6.3 (e) and 7.5 (c).

The following data are fixed in [Ni09, 1.4.1]. A small $\varepsilon > 0$ is chosen. Only x in the sector

$$S^{NI} := \{x \in \mathbb{C}^* \mid \arg x \in (-\tfrac{\pi}{4} + \varepsilon, \tfrac{\pi}{4} - \varepsilon) \bmod 2\pi\} \tag{11.13}$$

is considered. The four sectors for $k = 1, 2$,

$$\Omega_k^{(0)} := \{z \in \mathbb{C}^* \mid \arg z \in (\pi(k-2), \pi k) \bmod 2\pi\}, \tag{11.14}$$

$$\Omega_k^{(\infty)} := \{z \in \mathbb{C}^* \mid \arg z \in (\pi(k-\tfrac{3}{2})-2\varepsilon, \pi(k-\tfrac{1}{2}) + 2\varepsilon) \bmod 2\pi\}, \tag{11.15}$$

are used, and in each of them a solution $\Psi_k^{(0)}$ or $\Psi_k^{(\infty)}$ of (11.6) is considered.

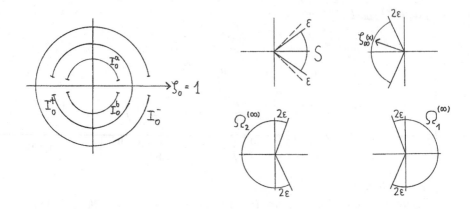

These solutions are fixed by their asymptotics near $z = 0$ respectively $z = \infty$,

$$\Psi_k^{(0)}(z) := A_k^{(0)}(z) \begin{pmatrix} e^{-i/z} & 0 \\ 0 & e^{i/z} \end{pmatrix} \tag{11.16}$$

with $A_k^{(0)} \in GL(2, \mathcal{A}|_{\Omega_k^{(0)} \cap S^1}), A_k^{(0)}(0) = P_0$,

$$\Psi_k^{(\infty)}(z) := A_k^{(\infty)}(z) \begin{pmatrix} e^{-ix^2 z} & 0 \\ 0 & e^{ix^2 z} \end{pmatrix} \tag{11.17}$$

with $A_k^{(\infty)} \in GL(2, (\rho_1^* \mathcal{A})|_{\Omega_k^{(\infty)} \cap S^1}), A_k^{(\infty)}(\infty) = \mathbf{1}_2$.

Then four Stokes matrices $S_k^{(0)}$ and $S_k^{(\infty)}$, $k = 1, 2$, are defined by

$$S_k^{(0)}(z) := \Psi_k^{(0)}(z)^{-1} \Psi_{k\pm1}^{(0)}(z) \quad \text{for } z \text{ with}$$
$$\arg z \in (\pi(k-1), \pi k) \bmod 2\pi,$$
$$S_k^{(\infty)}(z) := \Psi_k^{(\infty)}(z)^{-1} \Psi_{k\pm1}^{(\infty)}(z) \quad \text{for } z \text{ with} \tag{11.18}$$
$$\arg z \in (\pi(k-\tfrac{1}{2}) - 2\varepsilon, \pi(k-\tfrac{1}{2}) + 2\varepsilon) \bmod 2\pi,$$

and a connection matrix by

$$E^{NI} := \Psi_1^{(0)}(z)^{-1} \Psi_1^{(\infty)}(z) \quad \text{for } z \in \Omega_1^{(\infty)} (\subset \Omega_1^{(0)}). \tag{11.19}$$

The Stokes matrices satisfy

$$S_1^{(\infty)} = S_2^{(0)} = \begin{pmatrix} 1 & 0 \\ \zeta^{NI} & 1 \end{pmatrix}, \quad S_2^{(\infty)} = S_1^{(0)} = \begin{pmatrix} 1 & \zeta^{NI} \\ 0 & 1 \end{pmatrix} \tag{11.20}$$

for some $\zeta^{NI} \in \mathbb{C}$. For the connection matrix E^{NI}, two cases are distinguished, the special case (also called the *separatrix case* in [Ni09]):

$$E^{NI} = \pm i \begin{pmatrix} 0 & 1 \\ 1 & 0 \end{pmatrix}, \tag{11.21}$$

and the general case:

$$E^{NI} = \frac{\pm 1}{\sqrt{1 + pq}} \begin{pmatrix} 1 & p \\ -q & 1 \end{pmatrix} \tag{11.22}$$

for some $p, q \in \mathbb{C}$ with $pq \neq -1$ and $\zeta^{NI} = p + q$.

Lemma 11.4

(a) *For $u_0^1, u_0^2, u_\infty^1(x), u_\infty^2(x)$ as in (11.10) and $x \in S^{NI}$, the data in (2.1) and (2.2) are related as follows to the sectors $\Omega_k^{(0)}, \Omega_k^{(\infty)}$ in [Ni09, 1.4.1].*

$$\zeta_0 = 1, \quad \zeta_\infty(x) = -|x^2|/x^2, \tag{11.23}$$

$$\widehat{I}_0^+ = \mathbb{C}^* - \mathbb{R}_{>0}(-\zeta_0) = \Omega_1^{(0)}, \quad \widehat{I}_0^- = \mathbb{C}^* - \mathbb{R}_{>0}\zeta_0 = \Omega_2^{(0)}, \tag{11.24}$$

$$\widehat{I}_\infty^+(x) = \mathbb{C}^* - \mathbb{R}_{>0}(-\zeta_\infty(x)), \quad \widehat{I}_\infty^-(x) = \mathbb{C}^* - \mathbb{R}_{>0}\zeta_\infty(x), \tag{11.25}$$

$$\bigcap_{x \in S^{NI}} \widehat{I}_\infty^+(x) = \overline{\Omega_2^{(\infty)}} - \{0\}, \quad \bigcap_{x \in S^{NI}} \widehat{I}_\infty^-(x) = \overline{\Omega_1^{(\infty)}} - \{0\}. \tag{11.26}$$

(b) *The P_{3D6}-TEJPA bundle associated in Lemma 11.3 (b) to (11.6) comes equipped with the canonical 4-tuple $\underline{e}_0^\pm, \underline{e}_\infty^\pm$ of flat bases from Theorem 7.3 (c). For an isomonodromic family with $x \in S^{NI}$ the bases $\underline{e}_0^+, \underline{e}_0^-, \underline{e}_\infty^+, \underline{e}_\infty^-$ are constant (i.e. they depend flatly on x) on $\Omega_1^{(0)}, \Omega_2^{(0)}, \Omega_2^{(\infty)}, \Omega_1^{(\infty)}$, respectively, in view of (11.24) and (11.26). We have*

$$\underline{e}_0^+ = \underline{\rho}\,\Psi_1^{(0)} \begin{pmatrix} 1 & 0 \\ 0 & i \end{pmatrix}, \quad \underline{e}_0^- = \underline{\rho}\,\Psi_2^{(0)} \begin{pmatrix} 1 & 0 \\ 0 & i \end{pmatrix}, \tag{11.27}$$

$$\underline{e}_\infty^+ = \underline{\rho}\,\Psi_2^{(\infty)} \begin{pmatrix} 0 & 1 \\ -i & 0 \end{pmatrix}, \quad \underline{e}_\infty^- = \underline{\rho}\,\Psi_1^{(\infty)} \begin{pmatrix} 0 & 1 \\ -i & 0 \end{pmatrix}. \tag{11.28}$$

Let s and $B(\beta)$ be the Stokes parameter and the central connection matrix from (6.13) (for any β satisfying (2.23), i.e. $e^{-\beta} = 4x^2$) for $\underline{e}_0^\pm, \underline{e}_\infty^\pm$. We have

$$s = i\zeta^{NI}, \tag{11.29}$$

and, for the unique β with $\Im(\beta) \in (-\frac{\pi}{2} + 2\varepsilon, \frac{\pi}{2} - 2\varepsilon)$,

$$B(\beta) = \begin{pmatrix} b_1 & b_2 \\ -b_2 & b_1 + sb_2 \end{pmatrix} = \begin{pmatrix} 1 & 0 \\ 0 & -i \end{pmatrix} E^{NI} \begin{pmatrix} 0 & 1 \\ -i & 0 \end{pmatrix}, \tag{11.30}$$

$$E^{NI} = \begin{pmatrix} 1 & 0 \\ 0 & i \end{pmatrix} B(\beta) \begin{pmatrix} 0 & i \\ 1 & 0 \end{pmatrix} = \begin{pmatrix} b_2 & ib_1 \\ i(b_1 + sb_2) & b_2 \end{pmatrix}. \tag{11.31}$$

(c) For this β the special case (11.21) is the case

$$b_2 = 0, \quad b_1 \in \{\pm 1\}, \quad E^{NI} = ib_1 \begin{pmatrix} 0 & 1 \\ 1 & 0 \end{pmatrix}, \tag{11.32}$$

and the general case (11.22) is the case

$$b_2 \neq 0, \quad p = \frac{ib_1}{b_2}, \quad q = \frac{-i(b_1 + sb_2)}{b_2}, \quad E^{NI} = b_2 \begin{pmatrix} 1 & p \\ -q & 1 \end{pmatrix}. \tag{11.33}$$

For fixed x and fixed u_0^1, u_∞^1, β with $e^{-\beta} = 4x^2, \Im(\beta) \in (-\frac{\pi}{2} + 2\varepsilon, \frac{\pi}{2} - 2\varepsilon)$, $M_{3TJ}^{mon}(-i, ix^2)$ and V^{mat} can be identified with the manifold obtained by gluing two copies of the manifold

$$\{\zeta^{NI} \in \mathbb{C}\} \cong \mathbb{C},$$

(for the special case (11.21), taking both signs) and the double cover of the manifold

$$\{(\zeta^{NI}, p, q) \in \mathbb{C}^3 \mid pq \neq -1, \zeta^{NI} = p + q\},$$

on which $\sqrt{1 + pq}$ becomes a well defined function (for the general case (11.22), taking both signs).

For fixed (ζ^{NI}, p, q), the two choices (s, b_1, b_2) and $(s, -b_1, -b_2)$ correspond to two solutions f_0 and $-f_0$ of $P_{III}(0, 0, 4, -4)$.

Proof

(a) This follows from the definitions of ζ_0, ζ_∞ and $\hat{I}_0^\pm, \hat{I}_\infty^\pm(x), \Omega_k^{(0)}, \Omega_k^{(\infty)}$.

(b) First (11.27) will be proved. The global holomorphic bases $\underline{\sigma}_0$ and $\underline{p}$ are related by (11.11). The basis $\underline{\sigma}_0$ and the flat basis $\underline{e}_0^\pm$ on the sector $\hat{I}_0^\pm$ are related by

$$\underline{\sigma}_0 = \underline{e}_0^{\pm} \begin{pmatrix} e^{i/z} & 0 \\ 0 & e^{-i/z} \end{pmatrix} A_0^{\pm} C \tag{11.34}$$

with $A_0^{\pm} \in GL(2, \mathcal{A}|_{I_0^{\pm}})$, $\widehat{A}_0^{+} = \widehat{A}_0^{-}$, $\widehat{A}_0^{\pm}(0) = \mathbb{1}_2$. The basis $\underline{\rho}$ and the flat basis $\underline{\rho} \, \Psi_k^{(0)}$ on $\Omega_k^{(0)}$ are related by

$$\underline{\rho} = \underline{\rho} \, \Psi_k^{(0)} (\Psi_k^{(0)})^{-1} = \underline{\rho} \, \Psi_k^{(0)} \begin{pmatrix} e^{i/z} & 0 \\ 0 & e^{-i/z} \end{pmatrix} (A_k^{(0)})^{-1} \tag{11.35}$$

with $A_k^{(0)} \in GL(2, \mathcal{A}|_{\Omega_k^{(0)} \cap S^1})$, $A_k^{(0)}(0) = P_0$. Therefore (with $+ \leftrightarrow 1, - \leftrightarrow 2$)

$$\underline{e}_0^{\pm} = \underline{\rho} \, \Psi_{1/2}^{(0)} \begin{pmatrix} e^{i/z} & 0 \\ 0 & e^{-i/z} \end{pmatrix} (A_{1/2}^{(0)})^{-1} P_0 \begin{pmatrix} 1 & 0 \\ 0 & i \end{pmatrix} C$$

$$\times C^{-1} (A_0^{\pm})^{-1} \begin{pmatrix} e^{-i/z} & 0 \\ 0 & e^{i/z} \end{pmatrix} = \underline{\rho} \, \Psi_{1/2}^{(0)} \begin{pmatrix} 1 & 0 \\ 0 & i \end{pmatrix}.$$

Now (11.28) will be proved. The basis $\underline{\sigma}_0$ and the flat basis $\underline{e}_\infty^{\pm}$ on the sector $\widehat{I}_\infty^{\pm}(x)$ are related by

$$J(\underline{e}_0^{\pm}(z)) = \underline{e}_\infty^{\pm}(\rho_c(z)) \begin{pmatrix} 1 & 0 \\ 0 & -1 \end{pmatrix},$$

$$J(\underline{\sigma}_0(z)) = \underline{\sigma}_0(\rho_c(z)) \begin{pmatrix} 0 & f_0^{-1} \\ f_0 & 0 \end{pmatrix} = \underline{\sigma}_0(\rho_c(z)) \begin{pmatrix} 0 & e^{-\varphi/2} \\ e^{\varphi/2} & 0 \end{pmatrix},$$

so $\quad \underline{\sigma}_0(z) \begin{pmatrix} 0 & e^{-\varphi/2} \\ e^{\varphi/2} & 0 \end{pmatrix} = J(\underline{\sigma}_0(\rho_c(z)))$

$$= \underline{e}_\infty^{\pm}(z) \begin{pmatrix} 1 & 0 \\ 0 & -1 \end{pmatrix} \begin{pmatrix} e^{i/\rho_c(z)} & 0 \\ 0 & e^{-i/\rho_c(z)} \end{pmatrix} A_0^{\pm}(\rho_c(z)) C$$

$$= \underline{e}_\infty^{\pm}(z) \begin{pmatrix} 1 & 0 \\ 0 & -1 \end{pmatrix} \begin{pmatrix} e^{-ix^2 z} & 0 \\ 0 & e^{ix^2 z} \end{pmatrix} A_0^{\pm}(-1/x^2 z) C.$$

The basis $\underline{\rho}$ and the flat basis $\underline{\rho} \, \Psi_k^{(\infty)}$ are related by

$$\underline{\rho} = \underline{\rho} \, \Psi_k^{(\infty)} (\Psi_k^{(\infty)})^{-1} = \underline{\rho} \, \Psi_k^{(\infty)} \begin{pmatrix} e^{ix^2 z} & 0 \\ 0 & e^{-ix^2 z} \end{pmatrix} A_k^{(\infty)}(z)^{-1}$$

with $A_k^{(\infty)} \in GL(2, \mathcal{A}|_{\Omega_k^{(\infty)} \cap S^1})$, $A_k^{(\infty)}(\infty) = 1_2$. Therefore (with $+ \leftrightarrow 2, - \leftrightarrow 1$)

$$\underline{e}_\infty^\pm = \underline{\rho}\, \Psi_{2/1}^{(k)} \begin{pmatrix} e^{ix^2 z} & 0 \\ 0 & e^{-x^2 z} \end{pmatrix} A_k^{(\infty)}(z)^{-1} (-i) \begin{pmatrix} e^{\varphi/2} & -e^{-\varphi/2} \\ e^{\varphi/2} & e^{-\varphi/2} \end{pmatrix}$$

$$\times \begin{pmatrix} 0 & e^{-\varphi/2} \\ e^{\varphi/2} & 0 \end{pmatrix} C^{-1} A_0^\pm (-1/x^2 z)^{-1} \begin{pmatrix} e^{ix^2 z} & 0 \\ 0 & e^{-ix^2 z} \end{pmatrix} \begin{pmatrix} 1 & 0 \\ 0 & -1 \end{pmatrix}$$

$$= \underline{\rho}\, \Psi_{2/1}^{(k)} \begin{pmatrix} e^{ix^2 z} & 0 \\ 0 & e^{-x^2 z} \end{pmatrix} A_k^{(\infty)}(z)^{-1} \begin{pmatrix} 0 & -1 \\ -i & 0 \end{pmatrix}$$

$$\times A_0^\pm (-1/x^2 z)^{-1} \begin{pmatrix} e^{ix^2 z} & 0 \\ 0 & e^{-ix^2 z} \end{pmatrix} \begin{pmatrix} 1 & 0 \\ 0 & -1 \end{pmatrix}$$

$$= \underline{\rho}\, \Psi_{2/1}^{(k)} \begin{pmatrix} 0 & -1 \\ -i & 0 \end{pmatrix} \begin{pmatrix} 1 & 0 \\ 0 & -1 \end{pmatrix} = \underline{\rho}\, \Psi_{2/1}^{(k)} \begin{pmatrix} 0 & 1 \\ -i & 0 \end{pmatrix}.$$

Now, $S_1^{(0)} = \begin{pmatrix} 1 & \zeta^{NI} \\ 0 & 1 \end{pmatrix}$ is the Stokes matrix with

$$(\underline{\rho}\, \Psi_1^{(0)})|_{I_0^a} S_1^{(0)} = (\underline{\rho}\, \Psi_2^{(0)})|_{I_0^a},$$

and $S_0^a = \begin{pmatrix} 1 & s \\ 0 & 1 \end{pmatrix}$ is the Stokes matrix with

$$\underline{e}_0^+ |_{I_0^a} S_0^a = \underline{e}_0^- |_{I_0^a}.$$

Thus

$$\begin{pmatrix} 1 & -is \\ 0 & 1 \end{pmatrix} = \begin{pmatrix} 1 & 0 \\ 0 & i \end{pmatrix} S_0^a \begin{pmatrix} 1 & 0 \\ 0 & -i \end{pmatrix} = S_1^{(0)} = \begin{pmatrix} 1 & \zeta^{NI} \\ 0 & 1 \end{pmatrix}, \quad \text{so } s = i\zeta^{NI}.$$

For $x \in S^{NI}$, a β with $e^{-\beta} = 4x^2$ (2.23) and $\Im(\beta) \in (-\frac{\pi}{2} + 2\varepsilon, \frac{\pi}{2} - 2\varepsilon)$ encodes a path $[\beta]$ from 1 to e^β which stays within $\Omega_1^{(\infty)} \subset \widehat{I}_0^+ \cap \widehat{I}_\infty^-(x)$. Therefore $B(\beta)$ is the matrix with

$$\underline{e}_\infty^- |_{\Omega_1^{(\infty)}} = \underline{e}_0^+ |_{\Omega_1^{(\infty)}} B(\beta).$$

On the other hand

$$\underline{\rho}\, \Psi_1^{(\infty)} = \underline{\rho}\, \Psi_1^{(0)} E^{NI}.$$

This establishes (11.30) and (11.31).

(c) Equations (11.32) and (11.33) follow immediately from (11.31). They induce a natural bijection between the two copies of $\{\zeta^{NI} \in \mathbb{C}\}$ and the double cover for $\sqrt{1 + pq}$ of the manifold $\{(\zeta^{NI}, p, q) \in \mathbb{C}^3 \mid pq \neq -1, \zeta^{NI} = p + q\}$ on one side and $M_{3TJ}^{mon}(-i, ix^2)$ on the other side. Here two points (s, b_1, b_2) and $(s, -b_1, -b_2)$ are mapped to the same point (ζ, p, q) by (11.33). The symmetry R_2 from (7.30) and Remark 8.3 maps (s, b_1, b_2, f_0) to $(s, -b_1, -b_2, -f_0)$. □

Remarks 11.5

(i) In [IN86] and [FIKN06] the special case (11.21) is mentioned, but its role is not emphasized.

(ii) Chapters 8 and 11 of [IN86] contain rich results on the solutions on $\mathbb{R}_{>0}$ of (11.1). Using Lemma 11.1 they can be translated into results on the solutions of $P_{III}(0, 0, 4, -4)$ on $\mathbb{R}_{>0}$.

(iii) Chapter 8 of [IN86] studies first the asymptotics as $x \to \infty$ of solutions f of $P_{III}(0, 0, 4, -4)$ on $\mathbb{R}_{>0}$ which satisfy

$$\log f = O(x^{-1/2}), \quad \partial_x(\log f) = O(x^{-1/2}) \quad \text{for } x \to \infty. \quad (11.36)$$

In particular we have $f \to 1$ as $x \to \infty$ for these solutions. Reference [IN86, Theorem 8.1 and remark 8.2] state that a solution $f_{mult}(., s, B)|_{\mathbb{R}_{>0}}$ satisfies (11.36) in the case (11.22) with $pq > -1$. This is equivalent to $b_2 \in \mathbb{R}^*$. The case $b_2 = 0$ is the case (11.21) and is not considered in [IN86, ch. 8]. We expect that (11.36) holds there, too.

In [IN86, ch. 8] the condition $pq > -1$ is split into four cases, and two subcases of two of the cases are also given. The following table lists their conditions and notation (the manifolds $M_r, M_i, M_p, M_q, M_{\mathbb{R}}, M_I$) and the translation into conditions for s, b_1, b_2:

M_r	$pq > 0$	$b_2 \in]-1, 1[-\{0\}$
M_i	$-1 < pq < 0$	$b_2 \in \mathbb{R}_{<-1} \cup \mathbb{R}_{>1}$
M_p	$q = 0$	$b_2 = \pm 1, b_1 + sb_2 = 0$
M_q	$p = 0$	$b_2 = \pm 1, b_1 = 0$
$M_{\mathbb{R}} \subset M_r$	$pq > 0, |p| = |q|$	$s \in i\mathbb{R}, b_1 + \frac{s}{2}b_2 \in \mathbb{R}, b_2 \in \mathbb{R}^*$
$M_I \subset M_i$	$-1 < pq < 0, |p| = |q|$	$s \in (-2, 2), b_1 + \frac{1}{2}sb_2 \in i\mathbb{R}, \pm b_2 \in \mathbb{R}_{>1}$

The manifold $V^{mat, S^1} - \{(s, B) \mid b_2 = 0\}$ from (15.15) is a double cover of $M_{\mathbb{R}}$, and $V^{mat, i\mathbb{R}_{>0}}$ from (15.12) is isomorphic to M_I. For $(s, B) \in V^{mat, S^1}$, $f_{mult}(., s, B)|_{\mathbb{R}_{>0}}$ takes values in S^1 and u^{NI} has real values. For $(s, B) \in V^{mat, i\mathbb{R}_{>0}}$, $f_{mult}(., s, B)|_{\mathbb{R}_{>0}}$ takes values in $i\mathbb{R}_{>0}$ and u^{NI} takes values in $i\mathbb{R}_{>0}$; see Lemma 15.2 and Theorem 15.5 below and also [IN86, ch. 8].

For the four cases M_r, M_i, M_p, M_q [IN86, Theorem 8.1] gives precise asymptotic formulae for u^{NI} (and thus for $\log f$) as $x \to \infty$.

In the later part of [IN86, ch. 8], the monodromy data (s, B) of a solution $f_{mult}(., s, B)|_{\mathbb{R}_{>0}}$ which is smooth near 0 is connected with the asymptotics near 0. This is generalized in [Ni09], and we reformulate it in Chap. 12 and reprove it in Chap. 13. Further comments on [IN86, ch. 8] are made in Remark 15.6 (iii) below.

(iv) Reference [IN86, ch. 11] studies the singularities of real solutions φ on $\mathbb{R}_{>0}$ of (9.5) and thus implicitly the zeros and poles of real solutions of $P_{III}(0,0,4,-4)$ on $\mathbb{R}_{>0}$. Again they restrict to the case (11.22), which is precisely the case of real solutions of $P_{III}(0,0,4,-4)$ on $\mathbb{R}_{>0}$ which are not smooth near ∞. We take this up in Chaps. 15 ($V^{mat,\mathbb{R}}$ in (15.6)) and Chap. 18 (i.e. which solutions have which sequences of zeros and poles). However we do not reformulate formula [IN86, (11.10)] for the positions of zeros and poles of a solution for large x.

Chapter 12
Asymptotics of All Solutions Near 0

In this chapter we shall rewrite and extend one of the two main results of [Ni09], the asymptotic formulae as $x \to 0$ for all solutions of $P_{III}(0, 0, 4, -4)$. In [Ni09, 1.4.2], Niles distinguishes three cases a, b and c and makes implicitly the following finer separation into five cases a, $b+$, $b-$, $c+$, $c-$. Let $\mathbb{C}^{[sto]}$ be the complex plane with coordinate s, and define

$$\mathbb{C}^{[sto,a]} := \mathbb{C} - (\mathbb{R}_{\leq -2} \cup \mathbb{R}_{\geq 2}),$$

$$\mathbb{C}^{[sto,b\pm]} := (\pm 1)\mathbb{R}_{>2}, \quad \mathbb{C}^{[sto,b]} := \mathbb{C}^{[sto,b+]} \cup \mathbb{C}^{[sto,b-]},$$

$$\mathbb{C}^{[sto,c\pm]} := \{\pm 2\}, \quad \mathbb{C}^{[sto,c]} := \mathbb{C}^{[sto,c+]} \cup \mathbb{C}^{[sto,c-]}, \tag{12.1}$$

$$\mathbb{C}^{[sto,J_1 \cup J_2 \cup \ldots]} := \mathbb{C}^{[sto,J_1]} \cup \mathbb{C}^{[sto,J_2]} \cup \ldots \quad (J_1, J_2, \ldots \in \{a, b\pm, b, c\pm, c\}),$$

and

$$V^{mat,J} := \{(s, B) \in V^{mat} \mid s \in \mathbb{C}^{[sto,J]}\}$$
$$\text{for } J \in \{a, b\pm, b, c\pm, c, a \cup b+, \ldots\}. \tag{12.2}$$

For each of cases a, b and c, Niles has one formula for the special case (11.21) and one formula for the general case (11.22). The formulae for the cases b and c contain a sign which distinguishes $b+$, $b-$ and $c+$, $c-$.

We shall give a refined version (12.20) of the formula for the case a. Our versions (12.24) and (12.28) of Niles' formulae for the cases $b+$ and $c+$ will follow from (12.20) by analytic continuation. Our versions comprise the special case (11.21) and the general case (11.22). For the proof we only need Niles' formula for the case a in the general case (11.22) and a basic property of his formula for the case $b+$ in the general case (11.22).

© Springer International Publishing AG 2017
M.A. Guest, C. Hertling, *Painlevé III: A Case Study in the Geometry of Meromorphic Connections*, Lecture Notes in Mathematics 2198,
DOI 10.1007/978-3-319-66526-9_12

Remarks 12.1–12.3 prepare the formulae in Theorem 12.4 and make them more transparent.

Remarks 12.1 Lemma 5.2 (b) gives analytic isomorphisms (real analytic with respect to s in the case b)

$$
\begin{aligned}
V^{mat,a} &\to \mathbb{C}^{[sto,a]} \times \mathbb{C}^*, \quad (s, B) \mapsto (s, b_-), \\
V^{mat,b} &\to \mathbb{C}^{[sto,b]} \times \mathbb{C}^*, \quad (s, B) \mapsto (s, b_-),
\end{aligned}
\tag{12.3}
$$

so (s, b_-) are global coordinates on $V^{mat,a}$ as an analytic manifold. The fibration $V^{mat,a\cup b} \to \mathbb{C}^{[sto,a\cup b]}$ is an analytically locally trivial fibre bundle with fibres isomorphic to $\mathbb{C}^*$. Recall the definition of $\sqrt{\frac{1}{4}s^2 - 1}$ in Chap. 5. The first row in the following table lists some holomorphic functions on $V^{mat,a}$. The functions on $V^{mat,a}$ in the second row are obtained from these by analytic continuation over $V^{mat,b+}$.

$$
\begin{array}{c|c|c|c|c}
s & \sqrt{\frac{1}{4}s^2 - 1} & \alpha_\pm & \lambda_\pm & b_\pm \\
\hline
s & -\sqrt{\frac{1}{4}s^2 - 1} & \alpha_\mp \mp 1 & \lambda_\mp & b_\mp
\end{array}
\tag{12.4}
$$

In future, we denote a holomorphic function on $\mathbb{C}^{[sto,a]}$ by $\kappa(s, \sqrt{\frac{1}{4}s^2 - 1})$, and then denote the function on $\mathbb{C}^{[sto,a]}$ obtained by analytic continuation over $\mathbb{C}^{[sto,b+]}$ by $\kappa(s, -\sqrt{\frac{1}{4}s^2 - 1})$

Remarks 12.2

(i) Recall the action $m_{[1]}$ on $\mathbb{C} \times V^{mat}$ (with coordinates (ξ, s, B) with $-\frac{\beta}{2} = \xi + \log 2$) in (10.11):

$$
m_{[1]} : \mathbb{C} \times V^{mat} \to \mathbb{C} \times V^{mat}, \quad (\xi, s, B) \mapsto (\xi - i\pi, s, (\mathrm{Mon}_0^{mat})^{-1} B).
$$

The functions in the first row of the table (12.4) are with the exception of $b_\pm$ functions on $\mathbb{C} \times V^{mat,a}$ which are invariant under the action of $m_{[1]}$. As b_- is mapped under $m_{[1]}$ to $\lambda_-^{-1} b_-$, one obtains the function

$$
\widetilde{b}_- := e^{2(\xi - \log 2)\alpha_-} b_- = \left(\frac{x}{2}\right)^{2\alpha_-} b_-
\tag{12.5}
$$

which is invariant under $m_{[1]}$ (here x comes with the choice of $\xi = \log x$). The summand $-\log 2$ is inserted in order to simplify the formulae in Theorem 12.4.

We obtain analytic isomorphisms (real analytic with respect to s in case b)

$$\mathbb{C} \times V^{mat,a}/\langle m_{[1]}\rangle \to \mathbb{C}^* \times \mathbb{C}^{[sto,a]} \times \mathbb{C}^*, \quad [(\xi, s, B)] \mapsto (x^2, s, \widetilde{b}_-), \quad (12.6)$$

$$\mathbb{C} \times V^{mat,b}/\langle m_{[1]}\rangle \to \mathbb{C}^* \times \mathbb{C}^{[sto,b]} \times \mathbb{C}^*, \quad [(\xi, s, B)] \mapsto (x^2, s, \widetilde{b}_-),$$

so $(x^2, s, \widetilde{b}_-)$ are global coordinates on $\mathbb{C} \times V^{mat,a}/\langle m_{[1]}\rangle$. The fibration $\mathbb{C} \times V^{mat,a\cup b}/\langle m_{[1]}\rangle \to \mathbb{C}^{[sto,a\cup b]}$ is an analytically locally trivial fibre bundle with fibres isomorphic to $\mathbb{C}^* \times \mathbb{C}^*$.

Let us erase in table (12.4) the column with $b_\pm$ and extend the table by

$$
\begin{array}{c|c}
x^2 & \widetilde{b}_- \\
\hline
x^2 & \left(\frac{x}{2}\right)^2 \widetilde{b}_-^{-1}
\end{array}
\qquad (12.7)
$$

Then the first row lists some holomorphic functions on $\mathbb{C} \times V^{mat,a}/\langle m_{[1]}\rangle$, and the functions on $\mathbb{C} \times V^{mat,a}/\langle m_{[1]}\rangle$ in the second row are obtained from these by analytic continuation over $\mathbb{C} \times V^{mat,b+}/\langle m_{[1]}\rangle$.

If now an open neighbourhood U in $\mathbb{C} \times V^{mat,a\cup b+}/\langle m_{[1]}\rangle$ of a point in $\mathbb{C} \times V^{mat,c+}/\langle m_{[1]}\rangle$ and a holomorphic function in this neighbourhood are given, then the function has a convergent Laurent expansion

$$\sum_{(a_1,a_2)\in\mathbb{Z}^2} \kappa_{a_1,a_2}\left(s, \sqrt{\tfrac{1}{4}s^2 - 1}\right) \left(\tfrac{1}{2}x\right)^{2a_1} (\widetilde{b}_-)^{a_2}, \qquad (12.8)$$

where κ_{a_1,a_2} are coefficients holomorphic in $s \in \mathbb{C}^{[sto,a]} \cap \mathrm{pr}_s(U)$. The Laurent expansion must stay the same after continuation over $\mathbb{C} \times V^{mat,b+}/\langle m_{[1]}\rangle$. Therefore the coefficients must satisfy

$$\kappa_{a_1+a_2,-a_2}\left(s, \sqrt{\tfrac{1}{4}s^2 - 1}\right) = \kappa_{a_1,a_2}\left(s, -\sqrt{\tfrac{1}{4}s^2 - 1}\right). \qquad (12.9)$$

(ii) We also need coordinates on $\mathbb{C} \times V^{mat,c\pm}/\langle m_{[1]}\rangle$. Suppose $s \in \{\pm 2\}$. Then by Lemma 5.2 (b), $\widetilde{b}_1 := b_1 + \frac{1}{2}sb_2 \in \{\pm 1\}$ and $B \in V^{mat}$ is determined by $(\widetilde{b}_1, b_2) \in \{\pm 1\} \times \mathbb{C}$ and any such pair is realized by some matrix B. Since

$$(\mathrm{Mon}_0^{mat})^{-1}\begin{pmatrix} b_1 & b_2 \\ -b_2 & b_1 + sb_2 \end{pmatrix} = \begin{pmatrix} b_1 - s(b_2 + sb_1) & b_2 + sb_1 \\ -(b_2 + sb_1) & b_1 \end{pmatrix}, \qquad (12.10)$$

$m_{[1]}$ maps the functions in the first row to the functions in the second row,

$$
\begin{array}{c|c|c|c|c|c}
b_1 & b_2 & \widetilde{b}_1 & \xi & x\widetilde{b}_1 & \xi - \frac{i\pi}{4}s\widetilde{b}_1 b_2 =: \widetilde{b}_2 \\
\hline
b_1 - s(b_2 + sb_1) & s\widetilde{b}_1 - b_2 & -\widetilde{b}_1 & \xi - i\pi & x\widetilde{b}_1 & \xi - \frac{i\pi}{4}s\widetilde{b}_1 b_2 = \widetilde{b}_2
\end{array}
\qquad (12.11)
$$

In particular, $\widetilde{xb_1}$ and $\widetilde{b}_2$ are invariant functions and serve as global coordinates on $\mathbb{C} \times V^{mat,c\pm}/\langle m_{[1]} \rangle$. The map

$$\mathbb{C} \times V^{mat,c\pm}/\langle m_{[1]} \rangle \to \mathbb{C}^* \times \mathbb{C}, \quad [(\xi, s, B)] \mapsto (\widetilde{xb_1}, \widetilde{b}_2) \qquad (12.12)$$

is an analytic isomorphism.

Remark 12.3 The group $G^{mon} = \{id, R_1, R_2, R_3\}$ defined in (7.30) (with $R_3 = R_1 \circ R_2 = R_2 \circ R_1$) acts on $V^{mat} \cong V^{mon}$, $\mathbb{C} \times V^{mat}$, $\mathbb{C} \times V^{mat}/\langle m_{[1]} \rangle$ and on M^{mon}_{3FN} and respects the foliation there. Remark 8.3 fixes how it acts on the functions f_0, g_0 on M^{reg}_{3FN}. The action of $m_{[1]}$ on $\mathbb{C} \times V^{mat}$ is fixed in Remark 10.2 (iv). The following table lists some meromorphic functions on $\mathbb{C} \times V^{mat,a}$ and the actions of R_1, R_2 and $m_{[1]}$ on them.

	R_1	R_2	$m_{[1]}$
s	$-s$	s	s
$\lambda_\pm$	$\lambda_\mp$	$\lambda_\pm$	$\lambda_\pm$
$\alpha_\pm$	$-\alpha_\pm$	$\alpha_\pm$	$\alpha_\pm$
B	$\left(\begin{smallmatrix}1 & 0\\ 0 & -1\end{smallmatrix}\right) B \left(\begin{smallmatrix}1 & 0\\ 0 & -1\end{smallmatrix}\right)$	$-B$	$\mathrm{Mon}^{mat}_0(s)^{-1} B$
b_1	b_1	$-b_1$	$(1-s^2)b_1 - sb_2$
b_2	$-b_2$	$-b_2$	$b_2 + sb_1$
$b_1 + \frac{s}{2}b_2$	$b_1 + \frac{s}{2}b_2$	$-b_1 - \frac{s}{2}b_2$	$(1-\frac{s^2}{2})b_1 - \frac{s}{2}b_2$
$b_\pm$	$b_\mp$	$-b_\pm$	$\lambda_\mp^{-1}b_\pm$
ξ	ξ	ξ	$\xi - i\pi$
x	x	x	$-x$
f_{univ}	f_{univ}^{-1}	$-f_{univ}$	f_{univ}
g_{univ}	$-g_{univ}$	g_{univ}	g_{univ}

$$(12.13)$$

The core of Theorem 12.4 is the asymptotic formulae in [Ni09, 1.4.2 and ch. 3] for x near 0 of the solutions of the $P_{III}(0,0,4,-4)$ equation. In Theorem 12.4 they are extended, simplified and made more transparent.

Theorem 12.4 *There exist two continuous $m_{[1]}$-invariant and G^{mon}-invariant maps*

$$B_1 : i\mathbb{R} \times V^{mat} \to \mathbb{R}_{>0}, \quad B_2 : i\mathbb{R} \times V^{mat,a} \to \mathbb{R}_{>0} \qquad (12.14)$$

such that $B_2 \leq B_1$ (where both are defined) and such that the $m_{[1]}$-invariant and G^{mon}-invariant open subsets

$$U_1 := \{(\xi, s, B) \in \mathbb{C} \times V^{mat} \mid |e^{\xi}| < B_1(i\Im(\xi), s, B)\}$$
$$U_2 := \{(\xi, s, B) \in \mathbb{C} \times V^{mat,a} \mid |e^{\xi}| < B_2(i\Im(\xi), s, B)\} \tag{12.15}$$

satisfy the following:

(a) f_{univ} is holomorphic and invertible on $\mathbb{C} \times V^{mat,a} \cap U_2$ and holomorphic on $\{(\xi, s, B) \in \mathbb{C} \times V^{mat} \cap U_1 \mid \Re(s) > -1\}$. f_{univ}^{-1} is holomorphic on $\{(\xi, s, B) \in \mathbb{C} \times V^{mat} \cap U_1 \mid \Re(s) < 1\}$. The restriction of f_{univ} to $\mathbb{C} \times V^{mat,c} \cap U_1$ is holomorphic and invertible.

(b) $f_{mult}(., s, B)$ is holomorphic near 0 for $(s, B) \in V^{mat,a\cup b+\cup c}$ and invertible near 0 for $(s, B) \in V^{mat,a\cup c}$. $f_{mult}^{-1}(., s, B)$ is holomorphic near 0 for $(s, B) \in V^{mat,a\cup b-\cup c}$ ((b) rewrites (a) with $f_{mult}(., s, B)$, but is not precise about U_1 and U_2).

(c) f_{univ} is on $\mathbb{C} \times V^{mat,a\cup b+} \cap U_1$ a convergent sum

$$\sum_{(a_1,a_2)\in L} \kappa_{a_1,a_2}\left(s, \sqrt{\tfrac{1}{4}s^2 - 1}\right) \left(\tfrac{1}{2}x\right)^{2a_1} (\widetilde{b}_-)^{a_2}, \tag{12.16}$$

where $L \subset \mathbb{Z}^2$ is defined by the three inequalities

$$2a_1 - a_2 \geq -1, \ 2a_1 + a_2 \geq 1, \ 2a_1 + 3a_2 \geq -1, \tag{12.17}$$

and where κ_{a_1,a_2} are coefficients holomorphic in $s \in \mathbb{C}^{[sto,a]} \cap U$ which satisfy (12.9). In particular

$$\kappa_{0,1} = \frac{\Gamma(\tfrac{1}{2} - \alpha_-)}{\Gamma(\tfrac{1}{2} + \alpha_-)}, \quad \kappa_{1,-1} = \frac{\Gamma(\tfrac{1}{2} - (\alpha_+ + 1))}{\Gamma(\tfrac{1}{2} + (\alpha_+ + 1))}. \tag{12.18}$$

(d) Recall that as multi-valued functions in x

$$\widetilde{b}_- = \left(\tfrac{1}{2}x\right)^{2\alpha_-} b_-,$$
$$\left(\tfrac{1}{2}x\right)^2 (\widetilde{b}_-)^{-1} = \left(\tfrac{1}{2}x\right)^{2-2\alpha_-} b_-^{-1}. \tag{12.19}$$

If $(s, B) \in V^{mat,a}$ with $\Re(s) \geq 0$ ($\iff \Re(\alpha_-) \in [0, \tfrac{1}{2})$) then for small x

$$f_{mult}(x, s, B) = \kappa_{0,1} \widetilde{b}_- + \kappa_{1,-1} \left(\tfrac{1}{2}x\right)^2 (\widetilde{b}_-)^{-1} + O(|x|^2) \tag{12.20}$$

$$= \kappa_{0,1} \widetilde{b}_- + O(|x|^{2-2\Re(\alpha_-)}). \tag{12.21}$$

If $(s, B) \in V^{mat,a}$ with $\Re(s) \leq 0$ ($\Longleftrightarrow$ $\Re(\alpha_-) \in (-\frac{1}{2}, 0]$) then for small x

$$f_{mult}(x, s, B) = \kappa_{0,1} \widetilde{b}_- + O(|x|^{2+6\Re(\alpha_-)}). \tag{12.22}$$

If $(s, B) \in V^{mat,b+}$ ($\Longleftrightarrow$ $\Re(\alpha_-) = \frac{1}{2}, s \neq 2$) then for small x

$$f_{mult}(x, s, B) = \kappa_{0,1} \widetilde{b}_- + \kappa_{1,-1} \left(\frac{x}{2}\right)^2 (\widetilde{b}_-)^{-1} + O(|x|^2) \tag{12.23}$$

$$= -\frac{x}{t^{NI}} \sin\left(2t^{NI} \log \frac{x}{2} - 2 \arg \Gamma(1 + it^{NI}) + \delta^{NI}\right)$$

$$+ O(|x|^2), \tag{12.24}$$

where $e^{i\delta^{NI}} = b_-$ and $\Re(\delta^{NI}) \in [0, 2\pi)$, and t^{NI} is as in (5.17).

(e) *Part (a) and formula (12.24) show that $f_{mult}(., s, B)$ has zeros arbitrarily close to 0 within x with bounded argument $\arg(x) = \Im(\xi)$ if and only if $s \in \mathbb{R}_{>2}$, and that then the only such zeros are approximately given by*

$$x_k = 2 \exp\left(\frac{1}{2t^{NI}}(2 \arg \Gamma(1 + it^{NI}) - \delta^{NI})\right) \exp\left(\frac{-k\pi}{2t^{NI}}\right)$$

$$\times \left(1 + O\left(\exp\left(\frac{-k\pi}{2t^{NI}}\right)\right)\right) \quad \text{for } k \in \mathbb{Z}_{> a \text{ bound}}. \tag{12.25}$$

In particular, they all have approximately the same argument, which is

$$\frac{\Im(-\delta^{NI})}{2t^{NI}} = \frac{\log |b_-|}{2t^{NI}}. \tag{12.26}$$

(f) *Using R_1 in Remark 12.3, one can derive from (d) and (g) asymptotic formulae for $f_{mult}^{-1}(., s, B)$ for $(s, B) \in V^{mat,a \cup b - \cup c-}$. Using R_1 and (e), one finds that $f_{mult}(., s, B)$ has poles near 0 if and only if $s \in \mathbb{R}_{<-2}$ and that then there is only the sequence of poles tending to 0 in (12.25), where now $t^{NI} = t^{NI}(|s|)$ and $\delta^{NI} = \delta^{NI}(-s, b_1, -b_2) = -\delta^{NI}(s, b_1, b_2)$.*

(g) *f_{univ} is on $\mathbb{C} \times V^{mat,c+} \cap U_1$ holomorphic and invertible with leading term (leading with respect to $x \to 0$ with constant argument)*

$$- 2e^\xi \widetilde{b}_1 \left(\widetilde{b}_2 - \log 2 + \gamma_{Euler}\right). \tag{12.27}$$

Consider $(s, B) \in V^{mat,c+}$, i.e. $s = 2$. Then

$$f_{mult}(x, s, B) = -2x\widetilde{b}_1 \left(\log \frac{x}{2} - \frac{i\pi}{2}\widetilde{b}_1 b_2 + \gamma_{Euler}\right) + O(|x|^2). \tag{12.28}$$

Proof Niles has six formulae: for each of cases a, b and c, one in the special case (11.21) and one in the general case (11.22). We shall use only the formula for case a in the general case (11.22), and a basic property of the formula for case $b+$ in the general case (11.22). The basic property is that $f_{mult}(., s, B)$ for $(s, B) \in V^{mat,b+}$ is holomorphic for $x \in S^{NI}$ near 0. First we shall treat the cases a and b, that is, (a)–(f) without the statements in (a),(b) on the case c, then the case c, that is, (g) and the rest of (a),(b).

Two more remarks on Niles' formulae are appropriate. First, he restricts attention to $x \in S^{NI}$, but remarks at the end of [Ni09] that similar formulae should hold for all $x \in \mathbb{C}^*$ near 0. In fact, the same formulae hold for all $x \in \mathbb{C}^*$ near 0. This is contained in the proof below. Second, Niles' formulae start with the monodromy data (ζ^{NI}, p, q). Lemma (11.4) (c) shows that one such triple corresponds to two triples (s, b_1, b_2) and $(s, -b_1, -b_2) = R_2(s, b_1, b_2)$ and to two solutions f_0 and $-f_0$ of $P_{III}(0, 0, 4, -4)$. Thus his formulae are not specific about a global sign. We shall fix the sign using Remark 10.5.

Niles' formula for case a in the general case (11.22) is as follows [Ni09, 1.4.2 and 3.1 Theorem 7]: for $(s, B) \in V^{mat,a}$ and for $x \in S^{NI}$, any branch of the multi-valued function $u(x)$ with (11.5) $f_{mult}(x, s, B) = ie^{-iu(x)/2}$ (here a priori the distinguished branch of $f_{mult}(., s, B)$ should be taken) satisfies for $x \to 0$ (with constant argument)

$$u(x) = r^{NI} \log(4x) + s^{NI} + O(|x|^{2-|\Re(r^{NI})|}). \tag{12.29}$$

The constants r^{NI} and s^{NI} are as follows. r^{NI} is given by

$$r^{NI} = 4i\alpha^{NI}, \quad \text{with } \alpha^{NI} \text{ given by}$$
$$\zeta^{NI} = -2i\sin(\pi\alpha^{NI}), \ \Re(\alpha^{NI}) \in (-\tfrac{1}{2}, \tfrac{1}{2}).$$

Equation (11.29) $s = i\zeta^{NI}$ and (5.14) $\sin(\pi\alpha_-) = \tfrac{1}{2}s$ show $\alpha^{NI} = \alpha_-$. Thus

$$r^{NI} = 4i\alpha_-. \tag{12.30}$$

s^{NI} in (12.29) is given in [Ni09, 1.4.2 and 3.1 Theorem 7] by the first line of the following calculation, which make use of formulae in Lemmas 5.2 (b) and 11.4 (c),

$$e^{is^{NI}} = 2^{-3ir^{NI}} \frac{\Gamma^2(\tfrac{1}{2} - \tfrac{ir^{NI}}{4})}{\Gamma^2(\tfrac{1}{2} + \tfrac{ir^{NI}}{4})} \frac{1 + p\zeta^{NI} - p^2}{(e^{-\pi r^{NI}/4} - p)^{-2}} \tag{12.31}$$

$$= 2^{12\alpha_-} \frac{\Gamma^2(\tfrac{1}{2} + \alpha_-)}{\Gamma^2(\tfrac{1}{2} - \alpha_-)} \frac{1 + pq}{(e^{-\pi i\alpha_-} - p)^2}$$

$$= 2^{12\alpha_-} \frac{\Gamma^2(\tfrac{1}{2} + \alpha_-)}{\Gamma^2(\tfrac{1}{2} - \alpha_-)} \frac{1}{b_2^2(e^{-\pi i\alpha_-} - \tfrac{ib_1}{b_2})^2}$$

$$= 2^{12\alpha_-} \frac{\Gamma^2(\tfrac{1}{2} + \alpha_-)}{\Gamma^2(\tfrac{1}{2} - \alpha_-)} \frac{-1}{b_-^2}$$

We obtain for $(s, B) \in V^{mat,a}$ and $x \in S^{NI}$ and the distinguished branch of $f_{mult}(\cdot, s, B)$ for $x \to 0$ the asymptotic formula

$$
f_{mult}(x, s, B) = i e^{-iu(x)/2}
$$

$$
= (4x)^{2\alpha_-} i e^{-is^{NI}/2} + O(|x|^{2-4|\Re(\alpha_-)|+2\Re(\alpha_-)})
$$

$$
= \varepsilon \, \frac{\Gamma(\frac{1}{2} - \alpha_-)}{\Gamma(\frac{1}{2} + \alpha_-)} \left(\tfrac{1}{2}x \right)^{2\alpha_-} b_- \tag{12.32}
$$

$$
+ O(|x|^{2-4|\Re(\alpha_-)|+2\Re(\alpha_-)})
$$

with a sign $\varepsilon \in \{\pm 1\}$, which has yet to be determined and which is not fixed by the formulae in [Ni09].

Remark 10.5 gives $f_{mult}(x, 0, 1_2) = 1$. In the case $(s, B) = (0, 1_2)$ one has $\alpha_- = 0$, $b_- = 1$. Together with (12.32) this shows that the sign is $\varepsilon = 1$.

A priori the asymptotic formula (12.32) for f_{univ} is established only in the general case (11.22), but by continuity it holds also in the special case (11.21).

Now (12.21) and (12.22) with $\kappa_{0,1}$ as in (12.18) are established for $x \in S^{NI} \subset \mathbb{C}^*$. For some U_2 as in (12.15), f_{univ} restricted to $\{(\xi, s, B) \in \mathbb{C} \times V^{mat,a} \cap U_2 \mid e^{\xi} \in S^{NI}\}$ is a convergent sum

$$
\sum_{(a_1, a_2) \in \widetilde{L}} \kappa_{a_1, a_2}\left(s, \sqrt{\tfrac{1}{4}s^2 - 1}\right) \left(\tfrac{1}{2}x\right)^{2a_1} (\widetilde{b}_-)^{a_2}, \tag{12.33}
$$

where $\widetilde{L} \subset \mathbb{Z}^2$ is determined by the condition that

$$
\text{for } \Re(\alpha_-) \in (-\tfrac{1}{2}, \tfrac{1}{2}), \ 2a_1 + 2a_2 \Re(\alpha_-) \geq 2\Re(\alpha_-),
$$

$$
\text{equivalently: for } r \in \{-\tfrac{1}{2}, \tfrac{1}{2}\}, \ 2a_1 + 2a_2 r \geq 2r,
$$

$$
\text{equivalently: } 2a_1 - a_2 \geq -1, \ 2a_1 + a_2 \geq 1. \tag{12.34}
$$

But then it is convergent on the whole intersection $\mathbb{C} \times V^{mat,a} \cap U_2$ for some U_2 as in (12.15).

We do not need Niles' precise formula in [Ni09, 1.4.2 and 3.2 Theorem 8] for the case $b+$ in the general case (11.22), but just the following basic property, which is a consequence of that formula: f_{univ} is holomorphic (without poles) (or real analytic with respect to $s \in \mathbb{C}^{[sto,b+]}$) on $S^{NI} \times V^{mat,b+} \cap U_1$ for an open set U_1 as in (12.15).

Then the Laurent expansion is valid there, i.e. the Laurent expansion (12.33) must stay constant after continuation over $S^{NI} \times V^{mat,b+} \cap U_1$. Thus it satisfies (12.9). Therefore if $(a_1, a_2) \in \widetilde{L}$ then also $(a_1 + a_2, -a_2) \in \widetilde{L}$. Replacing (a_1, a_2) by $(a_1 + a_2, -a_2)$ in the two inequalities in (12.34) gives only the one new inequality $2a_1 + 3a_2 \geq -1$. Thus one can replace $\widetilde{L}$ by L in the Laurent expansion (12.33) to obtain the Laurent expansion (12.16). Now the convergence includes also $(\mathbb{C} - S^{NI}) \times V^{mat,b+} \cap U_1$ for a suitable U_1 as in (12.15). Part (c) is proved.

Thus f_{univ} is holomorphic on a neighbourhood of $\mathbb{C} \times V^{mat,b+} \cap U_1$ in $\mathbb{C} \times V^{mat,a\cup b+} \cap U_1$. This neighbourhood can be chosen for example as $\{(\xi, s, B) \in \mathbb{C} \times V^{mat,a\cup b+} \cap U_1 \mid \Re(s) > -1\}$, because (12.21) and (12.22) show that f_{univ} is holomorphic and invertible on $\mathbb{C} \times V^{mat,a} \cap U_2$ for a set U_2 as in (12.15). Part (a),(b) are proved up to the statements on the case c.

The following two pictures show the most relevant part of L and the graphs of the maps

$$(-\tfrac{1}{2}, \tfrac{1}{2}) \to \mathbb{R}, \quad r \mapsto a_1 + a_2 r$$

for the most important points in L.

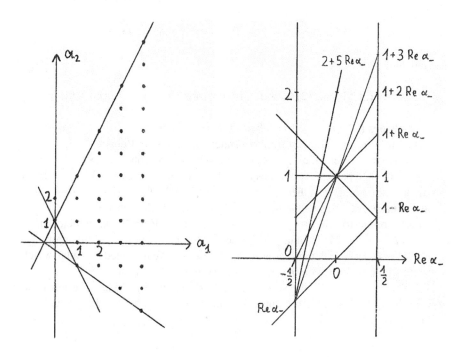

Equations (12.20) and (12.23) follow immediately. The next calculation derives (12.24) from (12.23). But of course (12.24) is equivalent to the formula for the case $b+$ in the general case (11.22) in [Ni09, 1.4.2 and 3.2 Theorem 8], up to the global sign, which is not fixed in [Ni09]. The calculation uses Lemma 5.1 (b).

$$\kappa_{0,1} \widetilde{b}_- + \kappa_{1,-1} \left(\tfrac{1}{2}x\right)^2 (\widetilde{b}_-)^{-1}$$

$$= \frac{\Gamma(-it^{NI})}{\Gamma(1 + it^{NI})} \left(\tfrac{1}{2}x\right)^{1+2it^{NI}} e^{i\delta^{NI}} + \frac{\Gamma(it^{NI})}{\Gamma(1 - it^{NI})} \left(\tfrac{1}{2}x\right)^{1-2it^{NI}} e^{-i\delta^{NI}}$$

$$= \frac{x}{2it^{NI}} \left(-\frac{\Gamma(1-it^{NI})}{\Gamma(1+it^{NI})} e^{2it^{NI}\log(x/2)} e^{i\delta^{NI}} \right.$$

$$+ \left. \frac{\Gamma(1+it^{NI})}{\Gamma(1-it^{NI})} e^{-2it^{NI}\log(x/2)} e^{-i\delta^{NI}} \right)$$

$$= -\frac{x}{t^{NI}} \sin\left(2t^{NI}\log(x/2) - 2\arg\Gamma(1+it^{NI}) + \delta^{NI}\right). \tag{12.35}$$

Part (d) is proved.

Now we turn to part (e). Denote by $x\, h(x, s, B)$ the remainder term of order $O(|x|^2)$ in (12.24), so $h(., s, B)$ has order $O(|x|)$. Then for k sufficiently large, x_k is a zero of $f_{mult}(., s, B)$ if

$$2t^{NI}\log(x_k/2) - 2\arg\Gamma(1+it^{NI}) + \delta^{NI} + k\pi$$

is small and

$$0 = \sin\left(2t^{NI}\log(x_k/2) - 2\arg\Gamma(1+it^{NI}) + \delta^{NI} + k\pi\right) + t^{NI}\, h(x_k, s, B).$$

This establishes (12.25). (12.26) is obvious. Part (e) is proved.

Part (f) follows immediately from the statements on R_1 in Remark 12.3.

We turn to part (g) and the case $c+$. f_{univ} must be holomorphic on $\mathbb{C} \times V^{mat,c+} \cap U_1$: by Lemma 8.5 a hypersurface of poles in $\mathbb{C} \times V^{mat}$ would intersect $\mathbb{C} \times V^{mat,c+} \cap U_1$ in a codimension 1 set. But as there are no poles in $\mathbb{C} \times V^{mat,a\cup b+} \cap U_1$ for $\Re(s) > -1$, the hypersurface of poles would have codimension 2 which is absurd.

An asymptotic formula for $f_{mult}(., s, B)$ for $(s, B) \in V^{mat,c+}$ can be determined from (12.20) or (12.24) by analytic (respectively, real analytic) continuation. We choose (12.24), as the calculations are slightly shorter. We just need formulae which connect some functions with the coordinates (s, b_2) on V^{mat} near $V^{mat,c+}$.

If $s \in \mathbb{C}^{[sto,b+]} = \mathbb{R}_{>2}$ is close to 2, then t^{NI} is close to 0, and the left hand sides in the following table are approximated by the right hand sides. Recall that $\Gamma'(1) = -\gamma_{Euler}$.

$\Gamma(1+it^{NI})$	$1 - \gamma_{Euler}\, it^{NI}$
$\arg\Gamma(1+it^{NI})$	$-\gamma_{Euler}\, t^{NI}$
$\sqrt{\frac{1}{4}s^2 - 1}$	$\sqrt{s-2}$
λ_-	$-1 - 2\sqrt{s-2}$
$t^{NI} = \frac{1}{2\pi}\log|\lambda_-|$	$\frac{1}{\pi}\sqrt{s-2}$
b_-	$\widetilde{b}_1(1 + \widetilde{b}_1\sqrt{\frac{1}{4}s^2 - 1}\, b_2)$
b_-	$\widetilde{b}_1(1 + \widetilde{b}_1 b_2\pi t^{NI})$
δ^{NI}	$(1-\widetilde{b}_1)\frac{\pi}{2} - i\widetilde{b}_1 b_2\pi t^{NI}$

Then (12.24) is close to

$$-\frac{x}{t^{NI}} \sin\left(2t^{NI}\log\frac{x}{2} + 2t^{NI}\gamma_{Euler} + (1-\widetilde{b}_1)\frac{\pi}{2} - i\widetilde{b}_1 b_2 \pi t^{NI}\right)$$

$$= -\frac{x\widetilde{b}_1}{t^{NI}} \sin\left(t^{NI}\left(2\log\frac{x}{2} + 2\gamma_{Euler} - i\pi\widetilde{b}_1 b_2\right)\right)$$

$$\approx -2x\widetilde{b}_1\left(\log\frac{x}{2} + \gamma_{Euler} - \frac{i\pi}{2}\widetilde{b}_1 b_2\right). \tag{12.36}$$

This proves (12.27) and part (g). Of course, (12.27) is equivalent to the formula for the case $c+$ in the general case (11.22) in [Ni09, 1.4.2 and 3.3 Theorem 9], up to the global sign, which is not fixed in [Ni09]. (12.27) shows that U_1 can be chosen such that also the statements on the case $c+$ in (a),(b) hold. □

Remarks 12.5

(i) By Theorem 10.3, f_0 on M_{3FN}^{mon} has simple zeros along the two hypersurfaces $M_{3FN}^{[0]}$ and $M_{3FN}^{[2]}$ and simple poles along the two hypersurfaces $M_{3FN}^{[1]}$ and $M_{3FN}^{[3]}$ and no other zeros or poles. The two types of zeros are distinguished by the value ∓ 2 of the first derivative with respect to x.

(ii) By Theorem 12.4 (a), f_{univ} is holomorphic on $\{(\xi, s, B) \in \mathbb{C} \times V^{mat} \mid \Re(s) > -1\} \cap U_1$. It has zeros of both types. More precisely, for any $(s, B) \in V^{mat,b+}$, the zero x_k of $f_{mult}(., s, B)$ satisfies $\partial_x f_{mult}(x_k, s, B) = (-1)^{k+1} 2$ because then the argument of the sine in (12.24) is approximately $k\pi \mod 2\pi$. So, x_k is a zero of type $M_{3FN}^{[0]}$ for even k, and a zero of type $M_{3FN}^{[2]}$ for odd k. Hence the two hypersurfaces $M_{3FN}^{[0]}$ and $M_{3FN}^{[2]}$ intersect the image in M_{3FN} of the set $(\{(\xi, s, B) \in \mathbb{C} \times V^{mat} \mid \Re(s) > -1\} \cap U_1) / \langle m_{[1]}\rangle$.

(iii) The two hypersurfaces $M_{3FN}^{[0]}$ and $M_{3FN}^{[2]}$ cannot restrict to the image in M_{3FN} of the set $(\{(\xi, s, B) \in \mathbb{C} \times V^{mat,b+} \mid \Re(s) > -1\} \cap U_1) / \langle m_{[1]}\rangle$, as then they would have real codimension 3. As for any $(s^0, B^0) \in V^{mat,b+}$ the x_k tend to zero for $k \to \infty$, also the functions $f_{mult}(., s, B)$ for $(s, B) \in V^{mat,a}$ close to (s^0, B^0) must have zeros rather close to 0. This implies the following: if $(s, B) \in V^{mat,a}$ approaches a point $(s^0, B^0) \in V^{mat,b+}$, then $B_2(s, B)$ must tend to 0, in contrast to $B_1(s, B)$ which tends to $B_1(s^0, B^0) > 0$.

(iv) Consider now a point $(2, B^0) \in V^{mat,c+}$ and $(s, B) \in V^{mat,b+}$ which tends to this point. Then t^{NI} tends to 0,

$$x_k = 2\exp\left(\frac{1}{2t^{NI}}(2\arg\Gamma(1 + it^{NI}) - \delta^{NI})\right)\exp\left(\frac{-k\pi}{2t^{NI}}\right)$$

tends by the proof of part (g) to

$$2\exp\left(-\gamma_{Euler} + \frac{i\pi\widetilde{b}_1 b_2}{2}\right)\exp\left(\frac{-(k + (1-\widetilde{b}_1)/2)\pi}{2t^{NI}}\right),$$

so its argument tends to

$$\tfrac{1}{2}\pi \widetilde{b}_1 b_2,$$

and x_k itself tends rapidly to 0. This helps to understand why $f_{mult}(.,2,B^0)$ has no zeros near 0, but $f_{mult}(.,s,B)$ for $(s,B) \in V^{mat,b+}$ close to $(2,B^0) \in V^{mat,c+}$ has zeros arbitrarily close to 0.

Chapter 13
Rank 2 TEPA Bundles with a Logarithmic Pole

This chapter offers an independent proof of the results in [Ni09] which are used above to prove Theorem 12.4, so together with the arguments in Chap. 12 it reproves Theorem 12.4. It uses the language of this monograph. But some crucial arguments are close to those in [Ni09] and [IN86, ch. 8]: the approximation of (sections of) the P_{3D6}-TEJPA bundles for small x by (sections of) a closely related bundle, and the explicit control of sections by Hankel functions.

Still, the details are quite different. We believe that the proof offered here is useful, as it shows clearly the origins of the different ingredients in the formulae in Theorem 12.4.

The function f_0 on M_{3FN}^{ini} (Lemma 10.1 (c)) induces the function $f_{univ} : \mathbb{C} \times V^{mat} \to \mathbb{C}$ which is the subject of Theorem 12.4. The function f_0 has simple zeros along the smooth hypersurfaces $M_{3FN}^{[0]}$ and $M_{3FN}^{[2]}$ and simple poles along $M_{3FN}^{[1]}$ and $M_{3FN}^{[3]}$. On M_{3FN}^{reg} the bases $\underline{\sigma}_0$ and $\underline{\sigma}_2$ of the normal forms in Theorem 8.2 (b) coincide, see (8.19). Here we shall call this basis $\underline{\sigma} = (\sigma_1, \sigma_2)$.

Formula (8.30) in Theorem 8.2 (g) shows: the holomorphic family of sections σ_1 for the pure P_{3D6}-TEJPA bundles in M_{3FN}^{reg} extends holomorphically to the hypersurfaces $M_{3FN}^{[0]}$ and $M_{3FN}^{[2]}$, so it is a holomorphic family of sections σ_1 on the union of the chart with $k = 0$ and the chart with $k = 2$ of M_{3FN}^{ini}. On the union of these two charts f_0 can now be calculated using σ_1 via

$$2f_0 = P(\sigma_1(z), J(\sigma_1(-\rho_1(z)))) \tag{13.1}$$

because $J(\sigma_1(-\rho_1(z)))) = f_0\,\sigma_2(-z)$ by (8.18) and $P(\sigma_1(z), \sigma_2(-z)) = 2$ by (8.16).

The idea of the proof of the crucial formulae (12.21) and (12.22) in Theorem 12.4 is to show that σ_1 is holomorphic on a suitable neighbourhood of $x = 0$ in M_{3FN} and to replace in (13.1) σ_1 by an approximation for x near 0 and calculate the right hand side of (13.1) for this approximation. The approximation will be a linear combination of *elementary sections*, which are defined after the following choices.

© Springer International Publishing AG 2017
M.A. Guest, C. Hertling, *Painlevé III: A Case Study in the Geometry of Meromorphic Connections*, Lecture Notes in Mathematics 2198, DOI 10.1007/978-3-319-66526-9_13

We shall restrict most of the time to $x \in \mathbb{C} - \mathbb{R}_{\leq 0}$. This is not a serious restriction, but it has the advantage that we can fix branches of $\log x$ on $\mathbb{C} - \mathbb{R}_{\leq 0}$ and of $\log z$ on $\widehat{I}_0^+$. We choose the branch ξ of $\log x$ with $\arg x = \Im(\xi) \in (-\pi, \pi)$ and $\beta \in \mathbb{C}$ with $\frac{1}{2} e^{-\beta/2} = x$ and $\Im(\beta) \in (-2\pi, 2\pi)$.

The choice $u_0^1 = u_\infty^1 = x = -u_0^2 = -u_\infty^2$ gives $c = u_\infty^1 / u_0^1 = 1$,

$$\zeta_0(x) = ix/|x|, \quad I_0^+(x) = S^1 - \{-\zeta_0\},$$
$$\zeta_\infty(x) = |x|/ix, \quad I_\infty^+(x) = S^1 - \{-\zeta_\infty\}. \tag{13.2}$$

We choose on $\widehat{I}_0^+$ the branch of $\log z$ with

$$\arg z = \Im(\log z) \in (-\tfrac{1}{2}\pi + \arg x, \tfrac{3}{2}\pi + \arg x).$$

Now let $(H, \nabla, x, x, P, A, J)$ be a P_{3D6}-TEJPA bundle with $x \in \mathbb{C} - \mathbb{R}_{\leq 0}$. The Φ^{mon} in (10.4) associates to it a unique tuple (β, s, B) with β as above, namely $\frac{1}{2} e^{-\beta/2} = x, \Im(\beta) \in (-2\pi, 2\pi)$. Throughout this chapter we shall assume that $s \in \mathbb{C}^{[sto, a \cup b]} = \mathbb{C} - \{\pm 2\}$. This is equivalent to the monodromy being semisimple.

Let $\underline{e}_0^\pm, \underline{e}_\infty^\pm$ be the (up to a global sign) unique 4-tuple of bases in Theorem 7.3 (c). Then (2.22), (5.1) and (6.13) show

$$\mathrm{Mon}(\underline{e}_0^+) = \underline{e}_0^+ \, \mathrm{Mon}_0^{mat}(s) \tag{13.3}$$

$$\text{with } \mathrm{Mon}_0^{mat}(s) = S^t S^{-1}, \quad S = \begin{pmatrix} 1 & s \\ 0 & 1 \end{pmatrix}.$$

Equations (5.4) and (5.15) give

$$\mathrm{Mon}_0^{mat}(s) \, v_\pm = \lambda_\pm \, v_\pm \tag{13.4}$$

$$\text{for } v_\pm = \begin{pmatrix} 1 \\ \mp \sqrt{\frac{1}{4}s^2 - 1} + \frac{s}{2} \end{pmatrix} = \begin{pmatrix} 1 \\ \mp i e^{-\pi i \alpha_\pm} \end{pmatrix},$$

so

$$\mathrm{Mon}(\underline{f}_0^+) = \underline{f}_0^+ \begin{pmatrix} \lambda_+ & 0 \\ 0 & \lambda_- \end{pmatrix} \quad \text{for } \underline{f}_0^+ := \underline{e}_0^+ (v_+ \, v_-). \tag{13.5}$$

$\underline{f}_0^+$ is a flat basis on $\widehat{I}_0^+$ of eigenvectors of the monodromy. Here $z \in \widehat{I}_0^+$ if $\arg z \in (-\tfrac{1}{2}\pi + \arg x, \tfrac{3}{2}\pi + \arg x)$. For z with different argument $\arg z$, $\underline{f}_0^+(z)$ is the flat extension of $\underline{f}_0^+$ on $\widehat{I}_0^+$.

Now we define for $k \in \mathbb{Z}$ the *elementary sections* on $\widehat{I}_0^+$

$$es_1^{(k)}(z) := f_0^{+1}(z)\, z^{\alpha_+ + k} = f_0^{+1}(z)\, exp((\alpha_+ + k)\log z),$$
$$es_2^{(k)}(z) := f_0^{+2}(z)\, z^{\alpha_- + k} = f_0^{+2}(z)\, exp((\alpha_- + k)\log z),$$
(13.6)

and extend them to (a priori) multi-valued sections on $\mathbb{C}^*$ by extending $f_0^{+1/2}$ flatly and $z^{\alpha_\pm + k}$ holomorphically (on $\widehat{I}_0^+$ $\log z$ is the branch chosen above with $\arg z = \Im(\log z) \in (-\tfrac{1}{2}\pi + \arg x, \tfrac{3}{2}\pi + \arg x)$). The elementary sections are single-valued holomorphic sections on $H|_{\mathbb{C}^*}$ because (in the formula (13.7) $e_{1/2}^{(k)}$ means either always $e_1^{(k)}$ or always $e_2^{(k)}$).

$$es_{1/2}^{(k)}(z\,e^{2\pi i}) = f_0^{+1/2}(z\,e^{2\pi i})\,(z\,e^{2\pi i})^{\alpha_\pm + k}$$
$$= Mon(f_0^{+1/2}(z))\,\lambda_{\mp}\,z^{\alpha_\pm + k} = es_{1/2}^{(k)}(z).$$
(13.7)

They form a holomorphic basis $\underline{es}^{(k)} = (es_1^{(k)}, es_2^{(k)})$ of $H|_{\mathbb{C}^*}$ with

$$\nabla_{z\partial_z}\underline{es}^{(k)}(z) = \underline{es}^{(k)}(z)\begin{pmatrix} \alpha_+ + k & 0 \\ 0 & \alpha_- + k \end{pmatrix}.$$
(13.8)

The following lemma says how the elementary sections behave with respect to A, J, P.

Lemma 13.1 *For $z \in \widehat{I}_0^+$ the following holds.*

(a)

$$A(f_{-0}^+(z)) = f_{-0}^+(z\,e^{\pi i})\begin{pmatrix} i\,e^{-\pi i\alpha_-} & 0 \\ 0 & -i\,e^{-\pi i\alpha_+} \end{pmatrix},$$
(13.9)

$$A(\underline{es}^{(k)}(z)) = \underline{es}^{(k)}(-z)\begin{pmatrix} i & 0 \\ 0 & -i \end{pmatrix}(-1)^k.$$
(13.10)

(b) In (13.11) $f_{-0}^+(1/z)$ is the flat extension of f_{-0}^+ on $\widehat{I}_0^+$ with $\arg 1/z = -\arg z$.

$$J(f_{-0}^+(z)) = f_{-0}^+(1/z)\begin{pmatrix} 0 & b_+ \\ b_- & 0 \end{pmatrix},$$
(13.11)

$$J(\underline{es}^{(k)}(z)) = \underline{es}^{(-k)}(1/z)\begin{pmatrix} 0 & b_+ \\ b_- & 0 \end{pmatrix}.$$
(13.12)

(c)

$$P(\underline{f}_0^+(z)^t, \underline{f}_0^+(z\,e^{\pi i})) = \begin{pmatrix} 0 & \lambda_- + 1 \\ \lambda_+ + 1 & 0 \end{pmatrix}, \qquad (13.13)$$

$$P(\underline{es}^{(k)}(z)^t, \underline{es}^{(l)}(-z)) = (-1)^l z^{k+l} \, 2\cos(\pi\alpha_+) \begin{pmatrix} 0 & 1 \\ 1 & 0 \end{pmatrix}. \qquad (13.14)$$

Proof We can assume $x = \frac{1}{2}$, because for any $x \in \mathbb{C} - \mathbb{R}_{\leq 0}$ a P_{3D6}-TEJPA bundle is connected by an isomonodromic family with a P_{3D6}-TEJPA bundle with $x = \frac{1}{2}$, and $\underline{f}_0^+(z)$, $\underline{es}^{(k)}(z)$, A, J and P (for fixed z) vary flatly with respect to x in this isomonodromic family.

(a)

$$\begin{aligned}
A(\underline{f}_0^+(z)) &= A(\underline{e}_0^+(z))(v_+ v_-) \\
&\overset{(7.11)}{=} \underline{e}_0^-(-z)\begin{pmatrix} 0 & -1 \\ 1 & 0 \end{pmatrix}(v_+ v_-) \\
&\overset{(2.21)}{=} \underline{e}_0^+(z\,e^{\pi i})\begin{pmatrix} 1 & s \\ 0 & 1 \end{pmatrix}\begin{pmatrix} 0 & -1 \\ 1 & 0 \end{pmatrix}(v_+ v_-) \\
&= \underline{e}_0^+(z\,e^{\pi i})\,T(s)^{-1}(v_+ v_-) \\
&\overset{(5.29)}{=} \underline{e}_0^+(z\,e^{\pi i})(v_+ v_-)\begin{pmatrix} ie^{-\pi i\alpha_-} & 0 \\ 0 & -ie^{-\pi i\alpha_+} \end{pmatrix} \\
&= \underline{f}_0^+(z\,e^{\pi i})\begin{pmatrix} ie^{-\pi i\alpha_-} & 0 \\ 0 & -ie^{-\pi i\alpha_+} \end{pmatrix}. \\
A(\underline{es}^{(k)}_{1/2}(z)) &= A(\underline{f}_0^{+1/2}(z)\,z^{\alpha_\pm + k}) \\
&\overset{(13.9)}{=} \underline{f}_0^{+1/2}(z\,e^{\pi i})(\pm i)\,e^{-\pi i\alpha_\mp}(z\,e^{\pi i})^{\alpha_\pm + k}(-1)^k\,e^{\pi i\alpha_\mp} \\
&= \underline{es}^{(k)}_{1/2}(z\,e^{\pi i})(\pm i)(-1)^k.
\end{aligned}$$

(b) $x = \frac{1}{2}$ implies $\zeta_0 = i = -\zeta_\infty$, $\widehat{I}_0^+ = \widehat{I}_\infty^-$, and $z \in \widehat{I}_0^+$ satisfies $\arg(z) \in (-\frac{1}{2}\pi, \frac{3}{2}\pi)$, $\arg(1/z) \in (-\frac{3}{2}\pi, \frac{1}{2}\pi)$. Thus $\underline{e}_\infty^-(1/z)$ is the flat extension of $\underline{e}_\infty^-$ on $\widehat{I}_\infty^-$ clockwise, and

$$\underline{e}_\infty^-(1/z) = \underline{e}_\infty^+(1/z)\,S_\infty^b = \underline{e}_\infty^+(1/z)(S^{-1})^t$$

by (2.21). Furthermore, $x = \frac{1}{2}$ implies $\beta = 0$, and (2.24) gives

$$\underline{e}_\infty^-(1/z) = \underline{e}_0^+(1/z)\,B.$$

Finally, observe that

$$S^t \begin{pmatrix} 1 & 0 \\ 0 & -1 \end{pmatrix} (v_+ \; v_-) = \begin{pmatrix} 1 & 0 \\ s & -1 \end{pmatrix} \begin{pmatrix} 1 & 1 \\ -\sqrt{\frac{1}{4}s^2 - 1} + \frac{s}{2} & \sqrt{\frac{s^2}{4} - 1} + \frac{s}{2} \end{pmatrix}$$

$$= \begin{pmatrix} 1 & 1 \\ \sqrt{\frac{1}{4}s^2 - 1} + \frac{1}{2}s & -\sqrt{\frac{1}{4}s^2 - 1} + \frac{1}{2}s \end{pmatrix} = (v_- \; v_+).$$

Now one calculates

$$J(f_{-0}^+(z)) = J(e_0^+(z)) (v_+ \; v_-)$$

$$\overset{(7.12)}{=} e_\infty^+(1/z) \begin{pmatrix} 1 & 0 \\ 0 & -1 \end{pmatrix} (v_+ \; v_-)$$

$$= e_\infty^-(1/z) S^t \begin{pmatrix} 1 & 0 \\ 0 & -1 \end{pmatrix} (v_+ \; v_-)$$

$$= e_\infty^-(1/z) (v_- \; v_+)$$

$$= e_0^+(1/z) B (v_- \; v_+)$$

$$= e_0^+(1/z) (b_- v_- \; b_+ v_+)$$

$$= f_{-0}^+(1/z) \begin{pmatrix} 0 & b_+ \\ b_- & 0 \end{pmatrix}.$$

$$J(es_{1/2}^{(k)}(z)) = J(f_0^{+1/2}(z) z^{\alpha_\pm + k})$$

$$\overset{(13.11)}{=} f_0^{+2/1}(1/z) b_\mp (1/z)^{\alpha_\mp - k}$$

$$= b_\mp \, es_{2/1}^{(-k)}(1/z).$$

(c)

$$P(f_{-0}^+(z)^t, f_{-0}^+(z\,e^{\pi i}))$$

$$= (v_+ \; v_-)^t \, P(e_0^+(z)^t, e_0^+(z\,e^{\pi i})) (v_+ \; v_-)$$

$$\overset{(2.21)}{=} (v_+ \; v_-)^t \, P(e_0^+(z)^t, e_0^-(-z)) S^{-1} (v_+ \; v_-)$$

$$\overset{(7.10)}{=} (v_+ \; v_-)^t \, 1_2 \, S^{-1} (v_+ \; v_-)$$

$$= \begin{pmatrix} 1 & -i\,e^{-\pi i\alpha_+} \\ 1 & i\,e^{-\pi i\alpha_-} \end{pmatrix} \begin{pmatrix} 1 & -s \\ 0 & 1 \end{pmatrix} \begin{pmatrix} 1 & 1 \\ -i\,e^{-\pi i\alpha_+} & i\,e^{-\pi i\alpha_-} \end{pmatrix}$$

$$= \begin{pmatrix} 1 + is\, e^{-\pi i \alpha_+} - e^{-2\pi i \alpha_+} & 1 - is\, e^{-\pi i \alpha_-} + 1 \\ 1 + is\, e^{-\pi i \alpha_+} + 1 & 1 - is\, e^{-\pi i \alpha_-} - e^{-2\pi i \alpha_-} \end{pmatrix}$$

$$= \begin{pmatrix} 0 & \lambda_- + 1 \\ \lambda_+ + 1 & 0 \end{pmatrix},$$

$$P(es_1^{(k)}(z), es_2^{(l)}(-z))$$

$$= P(f_0^{+1}(z)\, z^{\alpha_+ + k}, f_0^{+2}(z\, e^{\pi i})\,(z\, e^{\pi i})^{\alpha_- + l})$$

$$\overset{(13.13)}{=} (\lambda_- + 1)\, z^{k+l}\,(-1)^l\, e^{\pi i \alpha_-}$$

$$= (-1)^l\, z^{k+l}\,(e^{-\pi i \alpha_-} + e^{\pi i \alpha_-})$$

$$= (-1)^l\, z^{k+l}\, 2\cos(\pi \alpha_-)$$

$$= (-1)^l\, z^{k+l}\, 2\cos(\pi \alpha_+).$$

□

We keep the P_{3D6}-TEJPA bundle $(H, \nabla, x, x, P, A, J)$ which was chosen above with $x \in \mathbb{C} - \mathbb{R}_{\leq 0}$ and its triple (β, s, B).

DEFINITION/LEMMA 13.2 *In the following,*

$$\text{either } (\widetilde{\alpha}_+, \widetilde{\alpha}_-) := (\alpha_+(s), \alpha_-(s))$$

$$\text{or } (\widetilde{\alpha}_+, \widetilde{\alpha}_-) := (\alpha_-(s) - 1, \alpha_+(s) + 1)$$

(Remark 13.4 motivates the second choice). In any case $\widetilde{\alpha}_+ + \widetilde{\alpha}_- = 0$.

(a) *(Definition) Let $\widetilde{H} \to \mathbb{P}^1$ be a flat holomorphic pure rank 2 vector bundle on $\mathbb{P}^1$ with global basis $\underline{\chi} = (\chi_1, \chi_2)$. Define a flat meromorphic connection $\widetilde{\nabla}$, a pairing $\widetilde{P}$ and an automorphism $\widetilde{A}$ on $\widetilde{H}$ by*

$$\widetilde{\nabla}_{z\partial_z}\underline{\chi}(z) = \underline{\chi}(z) \left[\frac{x}{z}\begin{pmatrix} 0 & 1 \\ 1 & 0 \end{pmatrix} + \begin{pmatrix} \widetilde{\alpha}_+ & 0 \\ 0 & \widetilde{\alpha}_- \end{pmatrix} \right], \qquad (13.15)$$

$$\widetilde{P}(\underline{\chi}(z)^t, \underline{\chi}(-z)) = \begin{pmatrix} 0 & 2 \\ 2 & 0 \end{pmatrix}, \qquad (13.16)$$

$$\widetilde{A}(\underline{\chi}(z)) = \underline{\chi}(-z) \begin{pmatrix} i & 0 \\ 0 & -i \end{pmatrix}. \qquad (13.17)$$

(b) *(Lemma) $(\widetilde{H} \to \mathbb{P}^1, \widetilde{\nabla}, \widetilde{P}, \widetilde{A})$ is a TEPA bundle (Definitions 6.1 (a) and 7.1 (a)).*

Proof (a) Definition. (b) Obviously $\widetilde{P}$ is symmetric and nondegenerate, $\widetilde{A}^2 = -\,\mathrm{id}$, and $\widetilde{P}(\widetilde{A}\,a, \widetilde{A}\,b) = \widetilde{P}(a,b)$. The flatness of $\widetilde{P}$ and $\widetilde{A}$ follows from

$$\widetilde{P}(\nabla_{z\partial_z}\underline{\chi}(z)^t, \underline{\chi}(-z)) + \widetilde{P}(\underline{\chi}(z)^t, \nabla_{z\partial_z}\underline{\chi}(-z))$$

$$= \left[\frac{x}{z}\begin{pmatrix} 0 & 1 \\ 1 & 0 \end{pmatrix} + \begin{pmatrix} \widetilde{\alpha}_+ & 0 \\ 0 & \widetilde{\alpha}_- \end{pmatrix}\right]\begin{pmatrix} 0 & 2 \\ 2 & 0 \end{pmatrix}$$

$$+ \begin{pmatrix} 0 & 2 \\ 2 & 0 \end{pmatrix}\left[\frac{x}{-z}\begin{pmatrix} 0 & 1 \\ 1 & 0 \end{pmatrix} + \begin{pmatrix} \widetilde{\alpha}_+ & 0 \\ 0 & \widetilde{\alpha}_- \end{pmatrix}\right]$$

$$= 0 = z\partial_z\widetilde{P}(\underline{\chi}(z)^t, \underline{\chi}(-z))$$

and

$$\nabla_{z\partial_z}(\widetilde{A}(\underline{\chi}(z))) = \underline{\chi}(-z)\left[\frac{x}{-z}\begin{pmatrix} 0 & 1 \\ 1 & 0 \end{pmatrix} + \begin{pmatrix} \widetilde{\alpha}_+ & 0 \\ 0 & \widetilde{\alpha}_- \end{pmatrix}\right]\begin{pmatrix} i & 0 \\ 0 & -i \end{pmatrix}$$

$$= \underline{\chi}(-z)\begin{pmatrix} i & 0 \\ 0 & -i \end{pmatrix}\left[\frac{x}{z}\begin{pmatrix} 0 & 1 \\ 1 & 0 \end{pmatrix} + \begin{pmatrix} \widetilde{\alpha}_+ & 0 \\ 0 & \widetilde{\alpha}_- \end{pmatrix}\right]$$

$$= \widetilde{A}(\nabla_{z\partial_z}\underline{\chi}(z)).$$

$\square$

Lemma 13.3 *Consider the above P_{3D6}-TEJPA bundle $(H, \nabla, x, x, P, A, J)$ and the TEPA bundle $(\widetilde{H}, \widetilde{\nabla}, \widetilde{P}, \widetilde{A})$ in Definition/Lemma 13.2. Their restrictions to $\mathbb{C}$ are isomorphic, and the isomorphism*

$$(\widetilde{H}, \widetilde{\nabla}, \widetilde{P}, \widetilde{A})|_{\mathbb{C}} \to (H, \nabla, P, A)|_{\mathbb{C}} \tag{13.18}$$

is unique up to a sign.

The restrictions to $\mathbb{C}$ of the bundles will be identified via this isomorphism (i.e. by fixing one of the two isomorphisms which differ only by a sign).

Proof The tuple $(\widetilde{H}, \widetilde{\nabla}, \widetilde{P}, \widetilde{A})|_{\mathbb{C}}$ satisfies all properties of a P_{3D6}-TEPA bundle which do not concern the pole at ∞, but only the restriction to $\mathbb{C}$. Therefore all parts of Theorems 2.3, 6.3 and 7.3 hold which concern only the restriction to $\mathbb{C}$. In particular, for $u_0^1 = u_\infty^1 = -u_0^2 = -u_\infty^2 = x$, $\alpha_0^1 = \alpha_0^2 = 0$, $\zeta_0 = ix/|x|$ and $I_0^a, I_0^b, I_0^\pm$ as in (2.1), and $\widetilde{L} := \widetilde{H}|_{\mathbb{C}^*}$, the pole at 0 induces flat rank 1 subbundles $\widetilde{L}_0^{\pm j}$ ($j = 1, 2$) as after the Remarks 2.4. Theorem 7.3 (c) provides 2 up to a global sign unique bases $\widetilde{\underline{e}}_0^\pm = (\widetilde{e}_0^{\pm 1}, \widetilde{e}_0^{\pm 2})$ of $\widetilde{L}|_{T_0^\pm}$ such that

$\widetilde{e}_0^{\pm j}$ are flat generating sections of $\widetilde{L}_0^{\pm j}$,

the first half of (2.21) holds with $s_0^a = s_0^b = \widetilde{s}$ for some $\widetilde{s} \in \mathbb{C}$,

$$\widetilde{P}(\widetilde{\underline{e}}_0^\pm(z)^t, \widetilde{\underline{e}}_0^\pm(-z)) = \mathbf{1}_2, \tag{13.19}$$

$$\widetilde{A}(\widetilde{\underline{e}}_0^\pm(z)) = \widetilde{\underline{e}}_0^\mp(-z) \begin{pmatrix} 0 & -1 \\ 1 & 0 \end{pmatrix}.$$

By (2.22) the monodromy matrix with respect to $\widetilde{\underline{e}}_0^\pm$ is $\mathrm{Mon}_0^{mat}(\widetilde{s})$, so its eigenvalues are $\lambda_\pm(\widetilde{s})$.

On the other hand, the logarithmic pole at ∞ of the bundle $(\widetilde{H}, \widetilde{\nabla}, \widetilde{P}, \widetilde{A})$ shows that the eigenvalues of the monodromy are $e^{-2\pi i \widetilde{\alpha}_\pm}$ which is either $\lambda_\pm(s)$ or $\lambda_\mp(s)$. Thus $\{\lambda_+(s), \lambda_-(s)\} = \{\lambda_+(\widetilde{s}), \lambda_-(\widetilde{s})\}$, so $\widetilde{s} = s$ or $\widetilde{s} = -s$.

We have to show $\widetilde{s} = s$. Then the bases $\widetilde{\underline{e}}_0^\pm$ coincide up to a global sign with the analogous bases $\underline{e}_0^\pm$ in Theorem 7.3 (c) of the P_{3D6}-TEJPA bundle $(H, \nabla, x, x, P, A, J)$ with tuple (β, s, B). This gives then the (unique up to a sign) isomorphism in (13.18).

Define

$$\begin{aligned} \widetilde{\underline{f}}_0^+(z) &:= \widetilde{\underline{e}}_0^+(z) \, (v_+(\widetilde{s}) \ v_-(\widetilde{s})) \\ (\widetilde{\alpha}_+(\widetilde{s}), \widetilde{\alpha}_-(\widetilde{s})) &:= (\alpha_+(\widetilde{s}), \alpha_-(\widetilde{s})) \end{aligned} \right\} \quad \text{if } \widetilde{\alpha}_\pm = \alpha_\pm(s),$$

$$\tag{13.20}$$

$$\begin{aligned} \widetilde{\underline{f}}_0^+(z) &:= \widetilde{\underline{e}}_0^+(z) \, (v_-(\widetilde{s}) \ v_+(\widetilde{s})) \\ (\widetilde{\alpha}_+(\widetilde{s}), \widetilde{\alpha}_-(\widetilde{s})) &:= (\alpha_-(\widetilde{s}) - 1, \alpha_+(\widetilde{s}) + 1) \end{aligned} \right\} \quad \text{if } \widetilde{\alpha}_\pm = \alpha_\mp(s) \mp 1,$$

and in both cases

$$\widetilde{\underline{es}}_{1/2}^{(k)}(z) := \widetilde{\underline{f}}_0^{+1/2}(z) \, z^{\widetilde{\alpha}_\pm(\widetilde{s})} \quad \text{(with } 1 \leftrightarrow +, \ 2 \leftrightarrow -).$$

The formulae (13.9), (13.10), (13.13) and (13.14) hold in both cases if one replaces $A, P, \underline{f}_0^+, \underline{es}^{(k)}, \alpha_\pm, \lambda_\pm$ by

$$\widetilde{A}, \widetilde{P}, \widetilde{\underline{f}}_0^+, \widetilde{\underline{es}}^{(k)} := (\widetilde{es}_1^{(k)}, \widetilde{es}_2^{(k)}), \widetilde{\alpha}_\pm, \widetilde{\lambda}_\pm := e^{-2\pi i \widetilde{\alpha}_\pm(\widetilde{s})}.$$

Equations (13.10) and (13.17) give for χ_1 and χ_2 the elementary section expansions in

$$\chi_1 = \sum_{k \in 2\mathbb{Z}} a_{k,1}^+ \, \widetilde{es}_1^{(k)} + \sum_{k \in 2\mathbb{Z}-1} a_{k,1}^- \, \widetilde{es}_2^{(k)},$$

$$\tag{13.21}$$

$$\chi_2 = \sum_{k \in 2\mathbb{Z}-1} a_{k,2}^+ \, \widetilde{es}_1^{(k)} + \sum_{k \in 2\mathbb{Z}} a_{k,2}^- \, \widetilde{es}_2^{(k)}$$

with suitable coefficients $a_{k,1/2}^{\pm} \in \mathbb{C}$. Now the logarithmic pole at ∞ in (13.15) shows

$$\widetilde{\alpha}_{\pm}(\widetilde{s}) = \widetilde{\alpha}_{\pm}, \quad \text{so } \alpha_{\pm}(\widetilde{s}) = \alpha_{\pm}(s),$$
$$\text{so } \widetilde{s} = s \text{ and } \widetilde{\underline{e}}_0^{\pm} = \underline{e}_0^{\pm}, \tag{13.22}$$

which is what we had to prove, and furthermore

$$a_{k,1}^+ = a_{k,1}^- = a_{k,2}^+ = a_{k,2}^- = 0 \quad \text{for } k \geq 1, \tag{13.23}$$
$$a_{0,1}^+ \neq 0, \quad a_{0,2}^- \neq 0, \tag{13.24}$$

and

$$\widetilde{\underline{es}}^{(k)} = \underline{es}^{(k)}(z) \quad \text{in the case} \quad \widetilde{\alpha}_{\pm} = \alpha_{\pm}(s),$$
$$\widetilde{\underline{es}}^{(k)} = \underline{es}^{(k)}(z) \begin{pmatrix} 0 & z \\ z^{-1} & 0 \end{pmatrix} \quad \text{in the case} \quad \widetilde{\alpha}_{\pm} = \alpha_{\mp}(s) \mp 1. \tag{13.25}$$

$\square$

Remark 13.4 In Definition/Lemma 13.2 and Lemma 13.3 we are primarily interested in the case $\widetilde{\alpha}_{\pm} = \alpha_{\pm}(s)$. We consider also the other case $\widetilde{\alpha}_{\pm} = \alpha_{\mp}(s) \mp 1$ because it arises by analytic continuation of s over $\mathbb{C}^{[sto,b+]}$ from the case $\widetilde{\alpha}_{\pm} = \alpha_{\pm}(s)$.

We chose $\widetilde{\alpha}_{\pm}$, $(\widetilde{H}, \widetilde{\nabla}, \widetilde{P}, \widetilde{A})$, $\widetilde{f}_0^+$ (in (13.20)) and $\widetilde{\underline{es}}^{(k)}$ such that they fit together in the two cases by analytic continuation of s over $\mathbb{C}^{[sto,b+]}$. This will allow us to prove statements on the case $s \in \mathbb{C}^{[sto,b+]}$ by holomorphic extension from both cases for $s \in \mathbb{C}^{[sto,a]}$.

It also allows us to consider in the proof of Lemma 13.5 after (13.40) only the case $\widetilde{\alpha}_{\pm} = \alpha_{\pm}(s)$. In fact, we need that in Lemma 13.5 the case $\widetilde{\alpha}_{\pm} = \alpha_{\mp}(s) \mp 1$ follows by analytic continuation from the case $\widetilde{\alpha}_{\pm} = \alpha_{\pm}(s)$, because the leading parts of χ_1 and χ_2 for x near ∞ in the case $\widetilde{\alpha}_{\pm} = \alpha_{\mp}(s) \mp 1$ determine the coefficients $a_{-1,1}^-(1)$ and $a_{0,2}^-(1)$, but not the coefficient $a_{0,1}^+(1)$.

Lemma 13.5 *In (13.21) the coefficients* $a_{-2k,1}^+$, $a_{-2k-1,1}^+$, $a_{-2k,2}^-$, $a_{-2k-1,2}^-$ *for* $k \in \mathbb{Z}_{\geq 0}$ *(all others are 0 because of (13.23)) are*

$$a_{-2k,1}^+(x) = \Gamma(\widetilde{\alpha}_+ - \tfrac{1}{2}(2k-1)) \frac{2^{\widetilde{\alpha}_+ - 2k}(-1)^k}{\sqrt{\pi}\, k!} x^{-\widetilde{\alpha}_+ + 2k},$$

$$a_{-2k-1,2}^+(x) = \Gamma(\widetilde{\alpha}_+ - \tfrac{1}{2}(2k+1)) \frac{2^{\widetilde{\alpha}_+ - 2k-1}(-1)^k}{\sqrt{\pi}\, k!} x^{-\widetilde{\alpha}_+ + 2k+1},$$

$$a_{-2k,2}^-(x) = \Gamma(\widetilde{\alpha}_- - \tfrac{1}{2}(2k-1)) \frac{2^{\widetilde{\alpha}_- - 2k}(-1)^k}{\sqrt{\pi}\, k!} x^{-\widetilde{\alpha}_- + 2k}, \tag{13.26}$$

$$a_{-2k-1,1}^-(x) = \Gamma(\widetilde{\alpha}_- - \tfrac{1}{2}(2k+1)) \frac{2^{\widetilde{\alpha}_- - 2k-1}(-1)^k}{\sqrt{\pi}\, k!} x^{-\widetilde{\alpha}_- + 2k+1}.$$

Proof We make the dependence on x and z explicit in the basis $\underline{\chi}$ of sections of $\widetilde{H}$ by writing $\underline{\chi}(x, z)$, and we use the fact that we have an isomonodromic family in x. Then the shape of (13.15) shows

$$\underline{\chi}(x, z) = \underline{\chi}(1, z/x), \tag{13.27}$$

so in (13.21)

$$a_{k,1}^+(x)\,\widetilde{es}_1^{(k)}(z) = a_{k,1}^+(1)\,\widetilde{es}_1^{(k)}(z/x), \tag{13.28}$$

and analogously for the other terms. This gives the dependence on x in (13.26). From now on we restrict to the case $x = 1$.

In that case (13.15) together with (13.21), (13.23), (13.24) and

$$\nabla_{z\partial_z}\widetilde{es}_{1/2}^{(k)}(z) = (\widetilde{\alpha}_\pm + k)\,\widetilde{es}_{1/2}^{(k)}(z)$$

gives the recursive relations

$$\begin{aligned}
(2\widetilde{\alpha}_+ + k)\,a_{k,2}^+(1) &= a_{k+1,1}^+(1) \quad \text{for } k \in 2\mathbb{Z}_{\leq 0} - 1, \\
k\,a_{k,1}^+(1) &= a_{k+1,2}^+(1) \quad \text{for } k \in 2\mathbb{Z}_{\leq 0} - 2, \\
(2\widetilde{\alpha}_- + k)\,a_{k,1}^-(1) &= a_{k+1,2}^-(1) \quad \text{for } k \in 2\mathbb{Z}_{\leq 0} - 1, \\
k\,a_{k,2}^-(1) &= a_{k+1,1}^-(1) \quad \text{for } k \in 2\mathbb{Z}_{\leq 0} - 2.
\end{aligned} \tag{13.29}$$

With these recursive relations, the equations in (13.26) for $x = 1$ reduce to the special cases for $a_{0,1}^+(1)$ and $a_{0,2}^-(1)$,

$$a_{0,1}^+(1) = \Gamma(\widetilde{\alpha}_+ + \frac{1}{2})\,\frac{2^{\widetilde{\alpha}_+}}{\sqrt{\pi}},$$

$$a_{0,2}^-(1) = \Gamma(\widetilde{\alpha}_- + \frac{1}{2})\,\frac{2^{\widetilde{\alpha}_-}}{\sqrt{\pi}}. \tag{13.30}$$

Equation (13.30) will now be proved as follows. Equation (13.15) for the connection will be rewritten as two second order linear differential equations, which turn out to be Bessel equations and are solved by Hankel functions. From the known asymptotics of the Hankel functions near 0 and near ∞ we can read off (13.30).

Equation (13.15) in the case $x = 1$ has at 0 a semisimple pole of order 2 with eigenvalues ± 1. As in the proof of Theorem 10.3 (a), formula (10.16), one sees that the solution $\underline{\chi}$ takes the form

$$\underline{\chi}(z) = \underline{e}_0^+(z) \begin{pmatrix} e^{-1/z} & 0 \\ 0 & e^{1/z} \end{pmatrix} A_0^+(z)\,C \tag{13.31}$$

with $C = \left(\begin{smallmatrix} 1 & 1 \\ -i & i \end{smallmatrix} \right)$, $A_0^+ \in GL(2, \mathcal{A}|_{I_0^\pm})$, $\widehat{A}_0^+ = \mathbf{1}_2$.

We write

$$\underline{\chi}(z) = \underline{e}_0^+(z)\, h(z) = \underline{e}_0^+(z) \begin{pmatrix} h_{11}(z) & h_{12}(z) \\ h_{21}(z) & h_{22}(z) \end{pmatrix}.$$

Then the leading parts of $h_{jk}(z)$ near 0 are

$$\begin{pmatrix} h_{11}(z) & h_{12}(z) \\ h_{21}(z) & h_{22}(z) \end{pmatrix} \sim \begin{pmatrix} e^{-1/z} & 0 \\ 0 & e^{1/z} \end{pmatrix} \mathbf{1}_2\, C. \tag{13.32}$$

The second order linear differential equations satisfied by the functions $h_{jk}(z)$ are calculated as follows.

$$0 = \nabla_{z\partial_z} \underline{\chi} - \underline{\chi} \left[\frac{1}{z} \begin{pmatrix} 0 & 1 \\ 1 & 0 \end{pmatrix} + \begin{pmatrix} \widetilde{\alpha}_+ & 0 \\ 0 & \widetilde{\alpha}_- \end{pmatrix} \right]$$

$$= \underline{e}_0^+(z) \left[z\partial_z h - h \frac{1}{z} \begin{pmatrix} 0 & 1 \\ 1 & 0 \end{pmatrix} - h \begin{pmatrix} \widetilde{\alpha}_+ & 0 \\ 0 & \widetilde{\alpha}_- \end{pmatrix} \right]$$

$$= \underline{e}_0^+(z) \begin{pmatrix} z\partial_z h_{11} - \frac{1}{z}h_{12} - \widetilde{\alpha}_+ h_{11} & z\partial_z h_{12} - \frac{1}{z}h_{11} - \widetilde{\alpha}_- h_{12} \\ z\partial_z h_{21} - \frac{1}{z}h_{22} - \widetilde{\alpha}_+ h_{21} & z\partial_z h_{22} - \frac{1}{z}h_{21} - \widetilde{\alpha}_- h_{22} \end{pmatrix},$$

$$0 = (z\partial_z)^2 h_{j1} - z\partial_z(\tfrac{1}{z}h_{j2} + \widetilde{\alpha}_+ h_{j1})$$

$$= (z\partial_z)^2 h_{j1} + \tfrac{1}{z}h_{j2} - \tfrac{1}{z}z\partial_z h_{j2} - \widetilde{\alpha}_+ z\partial_z h_{j1}$$

$$= (z\partial_z)^2 h_{j1} + \tfrac{1}{z}h_{j2} - \tfrac{1}{z}(\tfrac{1}{z}h_{j1} + \widetilde{\alpha}_- h_{j2}) - \widetilde{\alpha}_+ z\partial_z h_{j1}$$

$$= (z\partial_z)^2 h_{j1} - \widetilde{\alpha}_+ z\partial_z h_{j1} - \tfrac{1}{z^2}h_{j1} + (1 - \widetilde{\alpha}_-)(z\partial_z h_{j1} - \widetilde{\alpha}_+ h_{j1})$$

$$= (z\partial_z)^2 h_{j1} + z\partial_z h_{j1} - (\tfrac{1}{z^2} + (1 - \widetilde{\alpha}_-)\widetilde{\alpha}_+)h_{j1}.$$

Write

$$z = i/t, \quad h_{jk}(z) = h_{jk}(i/t) = \sqrt{t}\,\widetilde{h}_{jk}(t). \tag{13.33}$$

Then

$$0 = (t\partial_t)^2(\sqrt{t}\,\widetilde{h}_{j1}) - t\partial_t(\sqrt{t}\,\widetilde{h}_{j1}) + (t^2 - (1 - \widetilde{\alpha}_-)\widetilde{\alpha}_+)\sqrt{t}\,\widetilde{h}_{j1}$$

$$= (\tfrac{1}{4}\sqrt{t}\,\widetilde{h}_{j1} + \sqrt{t}\,t\partial_t\widetilde{h}_{j1} + \sqrt{t}\,(t\partial_t)^2\widetilde{h}_{j1})$$

$$\quad - (\tfrac{1}{2}\sqrt{t}\,\widetilde{h}_{j1} + \sqrt{t}\,t\partial_t\widetilde{h}_{j1}) + (t^2 - (1 - \widetilde{\alpha}_-)\widetilde{\alpha}_+)\sqrt{t}\,\widetilde{h}_{j1}$$

$$= \sqrt{t}\left[(t\partial_t)^2\widetilde{h}_{j1} + (t^2 - (1 - \widetilde{\alpha}_-)\widetilde{\alpha}_+ - \tfrac{1}{4})\widetilde{h}_{j1}\right].$$

Thus

$$0 = (t\partial_t)^2 \widetilde{h}_{j1} + (t^2 - (\tfrac{1}{2} + \widetilde{\alpha}_+)^2) \widetilde{h}_{j1}, \tag{13.34}$$

and analogously

$$0 = (t\partial_t)^2 \widetilde{h}_{j2} + (t^2 - (\tfrac{1}{2} + \widetilde{\alpha}_-)^2) \widetilde{h}_{j2}. \tag{13.35}$$

Equations (13.34) and (13.35) are the cases $v = \tfrac{1}{2} + \widetilde{\alpha}_\pm$ of the Bessel equation (one of many possible references is [Te96, (9.1)])

$$0 = (t\partial_t)^2 y(t) + (t^2 - v^2)\, y(t). \tag{13.36}$$

A basis of solutions are the Hankel functions $H_v^{(1)}(t)$ and $H_v^{(2)}(t)$ [Te96, (9.3)], which are multi-valued on $\mathbb{C}^*$. Another basis is

$$H_{-v}^{(1)}(t) = e^{iv\pi} H_v^{(1)}(t), \quad H_{-v}^{(2)}(t) = e^{-iv\pi} H_v^{(2)}(t). \tag{13.37}$$

Their asymptotics near the logarithmic pole at 0 and the irregular pole at ∞ are known. If $\Re(v) > 0$ the leading part near 0 is [Te96, (9.13)]

$$H_v^{(1)}(t) \sim \frac{1}{\pi i} \Gamma(v)\, (2/t)^v, \quad H_v^{(2)}(t) \sim \frac{-1}{\pi i} \Gamma(v)\, (2/t)^v. \tag{13.38}$$

The leading part near ∞ is [Te96, (9.50) and before (9.52)]

$$H_v^{(1)}(t) \sim \sqrt{2/\pi t}\, e^{i(t - \frac{1}{2}v\pi - \frac{1}{4}\pi)} \quad \text{for } -\pi < \arg t < 2\pi,$$

$$H_v^{(2)}(t) \sim \sqrt{2/\pi t}\, e^{-i(t - \frac{1}{2}v\pi - \frac{1}{4}\pi)} \quad \text{for } -2\pi < \arg t < \pi. \tag{13.39}$$

Then (13.32), (13.34), (13.35) and (13.39) show that

$$h_{11}(z) = \sqrt{\pi/2}\, e^{i\frac{1}{2}\pi(\widetilde{\alpha}_+ + 1)} H_{\alpha_+ + \frac{1}{2}}^{(1)}(i/z)\, \sqrt{i/z},$$

$$h_{21}(z) = \sqrt{\pi/2}\, e^{-i\frac{1}{2}\pi(\widetilde{\alpha}_+ + 1)} H_{\alpha_+ + \frac{1}{2}}^{(2)}(i/z)\, (-i)\, \sqrt{i/z},$$

$$h_{12}(z) = \sqrt{\pi/2}\, e^{i\frac{1}{2}\pi(\widetilde{\alpha}_- + 1)} H_{\alpha_- + \frac{1}{2}}^{(1)}(i/z)\, \sqrt{i/z},$$

$$h_{22}(z) = \sqrt{\pi/2}\, e^{-i\frac{1}{2}\pi(\widetilde{\alpha}_- + 1)} H_{\alpha_- + \frac{1}{2}}^{(2)}(i/z)\, i\, \sqrt{i/z}. \tag{13.40}$$

Now for a moment we restrict to the case $\widetilde{\alpha}_\pm = \alpha_\pm(s)$ and to $s \in \mathbb{C}^{[sto,a]}$, so we exclude the case $\widetilde{\alpha}_\pm = \alpha_\mp(s) \mp 1$ and also $s \in \mathbb{C}^{[st,b]}$. Then $v = \tfrac{1}{2} + \widetilde{\alpha}_\pm = \tfrac{1}{2} + \alpha_\pm(s)$ and $\Re(v) > 0$. By (13.38) and (13.40) the leading parts of $h_{11}(z), h_{21}(z), h_{12}(z), h_{22}(z)$ for z near 0 are

$$h_{11}(z) \sim \sqrt{\pi/2}\, e^{i\frac{1}{2}\pi(\alpha_+ + 1)} \frac{1}{\pi i} \Gamma(\alpha_+ + \tfrac{1}{2})\, (2z/i)^{\alpha_+ + \frac{1}{2}} \sqrt{i/z}$$

$$= \Gamma(\alpha_+ + \tfrac{1}{2}) \frac{1}{\sqrt{\pi}} (2z)^{\alpha_+},$$

$$h_{21}(z) \sim \Gamma(\alpha_+ + \tfrac{1}{2}) \frac{1}{\sqrt{\pi}} (-i) e^{-i\pi\alpha_+} (2z)^{\alpha_+}, \tag{13.41}$$

$$h_{12}(z) \sim \Gamma(\alpha_- + \tfrac{1}{2}) \frac{1}{\sqrt{\pi}} (2z)^{\alpha_-},$$

$$h_{22}(z) \sim \Gamma(\alpha_- + \tfrac{1}{2}) \frac{1}{\sqrt{\pi}} i e^{-i\pi\alpha_-} (2z)^{\alpha_-},$$

so the leading parts of $\chi_1(z)$ and $\chi_2(z)$ for z near ∞ are

$$\chi_1(z) = \underline{e}_0^+ v_+ \Gamma(\alpha_+ + \tfrac{1}{2}) \frac{1}{\sqrt{\pi}} (2z)^{\alpha_+},$$

$$\chi_2(z) = \underline{e}_0^+ v_- \Gamma(\alpha_- + \tfrac{1}{2}) \frac{1}{\sqrt{\pi}} (2z)^{\alpha_-}. \tag{13.42}$$

This establishes (13.30) in the case $\widetilde{\alpha}_\pm = \alpha_\pm(s)$ and $s \in \mathbb{C}^{[sto,a]}$.

In the case $\widetilde{\alpha}_\pm = \alpha_\mp(s) \mp 1$ and/or $s \in \mathbb{C}^{[sto,b]}$ (13.30) follows by analytic continuation over $\mathbb{C}^{[sto,b+]}$ from the case $\widetilde{\alpha}_\pm = \alpha_\pm(s)$ and $s \in \mathbb{C}^{[sto,a]}$, see Remark 13.4. □

Now we return to the holomorphic family of sections σ_1 for the pure P_{3D6}-TEJPA bundles in M_{3FN}^{reg} which was considered at the beginning of this section.

Lemma 13.6 *There exists a continuous $m_{[1]}$-invariant and G^{mon}-invariant map*

$$B_3 : i\mathbb{R} \times V^{mat,a\cup b+} \to \mathbb{R}_{>0} \tag{13.43}$$

such that the set

$$U_3 := \{(\xi, s, B) \in \mathbb{C} \times V^{mat,a\cup b+} \mid |e^\xi| < B_3(i\Im(\xi), s, B)\} \tag{13.44}$$

and the induced subset $V_3 := (\Phi^{mon})^{-1}(U_3/\langle m_{[1]}^2 \rangle)$ of M_{3FN}^{mon} (see Lemma 10.1 (b)) satisfy: σ_1 is a holomorphic family of sections for the P_{3D6}-TEJPA bundles in V_3, and for those bundles with $s \in \mathbb{C}^{[sto,a]}$ an approximation of σ_1 for x near 0 is the leading part of χ_1 for the case $\widetilde{\alpha}_\pm = \alpha_\pm(s)$, i.e.

$$a_{0,1}^+(1) x^{-\alpha_+} e s_1^{(0)}(z). \tag{13.45}$$

Before we prove this lemma, we show that it gives those results of [Ni09] which are used in Chap. 12 together with the other arguments there to prove Theorem 12.4. σ_1 is holomorphic on $V_3 \subset M_{3FN}$, therefore formula (13.1) $2f_0 = P(\sigma_1(z), J(\sigma_1(-\rho_1(z))))$ shows that f_0 is holomorphic on V_3. This implies the basic property for the case $b+$ that f_{univ} is holomorphic (real analytic with respect to $s \in \mathbb{C}^{[sto,b]}$) on $S^{NI} \times V^{mat,b+} \cap U_1$ for an open set U_1 as in (12.15). For $s \in \mathbb{C}^{[sto,a]}$, (13.1) and (13.45) give the leading term of f_{mult} for x near 0,

$$
\begin{aligned}
&f_{mult}(x, s, B) \\
&\sim \tfrac{1}{2} P(a_{0,1}^+(1) x^{-\alpha+} \, es_1^{(0)}(z), J(a_{0,1}^+ x^{-\alpha+} \, es_1^{(0)}(-1/z))) \\
&\stackrel{(13.12)}{=} \tfrac{1}{2} (a_{0,1}^+)^2 x^{-2\alpha+} P(es_1^{(0)}(z), b_- \, es_2^{(0)}(-z)) \\
&\stackrel{(13.14)}{=} \tfrac{1}{2} (a_{0,1}^+)^2 x^{-2\alpha+} b_- 2\cos(\pi\alpha_+) \\
&\stackrel{(13.30)}{=} \Gamma(\alpha_+ + \tfrac{1}{2})^2 \frac{1}{\pi} (x/2)^{-2\alpha+} b_- \cos(\pi\alpha_+) \\
&= \frac{\Gamma(\tfrac{1}{2} + \alpha_+)}{\Gamma(\tfrac{1}{2} - \alpha_-)} (x/2)^{-2\alpha+} b_-.
\end{aligned}
\tag{13.46}
$$

The last equality uses the well known formula $\Gamma(\tfrac{1}{2} + x)\Gamma(\tfrac{1}{2} - x) = \pi/\cos(\pi x)$. Equation (13.46) gives the leading parts in (12.21) and (12.22). That the next parts are of order $O(|x|^{2-2\Re(\alpha-)})$ (respectively, $O(|x|^{2+6\Re(\alpha-)})$), follows from the arguments in the proof of Theorem 12.4, (see in particular (12.16) and the two pictures in the proof of Theorem 12.4). This completes our approach to Theorem 12.4. It remains to prove Lemma 13.6.

Proof of Lemma 13.6: A P_{3D6}-TEJPA bundle is given by its restrictions to $\mathbb{C}$ and $\mathbb{P}^1 - \{0\}$ together with P, A, J. By Lemma 13.3 its restriction to $\mathbb{C}$ is $\mathcal{O}_{\mathbb{C}} \chi_1 + \mathcal{O}_{\mathbb{C}} \chi_2$, and its restriction to $\mathbb{P}^1 - \{0\}$ is $\mathcal{O}_{\mathbb{P}^1 - \{0\}} J(\chi_1) + \mathcal{O}_{\mathbb{P}^1 - \{0\}} J(\chi_2)$. Therefore the 2-dimensional space $\Gamma(\mathbb{P}^1, \mathcal{O}(H))$ of global sections of H is

$$\Gamma(\mathbb{C}, \mathcal{O}_{\mathbb{C}} \chi_1 + \mathcal{O}_{\mathbb{C}} \chi_2) \cap \Gamma(\mathbb{P}^1 - \{0\}, \mathcal{O}_{\mathbb{P}^1 - \{0\}} J(\chi_1) + \mathcal{O}_{\mathbb{P}^1 - \{0\}} J(\chi_2)).$$

It splits into the 1-dimensional eigenspaces of A with eigenvalues $\pm i$. We have to show that a generating section of the eigenspace with eigenvalue i is nonzero at $z = 0$. Then it is (up to a scalar) σ_1.

We consider the sections χ_1 and χ_2 in Definition/Lemma 13.2 in the case $\widetilde{\alpha}_\pm = \alpha_\pm(s)$. Write

$$f_1^+(t) := \sum_{k \in 2\mathbb{Z}_{\geq 0}} a_{-k,1}^+(1)\, t^k,$$

$$f_1^-(t) := \sum_{k \in 2\mathbb{Z}_{\geq 0}} a_{-k-1,1}^-(1)\, t^k,$$

$$f_2^+(t) := \sum_{k \in 2\mathbb{Z}_{\geq 0}} a_{-k-1,2}^+(1)\, t^k, \tag{13.47}$$

$$f_2^-(t) := \sum_{k \in 2\mathbb{Z}_{\geq 0}} a_{-k,2}^-(1)\, t^k$$

so that

$$f_1^+, f_1^-, f_2^+, f_2^- \in \mathbb{C}\{t^2\} \cap \Gamma(\mathbb{C}, \mathcal{O}_\mathbb{C}) \tag{13.48}$$

and

$$\chi_1(x, z) = f_1^+\left(\frac{x}{z}\right) es_1^{(0)}\left(\frac{z}{x}\right) + f_1^-\left(\frac{x}{z}\right) es_2^{(-1)}\left(\frac{z}{x}\right),$$

$$\chi_2(x, z) = f_2^+\left(\frac{x}{z}\right) es_1^{(-1)}\left(\frac{z}{x}\right) + f_2^-\left(\frac{x}{z}\right) es_2^{(0)}\left(\frac{z}{x}\right). \tag{13.49}$$

Then

$$J(\chi_1)(x, z) = b_- f_1^+(xz)\, es_2^{(0)}(xz) + b_+ f_1^-(xz)\, es_1^{(1)}(xz),$$

$$J(\chi_2)(x, z) = b_- f_2^+(xz)\, es_2^{(1)}(xz) + b_+ f_2^-(xz)\, es_1^{(0)}(xz). \tag{13.50}$$

For small x we ask about the existence of

$$g_1(x, z), g_2(x, z) \in \mathbb{C}\{z^2\}, \quad h_1(x, z), h_2(x, z) \in \mathbb{C}\{z^{-2}\} \tag{13.51}$$

with

$$g_1 x^{\alpha_+} \chi_1 + g_2 x^{\alpha_-} z \chi_2$$

$$= h_1 x^{-\alpha_-} z^{-1} J(\chi_1) + h_2 x^{-\alpha_+} J(\chi_2). \tag{13.52}$$

Define

$$\gamma := \alpha_+ - (\alpha_- - 1) = 2\alpha_+ + 1 = 2 - ((\alpha_- + 1) - \alpha_+). \tag{13.53}$$

In view of (13.49), (13.52) is equivalent to the following two equations, which collect the coefficients of $es_1^{(0)}(z)$ and $es_2^{(-1)}(z)$:

$$g_1 f_1^+ \left(\frac{x}{z}\right) + g_2 x^{2-\gamma} f_2^+ \left(\frac{x}{z}\right)$$
$$= h_1 x^\gamma b_+ f_1^- (xz) + h_2 b_+ f_2^- (xz), \tag{13.54}$$

$$g_1 x^\gamma f_1^- \left(\frac{x}{z}\right) + g_2 z^2 f_2^- \left(\frac{x}{z}\right)$$
$$= h_1 b_- f_1^+ (xz) + h_2 x^{2-\gamma} z^2 b_- f_2^+ (xz). \tag{13.55}$$

The system (13.54) & (13.55) can be solved inductively in powers of x^2 if we fix the normalization

$$\text{coefficient of } z^0 \text{ in } g_1 = 1. \tag{13.56}$$

Each step in the induction can be separated into four substeps:

(1) (13.54) determines the coefficient in $\mathbb{C}[x^{\pm\gamma}, z^2]$ of a power of x^2 in g_1.
(2) (13.55) determines the coefficient in $\mathbb{C}[x^{\pm\gamma}, z^{-2}]$ of the same power of x^2 in h_1.
(3) (13.54) determines the coefficient in $\mathbb{C}[x^{\pm\gamma}, z^{-2}]$ of the same power of x^2 in h_2.
(4) (13.55) determines the coefficient in $\mathbb{C}[x^{\pm\gamma}, z^2]$ of the same power of x^2 in g_2.

As a result, at most the following monomials turn up in g_1, g_2, h_1, h_2:

$$g_1 : (xz)^{2k} x^{2l+m\gamma} \text{ with} \qquad\qquad k \in \mathbb{Z}_{\geq 0}, l \in 2\mathbb{Z}_{\geq 0},$$
$$m \in \{-l, -l+2, \dots, l+2\},$$

$$g_2 : (xz)^{2k} x^{2l+m\gamma} \text{ with} \qquad\qquad k \in \mathbb{Z}_{\geq 0}, l \in 2\mathbb{Z}_{\geq 0}+1,$$
$$m \in \{-l, -l+2, \dots, l+2\},$$

$$h_1 : \left(\frac{x}{z}\right)^{2k} x^{2l+m\gamma} \text{ with} \qquad\qquad k \in \mathbb{Z}_{\geq 0}, l \in 2\mathbb{Z}_{\geq 0}, \tag{13.57}$$
$$m \in \{-l+1, -l+3, \dots, l+1\},$$

$$h_2 : \left(\frac{x}{z}\right)^{2k} x^{2l+m\gamma} \text{ with} \qquad\qquad k \in \mathbb{Z}_{\geq 0}, l \in 2\mathbb{Z}_{\geq 0},$$
$$m \in \{-l, -l+2, \dots, l+2\}.$$

We leave the details of the proof of (13.57) to the reader.

The number γ satisfies $0 \leq \Re(\gamma) < 2$. Therefore the coefficients of all powers of z^2 in g_1, g_2, h_1, h_2 tend to 0, if $x \to 0$, except the coefficient 1 of z^0 in g_1, and the coefficient of z^0 in h_2, which is also bounded if $x \to 0$. This indicates that the power series g_1, g_2, h_1, h_2 are convergent if $x \to 0$. Also the details of the proof of this convergence is left to the reader. The convergence implies the first part of Lemma 13.6.

For x near 0, σ_1 and χ_1 are dominated by

$$x^{-\alpha+} \left(a_{0,1}^+(1)\, es_1^{(0)}(z) + x^{\gamma}\, a_{-1,1}^-(1)\, es_2^{(-1)}(z) \right). \tag{13.58}$$

If $s \in \mathbb{C}^{[sto,a]}$ then $x^{\gamma} \to 0$ if $x \to 0$. This establishes the second part (13.45) of Lemma 13.6. $\square$

Remark 13.7 If $s \in \mathbb{C}^{[sto,b+]}$ then $\Re(\gamma) = 0$, and σ_1 and χ_1 are dominated by (13.58). If one replaces σ_1 in (13.1) by (13.58) one obtains directly the leading part in (12.24).

Chapter 14
Symmetries of the Universal Family of Solutions of $P_{III}(0, 0, 4, -4)$

Symmetries of Painlevé equations and their solutions have been quite useful, see e.g. [GLSh02] and [No04]. In this chapter we will study the symmetries of the solutions of our Painlevé equation.

In Remark 10.2 the holomorphic function $f_{univ} : \mathbb{C} \times V^{mat} \to \mathbb{C}$ was introduced. By Theorem 10.3, the associated multi-valued functions on $\mathbb{C}^*$ (with distinguished branch, see Remark 10.2 (ii))

$$f_{mult}(., s, B), \quad (s, B) \in V^{mat},$$

give all solutions of $P_{III}(0, 0, 4, -4)$. We have $f_{mult}(x, s, B) = f_{univ}(\log x, s, B)$.

Certain symmetries of the function f_{univ} are given by the group $G^{mon} \times \langle m_{[1]} \rangle \cong (\mathbb{Z}_2 \times \mathbb{Z}_2) \times \mathbb{Z}$ (see Remarks 12.3 and 10.2 (iv)):

$$f_{univ} \circ R_1 = f_{univ}^{-1}, \tag{14.1}$$

i.e. $f_{univ}(\xi, -s, \left(\begin{smallmatrix} 1 & 0 \\ 0 & -1 \end{smallmatrix}\right) B \left(\begin{smallmatrix} 1 & 0 \\ 0 & -1 \end{smallmatrix}\right)) = f_{univ}^{-1}(\xi, s, B),$

$$f_{univ} \circ R_2 = -f_{univ}, \tag{14.2}$$

i.e. $f_{univ}(\xi, s, -B) = -f_{univ}(\xi, s, B),$

$$f_{univ} \circ R_3 = -f_{univ}^{-1}, \tag{14.3}$$

i.e. $f_{univ}(\xi, -s, -\left(\begin{smallmatrix} 1 & 0 \\ 0 & -1 \end{smallmatrix}\right) B \left(\begin{smallmatrix} 1 & 0 \\ 0 & -1 \end{smallmatrix}\right)) = -f_{univ}^{-1}(\xi, s, B),$

and

$$f_{univ} \circ m_{[1]} = f_{univ}, \tag{14.4}$$

i.e. $f_{univ}(\xi - i\pi, s, B) = f_{univ}(\xi, s, \text{Mon}_0^{mat}(s) B).$

© Springer International Publishing AG 2017
M.A. Guest, C. Hertling, *Painlevé III: A Case Study in the Geometry of Meromorphic Connections*, Lecture Notes in Mathematics 2198, DOI 10.1007/978-3-319-66526-9_14

In Theorems 14.1 and 14.2 two more symmetries of f_{univ} will be given. The proofs use Theorem 12.4, which allows us to recover the monodromy data (s, B) from the asymptotic behaviour of $f_{mult}(., s, B)$ as $x \to 0$. The symmetry in Theorem 14.1 will not be used in the rest of the monograph, but it is too beautiful to be omitted. The symmetry in Theorem 14.2 will be used in Chap. 15.

Theorem 14.1

(a) Let

$$R_4 : \mathbb{C} \times V^{mat} \to \mathbb{C} \times V^{mat},$$

$$(\xi, s, B) \mapsto (\xi + \tfrac{1}{2}\pi i, s, \left(\begin{smallmatrix} 0 & 1 \\ -1 & s \end{smallmatrix}\right) B). \tag{14.5}$$

Then

$$f_{univ} \circ R_4 = i f_{univ}, \tag{14.6}$$

$$\text{i.e. } (-i) f_{univ}(\xi + \tfrac{1}{2}\pi i, s, B) = f_{univ}(\xi, s, \left(\begin{smallmatrix} s & -1 \\ 1 & 0 \end{smallmatrix}\right) B),$$

and

$$R_4^2 = R_2 \circ m_{[1]}^{-1} = m_{[1]}^{-1} \circ R_2, \ R_4 \circ R_2 = R_2 \circ R_4 \tag{14.7}$$

$$R_4 \circ R_1 = R_2 \circ R_1 \circ R_4, \ R_4 \circ m_{[1]} = m_{[1]} \circ R_4. \tag{14.8}$$

(b) For $(s, B) \in V^{mat}$

$$(-i) f_{mult}(x\, e^{\pi i/2}, s, B) = f_{mult}(x, s, \left(\begin{smallmatrix} s & -1 \\ 1 & 0 \end{smallmatrix}\right) B). \tag{14.9}$$

This means that the multi-valued function $(x \mapsto (-i) f_{mult}(x\, e^{\pi i/2}, s, B)$ is the solution $f_{mult}(., s, \left(\begin{smallmatrix} s & -1 \\ 1 & 0 \end{smallmatrix}\right) B)$ of $P_{III}(0, 0, 4, -4)$.

Proof (a) and (b) The form (9.1) of the equation $P_{III}(0, 0, 4, -4)$ shows immediately that for any $(s, B) \in V^{mat}$ the multi-valued function $(x \mapsto (-i) f_{mult}(x\, e^{\pi i/2}, s, B))$ is a solution of $P_{III}(0, 0, 4, -4)$. By Theorem 10.3 (c) there exist unique monodromy data $(\widetilde{s}, \widetilde{B}) \in V^{mat}$ such that this function is $f_{mult}(., \widetilde{s}, \widetilde{B})$. The map

$$V^{mat} \to V^{mat}, \quad (s, B) \mapsto (\widetilde{s}, \widetilde{B}),$$

is an automorphism of V^{mat} which is at least analytic, and possibly algebraic. We shall show that $(\widetilde{s}, \widetilde{B}) = (s, \left(\begin{smallmatrix} s & -1 \\ 1 & 0 \end{smallmatrix}\right) B)$. This will establish (14.6) and (14.9). The identities in (14.7) and (14.8) are easily checked.

It is sufficient to show $(\widetilde{s}, \widetilde{B}) = (s, \left(\begin{smallmatrix} s & -1 \\ 1 & 0 \end{smallmatrix}\right) B)$ for $(s, B) \in V^{mat,a}$. This is equivalent to $(\widetilde{s}, \widetilde{b}_-) = (s, (-i) e^{\pi i \alpha -} b_-)$, by Lemmas 5.2 (b) and 5.3. The asymptotic formula (12.21) looks as follows for the multi-valued function $(x \mapsto (-i)f_{mult}(x\,e^{\pi i/2}, s, B))$:

$$(-i)f_{mult}(x\,e^{\pi i/2}, s, B)$$

$$= (-i)\frac{\Gamma(\frac{1}{2} - \alpha_-)}{\Gamma(\frac{1}{2} + \alpha_-)} \left(\tfrac{1}{2}x\,e^{\pi i/2}\right)^{2\alpha_-} b_- + O(|x|^{2-2\Re(\alpha_-)})$$

$$= \frac{\Gamma(\frac{1}{2} - \alpha_-)}{\Gamma(\frac{1}{2} + \alpha_-)} \left(\tfrac{1}{2}x\right)^{2\alpha_-} ((-i)e^{\pi i\alpha_-} b_-) + O(|x|^{2-2\Re(\alpha_-)}).$$

This gives $(\widetilde{s}, \widetilde{b}_-) = (s, (-i)e^{\pi i\alpha_-} b_-)$. $\qquad\square$

Theorem 14.2

(a) Let

$$R_5 : \mathbb{C} \times V^{mat} \to \mathbb{C} \times V^{mat}, \quad (\xi, s, B) \mapsto (\bar{\xi}, \bar{s}, \bar{B}^{-1}). \tag{14.10}$$

Then

$$f_{univ} \circ R_5 = \overline{f_{univ}}, \tag{14.11}$$

$$\text{i.e. } \overline{f_{univ}(\bar{\xi}, s, B)} = f_{univ}(\xi, \bar{s}, \bar{B}^{-1}),$$

and

$$R_5^2 = \text{id}, \quad R_5 \circ R_2 = R_2 \circ R_5, \quad R_5 \circ R_1 = R_1 \circ R_5, \tag{14.12}$$

$$R_5 \circ R_4 = R_2 \circ m_{[1]} \circ R_4 \circ R_5. \tag{14.13}$$

(b) For $(s, B) \in V^{mat}$

$$\overline{f_{mult}(\bar{x}, s, B)} = f_{mult}(x, \bar{s}, \bar{B}^{-1}). \tag{14.14}$$

This means that the multi-valued function $(x \mapsto \overline{f_{mult}(\bar{x}, s, B)})$ is the solution $f_{mult}(., \bar{s}, \bar{B}^{-1})$ of $P_{III}(0, 0, 4, -4)$.
(c) (Addendum to Lemmas 5.1 and 5.2) If $s \in \mathbb{C}$ with $s^2 \notin \mathbb{R}_{\geq 4}$ then

$$\overline{\sqrt{\tfrac{1}{4}s^2 - 1}} = -\sqrt{\tfrac{1}{4}\bar{s}^2 - 1}, \quad \lambda_\pm(\bar{s}) = \overline{\lambda_\mp(s)}, \quad v_\pm(\bar{s}) = \overline{v_\mp(s)}, \tag{14.15}$$

$$\alpha_\pm(\bar{s}) = \overline{\alpha_\pm(s)}, \quad b_\pm(\bar{s}, \bar{B}^{-1}) = b_\mp(\bar{s}, \bar{B}) = \overline{b_\pm(s, B)}. \tag{14.16}$$

If $s \in \mathbb{C}$ with $s^2 \in \mathbb{R}_{\geq 4}$, then $\bar{s} = s$ and

$$\sqrt{\tfrac{1}{4}s^2 - 1} \in \mathbb{R}_{\geq 0}, \quad \lambda_{\pm}(s) \in \mathbb{R}_{<0}, \quad v_{\pm}(s) \text{ real}, \qquad (14.17)$$

$$\overline{\alpha_{\pm}(s)} = \alpha_{\mp}(s) \mp 1, \quad b_{\pm}(s, \overline{B}^{-1}) = b_{\mp}(s, \overline{B}) = \overline{b_{\mp}(s, B)}. \qquad (14.18)$$

If $s = \pm 2$ then

$$v_{1/2}(s) = \overline{v_{1/2}(s)}, \quad \widetilde{b}_1(s, \overline{B}^{-1}) = \overline{\widetilde{b}_1(s, B)}, \qquad (14.19)$$

$$b_2(s, \overline{B}^{-1}) = -b_2(s, \overline{B}) = -\overline{b_2(s, B)}. \qquad (14.20)$$

Proof (c) The proof consists of elementary calculations and is omitted.

(a) and (b) The form (9.1) of the equation $P_{III}(0,0,4,-4)$ shows immediately that for any $(s, B) \in V^{mat}$ the multi-valued function $x \mapsto \overline{f_{mult}(\bar{x}, s, B)}$ is a solution of $P_{III}(0,0,4,-4)$. By Theorem 10.3 (c) there exist unique monodromy data $(\widetilde{s}, \widetilde{B}) \in V^{mat}$ such that this function is $f_{mult}(., \widetilde{s}, \widetilde{B})$. The map

$$V^{mat} \to V^{mat}, \quad (s, B) \mapsto (\widetilde{s}, \widetilde{B}),$$

is an automorphism of V^{mat} which is at least real analytic, possibly real algebraic. We shall show that $(\widetilde{s}, \widetilde{B}) = (\bar{s}, \overline{B}^{-1})$. This will establish (14.11) and (14.14). The identities in (14.12) and (14.13) are easily checked. It is sufficient to show that $(\widetilde{s}, \widetilde{B}) = (\bar{s}, \overline{B}^{-1})$ for $(s, B) \in V^{mat,a}$. This is equivalent to $(\widetilde{s}, \widetilde{b}_-) = (\bar{s}, \overline{b_-})$, by (14.16). The asymptotic formula (12.21) looks as follows for the multi-valued function $\overline{f_{mult}(\bar{x}, s, B)}$:

$$\overline{f_{mult}(\bar{x}, s, B)} = \frac{\Gamma(\tfrac{1}{2} - \overline{\alpha_-})}{\Gamma(\tfrac{1}{2} + \overline{\alpha_-})} \left(\tfrac{1}{2}x\right)^{2\overline{\alpha_-}} \overline{b_-} + O(|x|^{2 - 2\Re(\overline{\alpha_-})}).$$

This gives $(\widetilde{s}, \widetilde{b}_-) = (\bar{s}, \overline{b_-})$. $\qquad\qquad \square$

Remark 14.3 The following table extends table (12.13) and lists some meromorphic functions on $\mathbb{C} \times V^{mat,a}$ (so $s^2 \notin \mathbb{R}_{\geq 4}$) and their images under the actions of $R_4, R_5, R_1 \circ R_5$ and $R_2 \circ R_5$:

	R_4	R_5	$R_1 \circ R_5$	$R_2 \circ R_5$
s	s	$\overline{s}$	$-\overline{s}$	$\overline{s}$
$\sqrt{\frac{1}{4}s^2 - 1}$	$\sqrt{\frac{1}{4}s^2 - 1}$	$-\overline{\sqrt{\frac{1}{4}s^2 - 1}}$	$-\overline{\sqrt{\frac{1}{4}s^2 - 1}}$	$-\overline{\sqrt{\frac{1}{4}s^2 - 1}}$
$\lambda_\pm$	$\lambda_\pm$	$\overline{\lambda_\mp}$	$\overline{\lambda_\pm}$	$\overline{\lambda_\mp}$
$\alpha_\pm$	$\alpha_\pm$	$\overline{\alpha_\pm}$	$-\overline{\alpha_\pm}$	$\overline{\alpha_\pm}$
B	$\begin{pmatrix} 0 & 1 \\ -1 & s \end{pmatrix} B$	$\overline{B}^{-1}$	$\begin{pmatrix} 1 & 0 \\ 0 & -1 \end{pmatrix} \overline{B}^{-1} \begin{pmatrix} 1 & 0 \\ 0 & -1 \end{pmatrix}$	$-\overline{B}^{-1}$
b_1	$-b_2$	$\overline{b_1 + \overline{s}b_2}$	$\overline{b_1 + \overline{s}b_2}$	$-\overline{b_1} - \overline{s}\overline{b_2}$
b_2	$b_1 + sb_2$	$-\overline{b_2}$	$\overline{b_2}$	$\overline{b_2}$
$b_1 + \frac{1}{2}sb_2$	$-(1-\frac{1}{2}s^2)b_2 + \frac{1}{2}sb_1$	$\overline{b_1} + \frac{1}{2}\overline{s}\overline{b_2}$	$\overline{b_1} + \frac{1}{2}\overline{s}\overline{b_2}$	$-\overline{b_1} - \frac{1}{2}\overline{s}\overline{b_2}$
$b_\pm$	$\mp ie^{-\pi i\alpha_\pm} b_\pm$	$\overline{b_\pm}$	$\overline{b_\mp}$	$-\overline{b_\pm}$
ξ	$\xi + \frac{1}{2}\pi i$	$\overline{\xi}$	$\overline{\xi}$	$\overline{\xi}$
x	$x e^{\pi i/2}$	$\overline{x}$	$\overline{x}$	$\overline{x}$
f_{univ}	$i f_{univ}$	$\overline{f_{univ}}$	$\overline{f_{univ}}^{-1}$	$-\overline{f_{univ}}$

Beware that for $s^2 \in \mathbb{R}_{\geq 4}$ the images of $\sqrt{\frac{1}{4}s^2 - 1}, \lambda_\pm, \alpha_\pm$ and $b_\pm$ under $R_5, R_1 \circ R_5, R_2 \circ R_5$ are different, in particular we have:

	R_5		$R_1 \circ R_5$	$R_2 \circ R_5$
$\lambda_\pm$	$\lambda_\pm = \overline{\lambda_\pm}$		$\lambda_\mp$	$\lambda_\pm$
$\alpha_\pm$	$\alpha_\pm$		$-\alpha_\pm$	$\alpha_\pm$
$b_\pm$	$\overline{b_\mp}$		$\overline{b_\pm}$	$-\overline{b_\mp}$

Chapter 15
Three Families of Solutions on $\mathbb{R}_{>0}$

In this chapter we are interested in restrictions to $\mathbb{R}_{>0}$ of solutions of $P_{III}(0,0,4,-4)$, which are related to real solutions (possibly with singularities) on $\mathbb{R}_{>0}$ of one of the following three differential equations:

$$\text{radial sinh-Gordon}\oplus : \quad (x\partial_x)^2\varphi = 16x^2 \sinh \varphi, \tag{15.1}$$

$$\text{radial sinh-Gordon}\ominus : \quad (x\partial_x)^2\psi = -16x^2 \sinh \psi, \tag{15.2}$$

$$\text{radial sine-Gordon}: \quad (x\partial_x)^2 u = -16x^2 \sin u. \tag{15.3}$$

Recall that $(x\partial_x)^2 = x^2(\partial_x^2 + \frac{1}{x}\partial_x)$.

Remark 15.1 Solutions of (15.2) lead to CMC surfaces in $\mathbb{R}^3$ with rotationally symmetric metric (cf. [FPT94]). Solutions of (15.1) lead to CMC surfaces in $\mathbb{R}^{2,1}$ with rotationally symmetric metric (cf. [DGR10]). Solutions of (15.3) lead to timelike surfaces of constant negative Gauss curvature in Minkowski space with rotationally symmetric metric (cf. [Ko11]).

The following lemma makes precise how real solutions of (15.1)–(15.3) are related to those solutions of $P_{III}(0,0,4,-4)$ which are either real or purely imaginary or have values in S^1.

Lemma 15.2

(a) (i) *f is a local solution of $P_{III}(0,0,4,-4)$ on $\mathbb{R}_{>0}$ with values in $\mathbb{R}_{>0}$ (respectively, $\mathbb{R}_{<0}$) $\iff$ $\varphi = 2\log f$ is a local solution of (15.1) on $\mathbb{R}_{>0}$ with values in $\mathbb{R}$ (mod $4\pi i$) (respectively, $2\pi i + \mathbb{R}$ (mod $4\pi i$)).*

(ii) *By Theorem 10.3 (b), a real solution of $P_{III}(0,0,4,-4)$ on $\mathbb{R}_{>0}$ can have four types of singularities, which are indexed by $(\varepsilon_1,\varepsilon_2) \in \{\pm 1\}^2$ (or equivalently by $k \in \{0,1,2,3\}$, see Theorem 8.2 (b)). These are listed in (10.13). If $x_0 \in \mathbb{R}_{>0}$ is a singularity of f of type $(\varepsilon_1,\varepsilon_2)$, then the corresponding singularity of $\varphi = 2\log f$ is as follows: for some $\widetilde{g}_k \in \mathbb{R}$*

© Springer International Publishing AG 2017
M.A. Guest, C. Hertling, *Painlevé III: A Case Study in the Geometry of Meromorphic Connections*, Lecture Notes in Mathematics 2198,
DOI 10.1007/978-3-319-66526-9_15

$$\varphi(x) = (1 - \varepsilon_2)\pi i \varepsilon_1 + 2\varepsilon_1 \log((-2)(x - x_0)) + \varepsilon_1 \frac{x - x_0}{x_0}$$

$$-\varepsilon_1 \left(\frac{7}{12x_0^2} + \frac{4}{3x_0} \widetilde{g}_k \right)(x - x_0)^2 + O((x - x_0)^3). \qquad (15.4)$$

(b) (i) f is a local solution of $P_{III}(0, 0, 4, -4)$ on $\mathbb{R}_{>0}$ with values in $i\mathbb{R}_{>0}$ $\iff$ $\psi = 2\log f - \pi i$ is a local solution of (15.2) on $\mathbb{R}_{>0}$ with values in $\mathbb{R}$ (mod $4\pi i$).

(ii) Any local solutions f and ψ as in (i) extend to real analytic global solutions without singularities on $\mathbb{R}_{>0}$.

(c) (i) f is a local solution of $P_{III}(0, 0, 4, -4)$ on $\mathbb{R}_{>0}$ with values in S^1 $\iff$ $u = 2i \log f + \pi$ is a local solution of (15.3) on $\mathbb{R}_{>0}$ with values in $\mathbb{R}$. u is unique up to addition of multiples of 4π.

(ii) Any local solutions f and u as in (i) extend to real analytic global solutions without singularities on $\mathbb{R}_{>0}$.

Proof

(a) (i) This was stated up to the restriction to $\mathbb{R}_{>0}$ and the reality condition already in (9.1), (9.2) and (9.5).

(ii) Equation (10.13) gives the Taylor expansion of $\varepsilon_2 f^{\varepsilon_1}$ at a singularity x_0 of type $(\varepsilon_1, \varepsilon_2)$:

$$\varepsilon_2 f^{\varepsilon_1} = (-2)(x - x_0) + \frac{-1}{x_0}(x - x_0)^2$$

$$+ \left(\frac{1}{3x_0^2} + \frac{4}{3x_0} \widetilde{g}_k \right)(x - x_0)^3 + O((x - x_0)^4)$$

$$= (-2)(x - x_0) \left[1 + \frac{1}{2x_0}(x - x_0) \right.$$

$$\left. - \left(\frac{1}{6x_0^2} + \frac{2}{3x_0} \widetilde{g}_k \right)(x - x_0)^2 + O((x - x_0)^3) \right].$$

Using $\log(1 + x) = x - \frac{x^2}{2} + O(x^3)$ and

$$2\log f = 2\varepsilon_1 \log \varepsilon_2 + 2\varepsilon_1 \log(\varepsilon_2 f^{\varepsilon_1}) = (1 - \varepsilon_2)\pi i \varepsilon_1 + 2\varepsilon_1 \log(\varepsilon_2 f^{\varepsilon_1}),$$

we obtain (15.4).

(b) (i) is obvious. (ii) f as in (i) extends uniquely to a real analytic global solution on $\mathbb{R}_{>0}$ with values in $i\mathbb{R}$ which might have singularities a priori. But for any $(\varepsilon_1, \varepsilon_2) \in \{\pm 1\}^2$ one has $\varepsilon_2 \partial_x f^{\varepsilon_1}(x_0) \in i\mathbb{R}$, thus $\neq -2$. This and (10.13) show that f has no singularities.

(c) (i) is Lemma 11.1 and is obvious. (ii) is also obvious, as any local solution with values in S^1 extends to a global solution with values in S^1. $\qquad \square$

Lemma 15.4 will describe the restrictions of the spaces $M_{3FN}^{ini}(x_0)$ of initial conditions to the three families of solutions on $\mathbb{R}_{>0}$ in Lemma 15.2. This will be an elementary consequence of Lemma 10.1 (c) and Theorem 10.3 (e).

Theorem 15.5 will describe the restrictions of the spaces $M_{3FN}^{mon}(x_0)$ of monodromy data to the three families. Theorem 15.5 will follow from the fact that the three families of solutions in Lemma 15.2 are the fixed points of the symmetries R_5, $R_2 \circ R_5$ and $R_1 \circ R_5$, respectively, in the table in Remark 14.3.

Definition 15.3 For $x_0 \in \mathbb{R}_{>0}$, the sets

$$M_{3FN,\mathbb{R}}(x_0), M_{3FN,i\mathbb{R}_{>0}}(x_0) \text{ and } M_{3FN,S^1}(x_0)$$

are the subsets of $M_{3FN}(x_0)$ of those P_{3D6}-TEJPA bundles which correspond by Theorem 10.3 (e) to the regular or singular initial conditions at x_0 of solutions of $P_{III}(0,0,4,-4)$ on $\mathbb{R}_{>0}$ with values in $\mathbb{R}$, $i\mathbb{R}_{>0}$ or S^1, respectively. Their unions over all $x_0 \in \mathbb{R}_{>0}$ are called

$$M_{3FN,\mathbb{R}}, M_{3FN,i\mathbb{R}_{>0}} \text{ and } M_{3FN,S^1}$$

respectively.

The upper index *ini* in $M_{3FN,\mathbb{R}}^{ini}(x_0)$, $M_{3FN,i\mathbb{R}_{>0}}^{ini}(x_0)$, $M_{3FN,S^1}^{ini}(x_0)$, $M_{3FN,\mathbb{R}}^{ini}$, $M_{3FN,i\mathbb{R}_{>0}}^{ini}$ and M_{3FN,S^1}^{ini} denotes the additional structure (real algebraic or semi-algebraic manifold, charts, natural functions) which the corresponding set inherits from $M_{3FN}^{ini}(x_0)$ or M_{3FN}^{ini}.

The upper index *mon* in $M_{3FN,\mathbb{R}}^{mon}(x_0)$, $M_{3FN,i\mathbb{R}_{>0}}^{mon}(x_0)$, $M_{3FN,S^1}^{mon}(x_0)$, $M_{3FN,\mathbb{R}}^{mon}$, $M_{3FN,i\mathbb{R}_{>0}}^{mon}$ and M_{3FN,S^1}^{mon} denotes the additional structure (real algebraic or semi-algebraic or real-analytic manifold, foliation) which the corresponding set inherits from $M_{3FN}^{mon}(x_0)$ or M_{3FN}^{mon}.

Lemma 15.4

(a) (i) $M_{3FN,\mathbb{R}}^{ini}(x_0)$ is a real algebraic manifold with four natural charts. The charts have coordinates $(f_k, \widetilde{g}_k)$ for $k = 0,1,2,3$ and are isomorphic to $\mathbb{R} \times \mathbb{R}$. Each chart intersects each other chart in $\mathbb{R}^* \times \mathbb{R}$. The coordinate changes are given by (8.20) and (10.7). The intersection of all four charts is called $M_{3FN,\mathbb{R}}^{reg}(x_0)$. Each chart consists of $M_{3FN,\mathbb{R}}^{reg}(x_0)$ and a hyperplane $M_{3FN,\mathbb{R}}^{[k]}(x_0)$ isomorphic to $\{0\} \times \mathbb{R}$.

   (ii) $M_{3FN,\mathbb{R}}^{reg}(x_0)$ with the coordinates (f_0, g_0) is the space of initial conditions (10.14) at x_0 for at x_0 regular solutions f of $P_{III}(0,0,4,-4)$ on $\mathbb{R}_{>0}$ with values in $\mathbb{R}$. For $k = 0,1,2,3$, the space $M_{3FN,\mathbb{R}}^{[k]}(x_0) \cong \mathbb{R}$ with the coordinate $\widetilde{g}_k$ is the space of the initial condition (10.15) for the at x_0 singular solutions f with $f^{\varepsilon_1}(x_0) = 0$, $\varepsilon_2 \partial_x f^{\varepsilon_1}(x_0) = -2$.

   (iii) Because $x_0 \in \mathbb{R}_{>0}$, $M_{3FN,\mathbb{R}}^{ini}$ is only a real semi-algebraic manifold with four charts, which are given by putting together the charts for all $x_0 \in \mathbb{R}_{>0}$.

(b) $M^{ini}_{3FN,i\mathbb{R}_{>0}}(x_0)$ is a real semi-algebraic manifold isomorphic to $i\mathbb{R}_{>0} \times \mathbb{R}$ with coordinates (f_0, g_0). It is the space of initial conditions (10.14) at x_0 for solutions f of $P_{III}(0, 0, 4, -4)$ on $\mathbb{R}_{>0}$ with values in $i\mathbb{R}_{>0}$.

(c) $M^{ini}_{3FN,S^1}(x_0)$ is a real algebraic manifold isomorphic to $S^1 \times i\mathbb{R}$ with coordinates (f_0, g_0). It is the space of initial conditions (10.14) at x_0 for solutions f of $P_{III}(0, 0, 4, -4)$ on $\mathbb{R}_{>0}$ with values in S^1.

Proof Everything follows easily from Lemma 15.2, Lemma 10.1 (c) and Theorem 10.3, in particular formula (10.14) for the initial conditions at regular points. For (c), observe that if f takes values in S^1, then $\partial_x f(x_0) \in f(x_0) i\mathbb{R}$. Conversely, if $f(x_0) \in S^1$ and $\partial_x f(x_0) \in f(x_0) i\mathbb{R}$, then $u = 2i\log f + \pi$ satisfies $u(x_0) \in \mathbb{R}$, $\partial_x u(x_0) = 2i\frac{\partial_x f(x_0)}{f(x_0)} \in \mathbb{R}$, so u is real on $\mathbb{R}_{>0}$, and f takes values in S^1 on $\mathbb{R}_{>0}$. $\square$

Recall that by Lemma 10.1 (b), there is a canonical isomorphism for $\beta \in \mathbb{R}$ with $\frac{1}{2}e^{-\beta/2} = x_0$,

$$M^{mon}_{3FN}(x_0) \cong V^{mat}$$
$$(H, \nabla, x_0, x_0, P, A, J) \mapsto (s, B(\beta)). \tag{15.5}$$

Here s and $B(\beta)$ are associated to $(H, \nabla, x_0, x_0, P, A, J)$ as in Theorem 7.5 (b).

Theorem 15.5

(a) (i) *Equation (15.5) restricts to an isomorphism of real algebraic manifolds*

$$M^{mon}_{3FN,\mathbb{R}}(x_0) = V^{mat,\mathbb{R}}$$

$$:= \{(s, B) \in V^{mat} \mid s \in \mathbb{R}, b_1 + \tfrac{1}{2}sb_2 \in \mathbb{R}, b_2 \in i\mathbb{R}\}$$

$$= \{(s, B) \in V^{mat} \mid (s, B) = (\overline{s}, \overline{B}^{-1})\} \tag{15.6}$$

$$\cong \{(s, b_5, b_6) \in \mathbb{R}^3 \mid b_5^2 + (\tfrac{1}{4}s^2 - 1)b_6^2 = 1\}$$

with $b_5 = b_1 + \tfrac{1}{2}sb_2, b_6 = ib_2$.

(ii) $V^{mat,\mathbb{R}}$ *decomposes into* $V^{mat,J,\mathbb{R}} := V^{mat,\mathbb{R}} \cap V^{mat,J}$ *for* $J \in \{a, b+, b-, c+, c-\}$. *As real analytic manifolds*

$$V^{mat,a,\mathbb{R}} \xrightarrow{\cong} (-2, 2) \times \mathbb{R}^*, \quad (s, B) \mapsto (s, b_-), \tag{15.7}$$

$$V^{mat,b\pm,\mathbb{R}} \xrightarrow{\cong} (\pm 1)\,\mathbb{R}_{>2} \times S^1, \quad (s, B) \mapsto (s, b_-), \tag{15.8}$$

$$V^{mat,c\pm,\mathbb{R}} \xrightarrow{\cong} \{\pm 2\} \times \{\pm 1\} \times \mathbb{R}, \quad (s, B) \mapsto (s, \widetilde{b}_1, ib_2). \tag{15.9}$$

(iii) *As a C^∞-manifold $V^{mat,\mathbb{R}}$ is a sphere with four holes. The symmetries R_1 and R_2 (and R_3) act on it—see the table in Remark 12.3. The following schematical picture shows $V^{mat,\mathbb{R}}$ with these symmetries.*

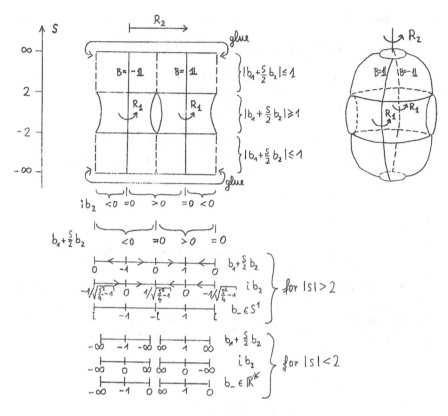

(iv) $M^{mon}_{3FN,\mathbb{R}}$ is a real semi-algebraic manifold, isomorphic to $\mathbb{R}_{>0} \times V^{mat,\mathbb{R}}$. The foliation which $M^{mon}_{3FN,\mathbb{R}}$ inherits from M^{mon}_{3FN}, corresponds to the foliation of $\mathbb{R}_{>0} \times V^{mat,\mathbb{R}}$ with leaves $\mathbb{R}_{>0} \times \{(s,B)\}$.

(v) The map

$$V^{mat,\mathbb{R}} \to \{\text{solutions of } P_{III}(0,0,4,-4) \text{ on } \mathbb{R}_{>0} \qquad (15.10)$$

$$\text{with values in } \mathbb{R}\}$$

$$(s,B) \mapsto f_{mult}(.,s,B)|_{\mathbb{R}_{>0}} \quad (\text{i.e. } x \text{ with } \arg x = 0)$$

is an isomorphism. $f_{mult}(.,s,B)|_{\mathbb{R}_{>0}}$ has a singularity (zero or pole) at x_0 if and only if (x_0,s,B) belongs to the image in $V^{mat,\mathbb{R}}(x_0)$ of $M^{[k]}_{3FN,\mathbb{R}}(x_0)$ for some $k \in \{0,1,2,3\}$. A pair $(\varepsilon_1,\varepsilon_2) \in \{\pm 1\}^2$ is associated with k as in Theorem 8.2 (b). The function has a zero at x_0 if $\varepsilon_1 = 1$ ($\iff$ $k = 0$ or 2) and a pole at x_0 if $\varepsilon_1 = -1$ ($\iff$ $k = 1$ or 3), and in any case

$$\partial_x f_{mult}(.,s,B)^{\varepsilon_1}_{|\mathbb{R}_{>0}}(x_0) = \varepsilon_2 (-2). \qquad (15.11)$$

(b) (i) Equation (15.5) restricts to an isomorphism of real semi-algebraic manifolds

$$M^{mon}_{3FN,i\mathbb{R}_{>0}}(x_0) = V^{mat,i\mathbb{R}_{>0}}$$

$$:= \{(s,B) \in V^{mat} \mid s \in (-2,2), b_1 + \tfrac{1}{2}sb_2 \in i\mathbb{R},$$

$$b_2 = \sqrt{(1 - (b_1 + \tfrac{1}{2}sb_2)^2)/(1 - \tfrac{1}{4}s^2)} \in \mathbb{R}_{>1}\}$$

$$= \{(s,B) \in V^{mat} \mid (s,B) = (\bar{s}, -\overline{B}^{-1}), b_2 \in \mathbb{R}_{>1}\} \qquad (15.12)$$

$$\cong \{(s,b_7,b_2) \in (-2,2) \times \mathbb{R} \times \mathbb{R}_{>1} \mid -b_7^2 + (1 - \tfrac{1}{4}s^2)b_2^2 = 1\}$$

with $b_7 = i(b_1 + \tfrac{1}{2}sb_2)$.

(ii) As a real analytic manifold

$$V^{mat,i\mathbb{R}_{>0}} \xrightarrow{\cong} (-2,2) \times \mathbb{R}_{>0}, \quad (s,B) \mapsto (s, -ib_-). \qquad (15.13)$$

(iii) The symmetry $R_3 = R_1 \circ R_2$ acts on $V^{mat,i\mathbb{R}_{>0}}$—see the table in Remark 12.3.

(iv) $M^{mon}_{3FN,i\mathbb{R}_{>0}}$ is a real semi-algebraic manifold, isomorphic to $\mathbb{R}_{>0} \times V^{mat,i\mathbb{R}_{>0}}$. The foliation which $M^{mon}_{3FN,i\mathbb{R}_{>0}}$ inherits from M^{mon}_{3FN}, corresponds to the foliation of $\mathbb{R}_{>0} \times V^{mat,i\mathbb{R}_{>0}}$ with leaves $\mathbb{R}_{>0} \times \{(s,B)\}$.

(v) The map

$$V^{mat,i\mathbb{R}_{>0}} \to \{\text{solutions of } P_{III}(0,0,4,-4) \text{ on } \mathbb{R}_{>0} \qquad (15.14)$$

$$\text{with values in } i\mathbb{R}_{>0}\}$$

$$(s,B) \mapsto f_{mult}(.,s,B)|_{\mathbb{R}_{>0}} \quad (\text{i.e. } x \text{ with } \arg x = 0)$$

is an isomorphism.

(c) (i) Equation (15.5) restricts to an isomorphism of real algebraic manifolds

$$M^{mon}_{3FN,S^1}(x_0) = V^{mat,S^1}$$

$$:= \{(s,B) \in V^{mat} \mid s \in i\mathbb{R}, b_1 + \tfrac{1}{2}sb_2 \in \mathbb{R}, b_2 \in \mathbb{R}\}$$

$$= \{(s,B) \in V^{mat} \mid (s,B) = (\bar{s}, \left(\begin{smallmatrix}1&0\\0&-1\end{smallmatrix}\right)\overline{B}^{-1}\left(\begin{smallmatrix}1&0\\0&-1\end{smallmatrix}\right))\} \qquad (15.15)$$

$$\cong \{(s,b_5,b_2) \in i\mathbb{R} \times \mathbb{R} \times \mathbb{R} \mid b_5^2 + (1 - \tfrac{1}{4}s^2)b_2^2 = 1\}$$

with $b_5 = b_1 + \tfrac{1}{2}sb_2$.

(ii) As a real analytic manifold

$$V^{mat,S^1} \xrightarrow{\cong} \mathbb{R} \times S^1, \quad (s,B) \mapsto (is, b_-). \qquad (15.16)$$

(iii) The symmetries R_1 and R_2 act on V^{mat,S^1}—see the table in Remark 12.3.

(iv) M^{mon}_{3FN,S^1} is a real semi-algebraic manifold, isomorphic to $\mathbb{R}_{>0} \times V^{mat,S^1}$. The foliation which M^{mon}_{3FN,S^1} inherits from M^{mon}_{3FN}, corresponds to the foliation of $\mathbb{R}_{>0} \times V^{mat,S^1}$ with leaves $\mathbb{R}_{>0} \times \{(s, B)\}$.

(v) The map

$$V^{mat,S^1} \to \{\text{solutions of } P_{III}(0, 0, 4, -4) \text{ on } \mathbb{R}_{>0} \tag{15.17}$$

$$\text{with values in } S^1\}$$

$$(s, B) \mapsto f_{mult}(., s, B)|_{\mathbb{R}_{>0}} \quad (\text{i.e. } x \text{ with } \arg x = 0)$$

is an isomorphism.

Proof Observe that for $(s, B) \in V^{mat}$

$$1 = b_1^2 + b_2^2 + sb_1 b_2 = (b_1 + \tfrac{1}{2}sb_2)^2 + (1 - \tfrac{1}{4}s^2)b_2^2. \tag{15.18}$$

(a) $f_{mult}(., s, B)|_{\mathbb{R}_{>0}}$ takes values in $\mathbb{R}$ if and only if it is a fixed point of the symmetry R_5 in Theorem 14.2 and Remark 14.3. And that holds if and only if (s, B) is a fixed point of the action of R_5 on V^{mat}, i.e.

$$(s, B) = (\bar{s}, \overline{B}^{-1}), \tag{15.19}$$

equivalently: $s \in \mathbb{R}, b_2 \in i\mathbb{R}, b_1 + \tfrac{1}{2}sb_2 \in \mathbb{R}$.

This establishes the isomorphisms (15.6) and (15.10). The table in Remark 14.3 shows furthermore that

$$(s, B) \in V^{mat,a} \Rightarrow (B = \overline{B}^{-1} \iff b_- \in \mathbb{R}),$$

$$(s, B) \in V^{mat,b\pm} \Rightarrow (B = \overline{B}^{-1} \iff b_- \in S^1),$$

$$(s, B) \in V^{mat,c\pm} \Rightarrow (B = \overline{B}^{-1} \iff b_2 \in i\mathbb{R}).$$

This gives the isomorphisms (15.7)–(15.9). The rest is clear from Lemma 10.1 and Theorem 10.3.

(b) $f_{mult}(., s, B)|_{\mathbb{R}_{>0}}$ takes values in $i\mathbb{R}$ if and only if it is a fixed point of the symmetry $R_2 \circ R_5$ in Remark 14.3. And that holds if and only if (s, B) is a fixed point of the action of $R_2 \circ R_5$ on V^{mat}, i.e.

$$(s, B) = (\bar{s}, -\overline{B}^{-1}), \tag{15.20}$$

equivalently: $s \in \mathbb{R}, b_2 \in \mathbb{R}, b_1 + \tfrac{1}{2}sb_2 \in i\mathbb{R}$,

equivalently: $s \in (-2, 2), b_2 \in \mathbb{R}, b_1 + \tfrac{1}{2}sb_2 \in i\mathbb{R}$ by (15.18),

equivalently: $s \in (-2, 2), b_1 + \tfrac{1}{2}sb_2 \in i\mathbb{R}$,

$$b_2 = \pm\sqrt{(1 - (b_1 + \tfrac{1}{2}sb_2)^2)/(1 - \tfrac{1}{4}s^2)} \in \mathbb{R}^*.$$

The set of these (s, b_1, b_2) has two components. By Lemma 15.2 (b) (ii) and the symmetry R_2 (or R_1), $f_{mult}(., s, B)|_{\mathbb{R}_{>0}}$ takes values in $i\mathbb{R}_{>0}$ on one component and values in $i\mathbb{R}_{<0}$ on the other component.

Because of (10.36) with $\varepsilon_2 = 1$, $f_{mult}(x, 0, \pm\left(\begin{smallmatrix} 0 & 1 \\ -1 & 0 \end{smallmatrix}\right)) = \pm i$, so $f_{mult}(., s, B)|_{\mathbb{R}_{>0}}$ takes values in $i\mathbb{R}_{>0}$ on the component $V^{mat, i\mathbb{R}_{>0}}$. This establishes the isomorphisms (15.12) and (15.14). The formulae in Lemma 5.2 (b)

$$b_{\pm} = (b_1 + \tfrac{1}{2} s b_2) \mp \sqrt{\tfrac{1}{4} s^2 - 1 b_2} \quad \text{and} \quad b_+ b_- = 1$$

give $b_- \in i\mathbb{R}_{>0}, b_+ \in i\mathbb{R}_{<0}$ and the isomorphism (15.13). The rest is clear from Lemma 10.1 and Theorem 10.3.

(c) $f_{mult}(., s, B)|_{\mathbb{R}_{>0}}$ takes values in S^1 if and only if it is a fixed point of the symmetry $R_1 \circ R_5$ in Remark 14.3. And that holds if and only if (s, B) is a fixed point of the action of $R_1 \circ R_5$ on V^{mat}, i.e.

$$(s, B) = (-\bar{s}, \left(\begin{smallmatrix} 1 & 0 \\ 0 & -1 \end{smallmatrix}\right) \overline{B}^{-1} \left(\begin{smallmatrix} 1 & 0 \\ 0 & -1 \end{smallmatrix}\right)), \tag{15.21}$$

equivalently: $s \in i\mathbb{R}, b_2 \in \mathbb{R}, b_1 + \tfrac{1}{2} s b_2 \in \mathbb{R}$.

This establishes the isomorphisms (15.15) and (15.17). The table in Remark 14.3 as well as (15.18) give $b_- \in S^1$ and the isomorphism (15.16). The rest is clear from Lemma 10.1 and Theorem 10.3.   □

Remarks 15.6

(i) The proof above of Theorem 15.5 uses the symmetry R_5 and Theorem 14.2. Theorem 15.5 can also be proved with the additional structures which are put onto the P_{3D6}-TEJPA bundles (for the three families) in Chap. 16.

(ii) $f_{mult}(., s, B)|_{\mathbb{R}_{>0}}$ has singularities (i.e. zeros or poles) arbitrarily close to 0 if and only if $(s, B) \in V^{mat, b, \mathbb{R}}$. This follows from (12.26) and the symmetry R_1.

(iii) References [IN86] and [FIKN06] study primarily the restriction of solutions of $P_{III}(0, 0, 4, -4)$ to $\mathbb{R}_{>0}$. In [IN86, ch. 11] the real solutions with asymptotic behaviour (12.21) as $x \to 0$ and with $b_2 \neq 0$, that means the solutions with $(s, B) \in V^{mat, a, \mathbb{R}}$ and $b_2 \neq 0$, and the distribution of their singularities (zeros or poles) as $x \to \infty$ are studied. We shall discuss this in Chaps. 16 and 18.

In [IN86, ch. 8] the solutions $f_{mult}(., s, B)|_{\mathbb{R}_{>0}}$ for $(s, B) \in V^{mat}$ with $b_2 \in \mathbb{R}^*$ and their asymptotic behaviour as $x \to \infty$ and $x \to 0$ are studied. These solutions have no singularities (zeros or poles) near $x = 0$ because of remark (ii) above and $\mathbb{R}^* \cap i\mathbb{R} = \emptyset$. Theorem 8.1 in [IN86] says that the solutions $f_{mult}(., s, B)$ with $b_2 \in \mathbb{R}^*$ have no singularities near $x = \infty$, and it gives asymptotic formulae as $x \to \infty$ for these solutions.

As these solutions are real analytic and have no singularities near 0 or near ∞, each of them can have only finitely many singularities. It is an interesting open question [IN86, p. 105] whether these solutions have any singularities at all. The solutions in the subfamilies $V^{mat,i\mathbb{R}_{>0}}$ and $V^{mat,S^1} - \{(s, B) \,|\, b_2 = 0\}$ have no singularities on $\mathbb{R}_{>0}$ by Lemma 15.2 (b)(ii) + (c)(ii).

Chapter 16
TERP Structures and P_{3D6}-TEP Bundles

The solutions of $P_{III}(0,0,4,-4)$ on $\mathbb{R}_{>0}$ which take values in $\mathbb{R}$ or in S^1 are related to the TERP structures which the second author had defined in [He03], motivated by [CV91, CV93, Du93], and which were studied subsequently in [HS07, HS10, HS11, Mo11b, Sa05a, Sa05b] and other papers. They generalize variations of (polarized) Hodge structures. The concept of TERP(0) bundle is defined below in Definition 16.1. It is a TEP bundle with an additional real structure. It can be pure or not, and if it is pure, it can be polarized or not. A pure polarized TERP(0) bundle generalizes a polarized Hodge structure.

A variation of polarized Hodge structures on a punctured disk can be approximated by a so-called nilpotent orbit of Hodge structures, and this gives rise to a limit polarized mixed Hodge structure [Sch73, CKS86]. This story was generalized to TERP structures in [HS07] (for the regular singular case, building on [Mo03]) and in [Mo11b] (for the irregular case, building on [Mo11a]).

The case of semisimple TERP structures of rank 2 is the first case which is far from the regular singular setting and the classical theory of variations of Hodge structures. It is closely related to the solutions of $P_{III}(0,0,4,-4)$ on $\mathbb{R}_{>0}$ with values in $\mathbb{R}$ or in S^1. A good understanding of these solutions gives a good understanding of this case and vice versa. In particular, singularity freeness or existence of singularities for real solutions on $\mathbb{R}_{>0}$ near $x = 0$ or $x = \infty$ can be derived from general results on certain 1-parameter orbits of TERP structures and limit mixed TERP structures. In Chap. 17 we shall recall these results in the case of semisimple TERP structures.

In this chapter, first TERP(0) bundles will be defined. Then it will be shown that, of the P_{3D6}-TEJPA bundles with $u_0^1 = u_\infty^1 = x \in \mathbb{R}_{>0}$, only those with $(s, B) \in V^{mat,\mathbb{R}} \cup V^{mat,S^1}$ can be enriched to TERP(0) bundles. Finally some general results on TERP structures will be cited and their implications for singularities of real solutions on $\mathbb{R}_{>0}$ of $P_{III}(0,0,4,-4)$ will be explained. The P_{3D6}-TEJPA bundles with

© Springer International Publishing AG 2017
M.A. Guest, C. Hertling, *Painlevé III: A Case Study in the Geometry of Meromorphic Connections*, Lecture Notes in Mathematics 2198,
DOI 10.1007/978-3-319-66526-9_16

$u_0^1 = u_\infty^1 = x \in \mathbb{R}_{>0}$ and $(s, B) \in V^{mat, i\mathbb{R}_{>0}}$ can be enriched by a quaternionic structure instead of a real structure. This is explained below.

Recall the antiholomorphic involution of $\mathbb{P}^1$ in (6.3),

$$\gamma : \mathbb{P}^1 \to \mathbb{P}^1, \quad z \mapsto 1/\bar{z}.$$

Definition 16.1

(a) A TERP(0) bundle is a TEP bundle $(H \to \mathbb{P}^1, \nabla, P)$ together with a $\mathbb{C}$-linear flat (on $H|_{\mathbb{C}^*}$) isomorphism

$$\text{(pointwise)} \quad \tau : H_z \to \overline{H_{\gamma(z)}}, \quad \text{for all } z \in \mathbb{P}^1, \tag{16.1}$$

$$\text{(for sections)} \quad \tau : \mathcal{O}(H) \to \overline{\gamma^* \mathcal{O}(H)}, \quad \mathcal{O}_{\mathbb{P}^1}\text{-linear}$$

with

$$\tau^2 = \text{id}, \tag{16.2}$$

$$P(\tau(a), \tau(b)) = \overline{P(a, b)} \text{ for } a \in H_z, b \in H_{-z}. \tag{16.3}$$

Here $\overline{H_{\gamma(z)}}$ is $H_{\gamma(z)}$ with the complex conjugate $\mathbb{C}$-linear structure, so $\tau : H_z \to H_{\gamma(z)}$ is $\mathbb{C}$-antilinear, and $\overline{\gamma^* \mathcal{O}(H)}$ is indeed a free $\mathcal{O}_{\mathbb{P}^1}$-module whose rank is rank H.

(b) A TERP(0) bundle is *pure* if H is pure, i.e., if H is a trivial holomorphic vector bundle (Remark 4.1 (iv)).

(c) (Lemma [HS10, Lemma 2.5]) The pairing $P(., \tau.) : H_z \times H_{-1/\bar{z}} \to \mathbb{C}$ is sesquilinear (linear $\times$ antilinear) and hermitian, in the sense that

$$P(b, \tau(a)) = \overline{P(a, \tau(b))} \quad \text{for } a \in H_z, b \in H_{-1/\bar{z}}. \tag{16.4}$$

This follows from (16.2), (16.3) and the symmetry of P. $P(., \tau.)$ has constant values on global sections of H. The induced pairing

$$S_{TERP} := P(., \tau.) : \Gamma(\mathbb{P}^1, \mathcal{O}(H)) \times \Gamma(\mathbb{P}^1, \mathcal{O}(H)) \to \mathbb{C} \tag{16.5}$$

is sesquilinear and hermitian. It is nondegenerate if and only if the TERP(0) bundle is pure.

(d) A pure TERP(0) bundle is *polarized* if S_{TERP} is positive definite.

Remarks 16.2

(i) τ restricted to $H|_{S^1}$ is fibrewise a $\mathbb{C}$-antilinear involution and defines a flat real subbundle $\ker(\tau - \text{id} : H|_{S^1} \to H|_{S^1})$ which extends to a flat real subbundle $H_{\mathbb{R}}'$ of $H' := H|_{\mathbb{C}^*}$ with $H_z = H_{\mathbb{R},z}' \oplus iH_{\mathbb{R},z}'$ for $z \in \mathbb{C}^*$.

(ii) One can recover (H, ∇, P, τ) from $((H, \nabla, P)|_{\mathbb{C}}, H_{\mathbb{R}}')$: τ restricted to $H|_{S^1}$ is the complex conjugation and is extended flatly to H'. Then $H|_{\mathbb{C}}$ and $\overline{\gamma^* H|_{\mathbb{C}}}$

are glued using τ to H. Now P extends holomorphically to ∞, ∇ extends meromorphically to ∞. In [He03] and [HS07] this observation was taken as starting point, there a TERP structure (in the case without parameters) was defined by the data $((H, \nabla, P)|_{\mathbb{C}}, H'_{\mathbb{R}})$.

(iii) This has the advantage that a TERP structure with parameters, i.e. a variation of TERP(0) bundles, can be defined by the "almost entirely holomorphic" object $((H, \nabla, P)|_{\mathbb{C} \times M}, H'_{\mathbb{R}})$. Here M is a complex manifold, $H|_{\mathbb{C} \times M}$ is a holomorphic vector bundle, ∇ is a flat connection on $H' := H|_{\mathbb{C}^* \times M}$ with pole of Poincaré rank ≤ 1 along $\{0\} \times M$, P is a flat holomorphic symmetric nondegenerate pairing, and $H'_{\mathbb{R}}$ is a flat real subbundle with $H'_{(z,t)} = H'_{\mathbb{R},(z,t)} \oplus iH'_{\mathbb{R},(z,t)}$ for $(z,t) \in \mathbb{C}^* \times M$ and such that P takes real values on $H'_{\mathbb{R}}$. The only non-holomorphic ingredient is $H'_{\mathbb{R}}$. Then τ is defined on H' as above, and $H|_{\mathbb{C} \times M}$ and $\overline{\gamma^* H}|_{\mathbb{C} \times M}$ are glued using τ to the bundle $H \to \mathbb{P}^1 \times M$. This bundle is holomorphic in $z \in \mathbb{P}^1$, but only real analytic with respect to the parameters $t \in M$.

(iv) This procedure, passing from the almost holomorphic object $((H, \nabla, P)|_{\mathbb{C} \times M}, H'_{\mathbb{R}})$ to the analytic $\times$ (real analytic) object $(H \to \mathbb{P}^1 \times M, \nabla, P, \tau)$, is related to the topological-antitopological fusion in [CV91, CV93] and to the Dorfmeister-Pedit-Wu method [DPW98, Do08] in the construction of CMC surfaces.

(v) In [He03, HS07] TERP structures of weight $w \in \mathbb{Z}$ are defined. These occur naturally, for example in singularity theory. They can be reduced (without losing information) to TERP structures of weight 0. TERP(0) bundles are TERP structures (without parameters) of weight 0.

(vi) Because of the involution τ, the pole at ∞ of a TERP(0) bundle is a "twin" of the pole at 0. In particular, if $u_0^1, \ldots, u_0^n$ are the eigenvalues of the pole part $[z\nabla_{z\partial_z}] : H_0 \to H_0$, then $\overline{u_0^1}, \ldots, \overline{u_0^n}$ are the eigenvalues of the pole part $[-\nabla_{\partial_z}] : H_\infty \to H_\infty$ at ∞, because $\overline{\gamma^*(-u_0^1/z)} = -\overline{u_0^1} z$.

(vii) If (H, ∇, P, τ) is a semisimple TERP(0) bundle of rank 2 with eigenvalues $u_0^1 \neq u_0^2$ of the pole part at 0, the eigenvalues of the pole part at ∞ are $u_\infty^1 = \overline{u_0^1}, u_\infty^2 = \overline{u_0^2}$. Tensoring the bundle with

$$\mathcal{O}_{\mathbb{P}^1} \, e^{(u_0^1 + u_0^2)/(2z) + (\overline{u_0^1} + \overline{u_0^2}) z/2}$$

is a mild twist and leads to a semisimple rank 2 TERP(0) bundle with eigenvalues $\widetilde{u_0^1} = \frac{u_0^1 - u_0^2}{2}$ and $\widetilde{u_0^2} = -\widetilde{u_0^1}$.

Theorem 16.3 *Let $(H, \nabla, x, x, P, A, J)$ be a P_{3D6}-TEJPA bundle with $x \in \mathbb{R}_{>0}$ and with canonically associated monodromy data $(s, B) \in V^{mat}$ (Lemma 10.1 (b), $B = B(\beta)$ for $\beta \in \mathbb{R}$ with $\frac{1}{2} e^{-\beta/2} = x$). Then*

$$\zeta_0 = i, \quad \zeta_\infty = -i, \quad c = u_\infty^1/u_0^1 = x/x = 1, \tag{16.6}$$
$$I_0^+ = I_\infty^- = S^1 - \{-i\}, \quad I_0^- = I_\infty^+ = S^1 - \{i\}, \quad I_0^a = I_\infty^a, \quad I_0^b = I_\infty^b.$$

(a) *The underlying TEP bundle can be enriched to a TERP(0) bundle if and only if $(s, B) \in V^{mat, \mathbb{R}} \cup V^{mat, S^1}$ (see Theorem 15.5 (a)(i)+(c)(i) for the definition of these subsets of V^{mat}).*

(b) (i) *Suppose $(s, B) \in V^{mat, \mathbb{R}}$. Consider for all $x \in \mathbb{R}_{>0}$ the P_{3D6}-TEJPA bundle $(H, \nabla, x, x, P, A, J)$ with monodromy data (s, B), and consider the associated real solution $f := f_{mult}(., s, B)|_{\mathbb{R}_{>0}}$ of $P_{III}(0, 0, 4, -4)$ on $\mathbb{R}_{>0}$. Then there are only two involutions which enrich the underlying TEP bundles to a TERP(0) bundles. One is the τ defined in (16.7), the other is $-\tau$. In (16.7) $\underline{e}_0^{\pm}, \underline{e}_\infty^{\pm}$ is the (unique up to a global sign) 4-tuple of bases associated to the P_{3D6}-TEJPA bundle in Theorem 7.3 (c).*

$$\tau(\underline{e}_0^{\pm}(z)) = \underline{e}_\infty^{\mp}(1/\bar{z}), \quad \tau(\underline{e}_\infty^{\pm}(z)) = \underline{e}_0^{\mp}(1/\bar{z}). \tag{16.7}$$

(ii) *(H, ∇, P, τ) is a pure TERP(0) bundle if and only if x is not a zero or a pole of f.*

(iii) *Suppose that the P_{3D6}-TEJPA bundle is pure. Then it has the normal form (8.15)–(8.18) for $k = 0$ with $f_0 = f(x)$. Then τ is the map with*

$$\underline{\tau}(\sigma_0(z)) = \underline{\sigma}_0(1/\bar{z}) \begin{pmatrix} 0 & f(x)^{-1} \\ f(x) & 0 \end{pmatrix} = J(\underline{\sigma}_0(\bar{z})). \tag{16.8}$$

The pairing $S = P(., \tau .)$ on $\Gamma(\mathbb{P}^1, \mathcal{O}(H))$ has the matrix

$$S(\underline{\sigma}_0(z)^t, \underline{\sigma}_0(-1/\bar{z})) = \begin{pmatrix} 2f(x) & 0 \\ 0 & 2f(x)^{-1} \end{pmatrix}. \tag{16.9}$$

So S is positive definite and (H, ∇, P, τ) is a pure and polarized TERP(0) bundle if $f(x) > 0$, and S is negative definite and (H, ∇, P, τ) is a pure, but not polarized TERP(0) bundle if $f(x) < 0$.

(iv)

$$A \circ \tau = \tau \circ A, \tag{16.10}$$

$$J \circ \tau = \tau \circ J. \tag{16.11}$$

(c) (i) *Suppose $(s, B) \in V^{mat, S^1}$, and let $f := f_{mult}(., s, B)|_{\mathbb{R}_{>0}}$ be the corresponding solution of $P_{III}(0, 0, 4, -4)$ on $\mathbb{R}_{>0}$ with values in S^1. Then there are only two involutions which enrich the underlying TEP bundle to a TERP(0) bundle. One is the τ defined in (16.12), the other is $-\tau$. In (16.12) $\underline{e}_0^{\pm}, \underline{e}_\infty^{\pm}$ is the (unique up to a global sign) 4-tuple of bases associated to the P_{3D6}-TEJPA* *bundle in Theorem 7.3 (c).*

$$\tau(\underline{e}_0^\pm(z)) = \underline{e}_\infty^\mp(1/\bar{z}) \begin{pmatrix} 1 & 0 \\ 0 & -1 \end{pmatrix},$$

$$\tau(\underline{e}_\infty^\pm(z)) = \underline{e}_0^\mp(1/\bar{z}) \begin{pmatrix} 1 & 0 \\ 0 & -1 \end{pmatrix}. \qquad (16.12)$$

(ii) (H, ∇, P, τ) *is a pure TERP(0) bundle.*

(iii) *It has the normal form (8.15)–(8.18) for* $k = 0$ *with* $f_0 = f(x) \in S^1$. *Then* τ *is the map with*

$$\underline{\tau}(\sigma_0(z)) = \underline{\sigma}_0(1/\bar{z}) \begin{pmatrix} f(x)^{-1} & 0 \\ 0 & f(x) \end{pmatrix}. \qquad (16.13)$$

The pairing $S = P(., \tau.)$ *on* $\Gamma(\mathbb{P}^1, \mathcal{O}(H))$ *has the matrix*

$$S(\underline{\sigma}_0(z)^t, \underline{\sigma}_0(-1/\bar{z})) = \begin{pmatrix} 0 & 2f(x) \\ 2f(x)^{-1} & 0 \end{pmatrix}. \qquad (16.14)$$

So S *is nondegenerate, but indefinite, and* (H, ∇, P, τ) *is a pure, but not polarized TERP(0) bundle.*

(iv)

$$A \circ \tau = -\tau \circ A, \qquad (16.15)$$

$$J \circ \tau = \tau \circ J. \qquad (16.16)$$

Proof Equation (16.6) is obvious. Let $\underline{e}_0^\pm, \underline{e}_\infty^\pm$ be the 4-tuple of bases associated to the P_{3D6}-TEJPA bundle in Theorem 7.3 (c). Then

$$\underline{e}_\infty^- = \underline{e}_0^+ B. \qquad (16.17)$$

(a) A map τ enriches the underlying TEP bundle to a TERP(0) bundle if and only if it is an isomorphism between (H, ∇, P) and $\overline{\gamma^*(H, \nabla, P)}$ with $\tau^2 = \mathrm{id}$.

We want to describe τ via its action on the 4-tuple of bases $\underline{e}_0^\pm, \underline{e}_\infty^\pm$. These bases generate the restrictions of H to the sectors $\widehat{I}_0^\pm, \widehat{I}_\infty^\pm$. We have to take care that the images under τ of these restrictions of H glue in the same way as the restrictions of H themselves, that τ respects P in the sense of (16.3) and that $\tau^2 = \mathrm{id}$ holds. We reorder these conditions as follows:

(α) τ respects the splittings of H on the sectors $\widehat{I}_0^\pm, \widehat{I}_\infty^\pm$ from the Stokes structure, and τ respects P.

(β) $\tau^2 = \mathrm{id}$.

(γ) τ respects the gluing of the restrictions to $\widehat{I_0^+}$ and $\widehat{I_0^-}$, that means, the Stokes structure at 0, and similarly the Stokes structure at ∞.

(δ) τ respects the gluing via the connection matrix of the bundles on $\mathbb{C}$ and on $\mathbb{P}^1 - \{0\}$.

(α) says that $\tau(e_\infty^\mp), \tau(e_0^\mp)$ is one of the eight 4-tuples of bases in (6.12),

$$\varepsilon_0 \, (e_0^{\pm 1}, \varepsilon_1 e_0^{\pm 2}), \varepsilon_0 \varepsilon_2 \, (e_\infty^{\pm 1}, \varepsilon_1 e_\infty^{\pm 2}) \quad \text{for} \quad \varepsilon_0, \varepsilon_1, \varepsilon_2 \in \{\pm\},$$

$$\text{so} \quad \tau(e_\infty^\mp) = \varepsilon_0 \, e_0^\pm \begin{pmatrix} 1 & 0 \\ 0 & \varepsilon_1 \end{pmatrix}, \quad \tau(e_0^\mp) = \varepsilon_0 \varepsilon_2 \, e_\infty^\pm \begin{pmatrix} 1 & 0 \\ 0 & \varepsilon_1 \end{pmatrix}. \tag{16.18}$$

(β) $\tau^2 = $ id says that $\varepsilon_2 = 1$.

Concerning (γ): recall from (6.13) that the 4-tuple of bases $e_0^\pm, e_\infty^\pm$ satisfies (2.21) with

$$S = S_0^a \quad \text{and} \quad S_\infty^a = S^{-1}, S_0^b = S^t, S_\infty^b = (S^t)^{-1},$$

and that any of the eight 4-tuples of bases in (6.12) satisfies (2.21) with S replaced by $\begin{pmatrix} 1 & 0 \\ 0 & \varepsilon_1 \end{pmatrix} S \begin{pmatrix} 1 & 0 \\ 0 & \varepsilon_1 \end{pmatrix}$. Now the following four calculations show that the 4-tuple $\tau(e_\infty^\mp), \tau(e_0^\mp)$ satisfies (2.21) with S replaced by $\overline{S}$. We give the details only for the first calculation. It uses (2.21) and (16.6).

$$\tau(e_\infty^+)|_{I_0^a} = \tau(e_\infty^+|_{I_\infty^a}) = \tau(e_\infty^-|_{I_\infty^a} \, (S_\infty^a)^{-1})$$

$$= \tau(e_\infty^-)|_{I_0^a} \, (\overline{S_\infty^a})^{-1} = \tau(e_\infty^-)|_{I_0^a} \, \overline{S},$$

$$\tau(e_\infty^+)|_{I_0^b} = \cdots = \tau(e_\infty^-)|_{I_0^b} \, \overline{S^t},$$

$$\tau(e_0^+)|_{I_\infty^a} = \cdots = \tau(e_0^-)|_{I_\infty^a} \, \overline{S}^{-1},$$

$$\tau(e_0^+)|_{I_\infty^b} = \cdots = \tau(e_0^-)|_{I_\infty^b} \, (\overline{S^t})^{-1}.$$

(γ) is equivalent to

$$\begin{pmatrix} 1 & 0 \\ 0 & \varepsilon_1 \end{pmatrix} S \begin{pmatrix} 1 & 0 \\ 0 & \varepsilon_1 \end{pmatrix} = \overline{S}, \quad \text{so} \quad \varepsilon_1 S = \overline{S}. \tag{16.19}$$

Concerning (δ): the analogue of (16.17) for $\varepsilon_0 e_\infty^- \begin{pmatrix} 1 & 0 \\ 0 & \varepsilon_1 \end{pmatrix}$ and $\varepsilon_0 e_0^+ \begin{pmatrix} 1 & 0 \\ 0 & \varepsilon_1 \end{pmatrix}$ is

$$\varepsilon_0 e_\infty^- \begin{pmatrix} 1 & 0 \\ 0 & \varepsilon_1 \end{pmatrix} = \varepsilon_0 e_0^+ \begin{pmatrix} 1 & 0 \\ 0 & \varepsilon_1 \end{pmatrix} \left[\begin{pmatrix} 1 & 0 \\ 0 & \varepsilon_1 \end{pmatrix} B \begin{pmatrix} 1 & 0 \\ 0 & \varepsilon_1 \end{pmatrix} \right].$$

The analogue of (16.17) for $\tau(\underline{e}_0^+)$ and $\tau(\underline{e}_\infty^-)$ is

$$\tau(\underline{e}_0^+) = \tau(\underline{e}_\infty^-)\,\overline{B}^{-1}.$$

(δ) is equivalent to

$$\begin{pmatrix} 1 & 0 \\ 0 & \varepsilon_1 \end{pmatrix} B \begin{pmatrix} 1 & 0 \\ 0 & \varepsilon_1 \end{pmatrix} = \overline{B}^{-1}. \tag{16.20}$$

Equations (16.19) and (16.20) give (15.18) for $\varepsilon_1 = 1$ and (15.21) for $\varepsilon_1 = -1$. This shows that τ with all desired properties exists if and only if $(s, B) \in V^{mat,\mathbb{R}} \cup V^{mat,S^1}$. It also shows that τ is unique up to sign and that it is given by (16.7) for $\varepsilon_1 = 1$ and by (16.12) for $\varepsilon_1 = -1$. This proves (b) (i) and (c) (i).

(b) (i)   is proved above.

(ii)   This follows from Theorem 10.3 (b).

(iii)   Recall Remark 8.1 (ii): by the correspondence (2.19), the bases $\underline{e}_0^\pm$ correspond to a basis $\underline{v}_0$ of H_0, the bases $\underline{e}_\infty^\pm$ correspond to a basis $\underline{v}_\infty$ of H_∞. Then $\tau(\underline{v}_0) = \underline{v}_\infty$. The construction of the basis σ_0 in the proof of Theorem 8.2 (b) shows that

$$\underline{\sigma}_0(0) = \underline{v}_0\,C, \quad \underline{\sigma}_0(\infty) = \underline{v}_\infty\,C \begin{pmatrix} f(x) & 0 \\ 0 & f(x)^{-1} \end{pmatrix}.$$

Then

$$\tau(\underline{\sigma}_0(0)) = \underline{v}_\infty\,\overline{C} = \underline{\sigma}_0(\infty) \begin{pmatrix} f(x)^{-1} & 0 \\ 0 & f(x) \end{pmatrix} C^{-1}\overline{C}$$

$$= \underline{\sigma}_0(\infty) \begin{pmatrix} 0 & f(x)^{-1} \\ f(x) & 0 \end{pmatrix}.$$

as $C^{-1}\overline{C} = \begin{pmatrix} 0 & 1 \\ 1 & 0 \end{pmatrix}$.

As τ acts on $\Gamma(\mathbb{P}^1, \mathcal{O}(H))$, this establishes (16.8). (16.9) and the conclusions for $f(x) > 0$ and for $f(x) < 0$ follow from (16.8) and (8.16).

(iv)   Compare (16.7) with (7.11) and (7.12).

(c) (i)   is proved above.

(ii)   This follows from Theorem 10.3 (b) and Lemma 15.2 (c) (ii).

(iii) This is analogous to the proof of (b) (iii). The only change is

$$\tau(\underline{v}_0) = \underline{v}_\infty \begin{pmatrix} 1 & 0 \\ 0 & -1 \end{pmatrix},$$

$$\tau(\underline{\sigma}_0(0)) = \underline{v}_\infty \begin{pmatrix} 1 & 0 \\ 0 & -1 \end{pmatrix} \overline{C}$$

$$= \underline{\sigma}_0(\infty) \begin{pmatrix} f(x)^{-1} & 0 \\ 0 & f(x) \end{pmatrix} C^{-1} \begin{pmatrix} 1 & 0 \\ 0 & -1 \end{pmatrix} \overline{C}$$

$$= \underline{\sigma}_0(\infty) \begin{pmatrix} f(x)^{-1} & 0 \\ 0 & f(x) \end{pmatrix}.$$

(iv) Compare (16.12) with (7.11) and (7.12).   □

Remark 16.4 Theorem 16.3 (c) (ii) says that $M_{3FN,S^1} \subset M_{3FN}^{reg}$. The proof above uses the fact that points in M_{3FN}^{sing} correspond to zeros or poles of solutions f of $P_{III}(0,0,4,-4)$ and that the solutions on $\mathbb{R}_{>0}$ with values in S^1 have no poles or zeros.

The following is an alternative proof which does not use isomonodromic families and solutions of $P_{III}(0,0,4,-4)$, but (16.15).

Suppose that $(H, \nabla, x_0, x_0, P, A, J)$ is a P_{3D6}-TEJPA bundle in $M_{3FN,S^1}(x_0) \cap M_{3FN}^{[k]}(x_0)$ for some $k \in \{0,1,2,3,\}$ (and the corresponding $(\varepsilon_1, \varepsilon_2) \in \{\pm 1\}^2$). Consider its normal form in (8.21)–(8.24). Then

$$\Gamma(\mathbb{P}^1, \mathcal{O}(H)) = \mathbb{C}\,\psi_1 \oplus \mathbb{C}\,z\psi_1,$$

$$A(\psi_1) = i\varepsilon_1\,\psi_1, \quad A(z\psi_1) = -i\varepsilon_1\,z\psi_1,$$

$$\tau(\psi_1) = \kappa\,z\psi_1 \quad \text{for some } \kappa \in \mathbb{C}^*.$$

This and (16.15) $A \circ \tau = -\tau \circ A$ give a contradiction by the following calculation,

$$-i\varepsilon_1\,\tau(\psi_1) = A(\tau(\psi_1)) = -\tau(A(\psi_1)) = i\varepsilon_1\,\tau(\psi_1).$$

One can ask about variants of the notion of TERP(0) bundles. In particular, one can ask: are there P_{3D6}-TEP bundles which can be enriched by a τ which satisfies all properties of a TERP(0) bundle except $\tau^2 = \mathrm{id}$? The question is justified by the positive answer which singles out the P_{3D6}-TEJPA bundles in $M_{3FN,i\mathbb{R}_{>0}} \cup M_{3FN,i\mathbb{R}_{<0}}$, where $M_{3FN,i\mathbb{R}_{<0}} := R_1(M_{3FN,i\mathbb{R}_{>0}}) = R_2(M_{3FN,i\mathbb{R}_{>0}})$. This is made precise in the following theorem.

Theorem 16.5

(i) A P_{3D6}-TEJPA bundle $(H, \nabla, x, x, P, A, J)$ with $x \in \mathbb{R}_{>0}$ can be enriched by a τ which satisfies all properties in Definition 16.1 (a) except that now $\tau^2 = \mathrm{id}$

is replaced by $\tau^2 \neq$ id, if and only if the P_{3D6}-TEJPA bundle is in $M_{3FN,i\mathbb{R}_{>0}} \cup M_{3FN,i\mathbb{R}_{<0}}$. Then τ is unique up to a sign. One choice is τ as in (16.21),

$$\tau(\underline{e}_0^\pm(z)) = \underline{e}_\infty^\mp(1/\bar{z}), \quad \tau(\underline{e}_\infty^\pm(z)) = -\underline{e}_0^\mp(1/\bar{z}). \tag{16.21}$$

Here $\underline{e}_0^\pm, \underline{e}_\infty^\pm$ is the 4-tuple of bases in Theorem 7.3 (c).

(ii) H is a pure twistor.

(iii) The P_{3D6}-TEJPA bundle has the normal form (8.15)–(8.18) for $k = 0$ with $f_0 = f(x)$. Then τ is the map with

$$\underline{\tau}(\sigma_0(z)) = \underline{\sigma}_0(1/\bar{z}) \begin{pmatrix} 0 & f(x)^{-1} \\ f(x) & 0 \end{pmatrix} = J(\underline{\sigma}_0(\bar{z})). \tag{16.22}$$

(iv)

$$\tau^2 = -\,\mathrm{id}, \tag{16.23}$$

$$A \circ \tau = \tau \circ A, \tag{16.24}$$

$$J \circ \tau = \tau \circ J. \tag{16.25}$$

$\tau^2 = -$ id says that τ enriches the complex structure on each fibre H_z to a quaternionic structure.

Proof One can follow the proof of Theorem 16.3 (a)+(b). (α) and (16.18) are unchanged. $\tau^2 \neq$ id requires $\varepsilon_2 = -1$. This implies $\tau^2 = -$ id.

(γ) and (16.19) are unchanged. But (δ) changes to

$$-\begin{pmatrix} 1 & 0 \\ 0 & \varepsilon_1 \end{pmatrix} B \begin{pmatrix} 1 & 0 \\ 0 & \varepsilon_1 \end{pmatrix} = \overline{B}^{-1}, \quad \text{together with } \varepsilon_1 s = \bar{s}. \tag{16.26}$$

In the case $\varepsilon_1 = -1$ this yields $s, b_2, b_1 + \frac{s}{2}b_2 \in i\mathbb{R}$, which is not possible because

$$1 = (b_1 + \tfrac{1}{2}sb_2)^2 + (1 - \tfrac{1}{4}s^2)\,b_2^2.$$

Thus $\varepsilon_1 = 1$. In this case (16.28) is (15.20), so $(s, B) \in M_{3FN,i\mathbb{R}_{>0}} \cup M_{3FN,i\mathbb{R}_{<0}}$. Equation (16.21) is proved.

(ii) This follows from Theorem 10.3 (b) and Lemma 15.2 (b) (ii).

(iii) This is analogous to the proof of Theorem 16.3 (b) (iii).

(iv) Compare (16.21) with (7.11) and (7.12). $\square$

Remarks 16.6

(i) The pairing $iP(.,\tau.) : H_z \times H_{-1/\bar{z}} \to \mathbb{C}$ is hermitian. It induces a hermitian, nondegenerate and indefinite pairing on $\Gamma(\mathbb{P}^1, \mathcal{O}(H))$ with matrix

$$iP(\underline{\sigma}_0^t(z), \tau(\underline{\sigma}_0(-1/\bar{z}))) = \begin{pmatrix} 2if(x) & 0 \\ 0 & 2if(x)^{-1} \end{pmatrix}. \qquad (16.27)$$

The pairing $P(.,\tau \circ A(.)) : H_z \times H_{1/\bar{z}} \to \mathbb{C}$ is hermitian. Its restriction to $H|_{S^1}$ and the induced pairing on $\Gamma(\mathbb{P}^1, \mathcal{O}(H))$ are hermitian, nondegenerate and positive definite in the case $M_{3FN, i\mathbb{R}_{>0}}(x_0)$, with matrix

$$iP(\underline{\sigma}_0^t(z), \tau \circ A(\underline{\sigma}_0(1/\bar{z}))) = \begin{pmatrix} -2if(x) & 0 \\ 0 & 2if(x)^{-1} \end{pmatrix}. \qquad (16.28)$$

(ii) Theorem 16.5 (ii) says that $M_{3FN, i\mathbb{R}_{>0}} \subset M_{3FN}^{reg}$. Its proof is similar to the proof of Theorem 16.3 (c) (ii). As in Remark 16.4, we can give an alternative proof which does not use isomonodromic families and solutions of $P_{III}(0, 0, 4, -4)$. But it is very different from the proof in Remark 16.4. It uses the pairing $P(.,\tau \circ A(.))$ on $H|_{S^1}$. The fact that this pairing is hermitian and positive definite permits application of the Iwasawa decomposition for loop groups to conclude that H is a pure twistor. We shall not give details here.

(iii) The appearance of a quaternionic structure on the vector bundle H in Theorem 16.5 is not a surprise. An isomonodromic family of such bundles can be related via equation (15.2) or more directly (via the Sym-Bobenko formula) to a CMC surface in $\mathbb{R}^3$ with rotationally symmetric metric. The appearance of a quaternionic structure in the construction of CMC surfaces in $\mathbb{R}^3$ from vector bundles with connections is well established [FLPP01].

Chapter 17
Orbits of TERP Structures and Mixed TERP Structures

The real solutions (possibly with zeros and/or poles) of $P_{III}(0, 0, 4, -4)$ on $\mathbb{R}_{>0}$ are by (15.10) the functions $f = f_{mult}(., s, B)|_{\mathbb{R}_{>0}}$ for $(s, B) \in V^{mat, \mathbb{R}}$. They will be studied in detail in Chap. 18.

By Theorem 10.3 any solution corresponds to an isomonodromic family of P_{3D6}-TEJPA bundles $(H(x), \nabla, x, x, P, A, J)$ for $x \in \mathbb{R}_{>0}$ with monodromy data (s, B). By Theorem 16.3 (b) with the τ in (16.9), this can be enriched to an isomonodromic family of TERP(0) bundles. Then $H(x)$ is pure if and only if x is not a zero or a pole of f, and then $H(x)$ is polarized if $f(x) > 0$, and it has negative definite pairing S if $f(x) < 0$.

The 1-parameter isomonodromic families $(H(x), \nabla, P, \tau)$ for $x \in \mathbb{R}_{>0}$ turn out to be special cases of 1-parameter isomonodromic families of TERP(0) bundles which had been studied in [HS07, Mo11b] and which are called Euler orbits in Definition 17.1 below. The situation where all members with large x or when all members with small x are pure and polarized was investigated in [HS07, Mo11b]. In our case this corresponds to smoothness and positivity of f near ∞ or 0.

The characterization for large x will be stated below in the semisimple case, and the characterisation for small x will be discussed informally. This will be preceded by the definition of Euler orbits of TERP(0) bundles and a discussion of semisimple TERP(0) bundles. At the end of the chapter a result from [HS11] will be explained in the special case of semisimple TERP(0) bundles. It provides semisimple TERP(0) bundles such that they and all their semisimple isomonodromic deformations are pure and polarized. This will give all solutions f of $P_{III}(0, 0, 4, -4)$ on $\mathbb{R}_{>0}$ which are smooth and positive on $\mathbb{R}_{>0}$.

Definition 17.1 Let (H, ∇, P, τ) be a TERP(0) bundle. Let

$$\pi_{orb} : \mathbb{C} \times \mathbb{C}^* \to \mathbb{C}, \quad (z, r) \mapsto z r^{-1}. \tag{17.1}$$

© Springer International Publishing AG 2017
M.A. Guest, C. Hertling, *Painlevé III: A Case Study in the Geometry of Meromorphic Connections*, Lecture Notes in Mathematics 2198, DOI 10.1007/978-3-319-66526-9_17

Then $(\pi_{orb}^*(H|_\mathbb{C}, \nabla, P), \pi_{orb}^*(H'_\mathbb{R}))$ is the almost entirely holomorphic object of a variation on $M := \mathbb{C}^*$ of TERP(0) bundles mentioned in Remark 16.2 (iii). τ on $\pi_{orb}^*(H|_\mathbb{C})|_{S^1 \times M}$ is defined as complex conjugation with respect to the real structure $\pi_{orb}^*(H'_\mathbb{R})|_{S^1 \times M}$, and it is extended flatly to $\mathbb{C}^* \times M$. Then $\pi_{orb}^*(H|_\mathbb{C})$ and $\overline{\gamma^*(\pi_{orb}^*(H|_\mathbb{C}))}$ are glued by τ to a bundle $G \to \mathbb{P}^1 \times M$, which is holomorphic in $z \in \mathbb{P}^1$, but only real analytic with respect to $r \in M = \mathbb{C}^*$.

(G, ∇, P, τ) is a special variation of TERP(0) bundles with parameter space $M = \mathbb{C}^*$ which is called here an *Euler orbit*. (See [He03, ch. 2] or Remark 16.2 (iii) for the general notion of a variation of TERP(0) bundles—in [He03] this is called a TERP structure.)

The single TERP(0) bundle for $r \in M = \mathbb{C}^*$ is $(G(r), \nabla, P, \tau)$ with $G(r) = G|_{\mathbb{P}^1 \times \{r\}}$. It is obtained by gluing $\pi_r^*(H|_\mathbb{C})$ and $\overline{\gamma^*(\pi_r^*(H|_\mathbb{C}))}$ via τ, here

$$\pi_r : \mathbb{C} \to \mathbb{C}, \quad z \to z r^{-1}. \tag{17.2}$$

Of course $(G(1), \nabla, P, \tau) = (H, \nabla, P, \tau)$.

Remarks 17.2

(i) If $u_0^1, \ldots, u_0^n$ are the eigenvalues of the pole part $[z\nabla_{z\partial_z}] : H_0 \to H_0$ at 0, then $ru_0^1, \ldots, ru_0^n$ are the eigenvalues of the pole part of $(G(r), \nabla)$ at 0.

(ii) In the semisimple case the Euler orbit of a TERP(0) bundle is simply the isomonodromic 1-parameter family of TERP(0) bundles $(G(r), \nabla, P, \tau)$ for $r \in \mathbb{C}^*$ such that for fixed r the eigenvalues of the pole part at 0 are $ru_0^1, \ldots, ru_0^n$ and the eigenvalues of the pole part $[-\nabla_{\partial_z}] : G(r)_\infty \to G(r)_\infty$ are $\overline{ru_0^1}, \ldots, \overline{ru_0^n}$.

(iii) Any member $(G(r), \nabla, P, \tau)$ of the Euler orbit of a TERP(0) bundle (H, ∇, P, τ) induces up to a rescaling of the parameter space the same Euler orbit.

(iv) The definition of the pull-back $\pi_{orb}^*(H|_\mathbb{C}) = G|_{\mathbb{C} \times \mathbb{C}^*}$ implies the existence of canonical isomorphisms

$$G(r_1)_{z_1} \cong H_{z_1 r_1^{-1}} = H_{z_2 r_2^{-1}} \cong G(r_2)_{z_2} \quad \text{if } z_1 r_1^{-1} = z_2 r_2^{-1}.$$

It is shown in [HS07, Lemma 4.4] that these isomorphisms extend to $z_1 = z_2 = \infty$ if $|r_1| = |r_2|$ and induce a bundle isomorphism $G(r_1) \cong G(r_2)$ over the automorphism $\mathbb{P}^1 \to \mathbb{P}^1$, $z \mapsto z\frac{r_2}{r_1}$.

Therefore the restriction of the variation (G, ∇, P, τ) on $\mathbb{C}^*$ to any circle $\{r \in \mathbb{C}^* \,|\, |r| = r_0\}$, $r_0 \in \mathbb{R}_{>0}$, is a (rather trivial) variation of TERP(0) bundles. All bundles are pure/polarized if one bundle is pure/polarized. The interesting part of the variation (G, ∇, P, τ) on $\mathbb{C}^*$ is its restriction to $\mathbb{R}_{>0} \subset \mathbb{C}^*$.

(v) Because of (iv) and Remark 16.2 (vi) and (vii), for the study of semisimple rank 2 TERP(0) bundles we can restrict to the case $u_0^1 = u_\infty^1 = x \in \mathbb{R}_{>0}$, $u_0^2 = u_\infty^2 = -x$. Such a TERP(0) bundle is a P_{3D6}-TEP bundle (H, ∇, x, x, P)

with additional real structure given by τ. Theorem 16.3 relates such a TERP(0) bundle to a value $f(x)$ of a solution $f = f_{mult}(.,s,B)|_{\mathbb{R}_{>0}}$ of $P_{III}(0,0,4,-4)$ on $\mathbb{R}_{>0}$ with $(s,B) \in V^{mat,\mathbb{R} \cup S^1}$.

By Theorem 16.3, in the case of $(s,B) \in V^{mat,S^1}$ the TERP(0) bundle is pure and has indefinite pairing S. In the case of $(s,B) \in V^{mat,\mathbb{R}}$ it is pure if and only if x is not a zero or pole of f, otherwise $\mathcal{O}(H) \cong \mathcal{O}_{\mathbb{P}^1}(1) \oplus \mathcal{O}_{\mathbb{P}^1}(-1)$. If $f(x) > 0$ it is pure and polarized, if $f(x) < 0$ it is pure with negative definite pairing S.

Theorem 16.3 and the results in Chap. 18 give a complete picture of the semisimple rank 2 TERP(0) bundles and their Euler orbits.

(vi) By (15.10) and Theorem 16.3 and Remark 17.2 (ii), the solutions f of $P_{III}(0,0,4,-4)$ on $\mathbb{R}_{>0}$ with values in $\mathbb{R}$ or in S^1 correspond to the restrictions to $\mathbb{R}_{>0}$ of the Euler orbits of the TERP(0) bundles with $u_0^1 = u_\infty^1 = x \in \mathbb{R}_{>0}$, $u_0^2 = u_\infty^2 = -x$. Results on their Euler orbits are equivalent to results on solutions f.

(vii) The TERP structures in [He03] give a framework for the data studied in [CV91, CV93, Du93], which are essentially certain variations of TERP bundles. The Euler orbits appear in [CV91, CV93] from the renormalisation group flow. There the limits $r \to \infty$ and $r \to 0$ are called *infrared limit* and *ultraviolet limit*.

Definition 17.3 ([HS07, def. 4.1]) Let (H, ∇, P, τ) be a TERP(0) bundle. Its Euler orbit $(G(r), \nabla, P, \tau)$, $r \in \mathbb{C}^*$, is called a *nilpotent orbit* of TERP(0) bundles if, for all large $|r|$, $(G(r), \nabla, P, \tau)$ is a pure and polarized TERP(0) bundle. Its Euler orbit is called a *Sabbah orbit* of TERP(0) bundles if, for all small $|r|$, $(G(r), \nabla, P, \tau)$ is a pure and polarized TERP(0) bundle.

Remarks 17.4

(i) A real solution (possibly with zeros and/or poles) on $\mathbb{R}_{>0}$ of $P_{III}(0,0,4,_4)$ is positive for large x (respectively, small x) if and only if its Euler orbit of TERP(0) bundles is a nilpotent orbit (respectively, a Sabbah orbit).

(ii) Reference [HS07, Theorem 7.3] gives a precise characterisation of Sabbah orbits:

The Euler orbit of a TERP(0) bundle is a Sabbah orbit if a certain candidate for a sum of two polarized mixed Hodge structures (briefly: PMHS) is indeed a sum of two PMHS.

The definition of the candidate starting from the TERP(0) bundle is lengthy. It uses a variant due to Sabbah of the Kashiwara-Malgrange V-filtration, and the definition of the polarizing form needs special care. Also the definition of a PMHS is nontrivial. All definitions can be found in [HS07], so we shall not reproduce them here.

(iii) Another reason for omitting the details in the characterization of Sabbah orbits is that the application here would be a characterization of those $(s,B) \in V^{mat,\mathbb{R}}$ for which $f_{mult}(.,s,B)|_{\mathbb{R}_{>0}}$ is positive for small x. But Chap. 12 settles this completely and provides richer information, which we cannot extract easily from [HS07, Theorem 7.3]. See Theorem 18.2 (c)+(d) for the details.

(iv) In fact, for $(s, B) \in V^{mat,\mathbb{R}}$ $f_{mult}(., s, B)$ is positive for small x if and only if $|s| \leq 2$ and $b_1 + \frac{s}{2}b_2 \in \mathbb{R}_{\geq 1}$. If $|s| < 2$ then $\mathrm{Mon}(s)$ is semisimple and the candidate in [HS07, Theorem 7.3] is a pure polarized Hodge structure. But if $|s| = 2$ then $\mathrm{Mon}(s)$ has a 2×2 Jordan block and the candidate is a polarized (and truly) *mixed* Hodge structure.

Now we shall review the notion of semisimple TERP(0) bundles and their monodromy data, following [HS07, ch. 8,10]. Then we shall define semisimple mixed TERP structures and formulate the characterization of nilpotent orbits. Finally a result of [HS11] will show that for certain semisimple mixed TERP(0) bundles they and all their semisimple isomonodromic deformations are pure and polarized TERP(0) bundles.

First we cite and explain a lemma from [HS07] on semisimple TEP bundles. Recall (Remark 6.2 (i)) that the TEP structures of weight 0 in [HS07] are the restrictions to $\mathbb{C}$ of the TEP bundles in Definition 6.1 (a) (which are defined on $\mathbb{P}^1$).

Let us call two matrices T and T' in $M(n \times n, \mathbb{C})$ *sign equivalent* if there is a matrix $B = \mathrm{diag}(\varepsilon_1, \dots, \varepsilon_n)$ with $\varepsilon_1, \dots, \varepsilon_n \in \{\pm 1\}$ such that $BTB = T'$.

Lemma 17.5 ([HS07, Lemma 10.1 (1)])   *Fix pairwise distinct values $u_1, \dots, u_n \in \mathbb{C}$ and $\xi \in S^1$ with $\Re(\frac{u_i - u_j}{\xi}) < 0$ for $i < j$. There is a natural $1{:}1$ correspondence between the set of restrictions to $\mathbb{C}$ of semisimple TEP bundles with pole part at 0 having eigenvalues $u_1, \dots, u_n$, and the set of sign equivalence classes of upper triangular matrices $T \in Gl(n, \mathbb{C})$ with diagonal entries equal to 1. The matrices T are the Stokes matrices of the pole at 0 of the TEP bundle.*

A similar statement for the case without pairing holds and is a special case of a Riemann-Hilbert correspondence between germs of holomorphic vector bundles with meromorphic connections and their Stokes data, see for example [Si67, Si90, BJL79, Ma83a], [Sa02, ch. II 5,6], [Bo01], [Mo11a, Theorem 4.3.1]. There one has two Stokes matrices and n *exponents* in $\mathbb{C}$ determining the regular singular rank 1 pieces in the formal isomorphism class. In the case of a TEP bundle, the n exponents are all equal to 0, and the second Stokes matrix is the transpose of the first.

Let us explain the $1{:}1$ correspondence in more detail. The proof can be found in [HS07].

The formal decomposition of Turrittin works in the semisimple case without ramification (e.g. [Sa02, II Theorem 5.7]) and gives a formal isomorphism

$$\Psi : (\mathcal{O}(H)_0[z^{-1}], \nabla) \otimes_{\mathbb{C}\{z\}[z^{-1}]} \mathbb{C}[[z]][z^{-1}] \tag{17.3}$$

$$\rightarrow \oplus_{j=1}^{n} e^{-u_j/z} \otimes (\mathcal{R}_j, \nabla_j) \otimes_{\mathbb{C}\{z\}[z^{-1}]} \mathbb{C}[[z]][z^{-1}]$$

where $\mathcal{R}_j$ is a $\mathbb{C}\{z\}[z^{-1}]$ vector space of dimension 1 and ∇_j is a meromorphic connection on it with a regular singular pole at 0.

With the general results in [Sa02, II 5. and III 2.1] it follows easily that Ψ extends to a formal isomorphism

$$\Psi : (\mathcal{O}(H)_0, \nabla, P) \otimes_{\mathbb{C}\{z\}} \mathbb{C}[[z]] \tag{17.4}$$

$$\to \oplus_{j=1}^n e^{-u_j/z} \otimes (\mathcal{O}(H_j)_0, \nabla_j, P_j) \otimes_{\mathbb{C}\{z\}} \mathbb{C}[[z]]$$

of germs at 0 of TEP bundles [HS07, Lemma 8.2]. Here $(\mathcal{O}(H_j)_0, \nabla_j, P_j)$ is the germ at 0 of a rank 1 TEP bundle with regular singular pole at 0 and with $\mathcal{R}_j = \mathcal{O}(H_j)_0[z^{-1}]$, and $e^{-u_j/z} \otimes \mathcal{O}(H_j)_0$ is the germ at 0 whose holomorphic sections are obtained by multiplying those in $\mathcal{O}(H_j)_0$ with $e^{-u_j/z}$. Then $e^{-u_j/z} \otimes (\mathcal{O}(H_j)_0, \nabla, P_j)$ is the germ at 0 of a rank 1 TEP bundle with eigenvalue u_j of the pole part at 0.

We remark that up to isomorphism there is only one germ at 0 of a rank 1 TEP bundle with regular singularity at 0: the regular singularity leads to a generating section $z^\alpha e_0$ where e_0 is a flat multi-valued section on $\mathbb{C}^*$, the pairing P implies $\alpha = 0$ and the single-valuedness of e_0 and $P(e_0(z), e_0(-z)) = $ constant $\in \mathbb{C}^*$.

Denote by f_j a flat generating section with $P(f_j(z), f_j(-z)) = 1$ of the germ $(\mathcal{O}(H_j)_0, \nabla_j, P_j)$. It is unique up to sign.

For any n distinct values $u_1, \ldots, u_n \in \mathbb{C}$ the set

$$\Sigma := \{\xi \in \mathbb{C} \mid \exists\, i, j \text{ with } i \neq j \text{ and } \Re(\tfrac{u_i - u_j}{\xi}) = 0\}$$

of *Stokes directions* is finite. For any $\xi \in S^1 - \Sigma$ one can renumber the n distinct values such that then $\Re(\tfrac{u_i - u_j}{\xi}) < 0$ for $i < j$ holds. The choice of a ξ and a numbering of $u_1, \ldots, u_n$ with this property is assumed in Lemma 17.5 and in the following. Let I_0^a (respectively, I_0^b) be the component of $S^1 - \Sigma$ which contains ξ (respectively, $-\xi$), and denote

$$I_0^\pm := I_0^a \cup I_0^b \cup \{z \in S^1 \mid \pm \Im(z/\xi) \leq 0\}.$$

This generalizes the notation in Chap. 2 for the rank 2 case. Each of the sets I_0^+ and I_0^- contains exactly one of the two Stokes directions $\pm\xi'$ for any $\xi' \in \Sigma$.

Denote by $\mathcal{A}$ the sheaf on S^1 of holomorphic functions in neighbourhoods of 0 in sectors which have an asymptotic expansion in $\mathbb{C}[[z]]$ in the sense of [Ma83a, ch. 3].

A result of Hukuhara and many others (e.g. [Ma83a, ch. 3-5], [Sa02, II 5.12]) says in our case that the formal isomorphism Ψ in (17.4) lifts in the sectors $\widehat{I_0^\pm}$ to unique isomorphisms $\Psi^\pm$ with coefficients in $\mathcal{A}|_{I_0^\pm}$,

$$\Psi^\pm : (\mathcal{O}(H)_0, \nabla) \otimes_{\mathbb{C}\{z\}} \mathcal{A}|_{I_0^\pm} \tag{17.5}$$

$$\to \oplus_{j=1}^n e^{-u_j/z} \otimes (\mathcal{O}(H_j)_0, \nabla_j) \otimes_{\mathbb{C}\{z\}} \mathcal{A}|_{I_0^\pm}$$

which together respect the pairing in the following sense: The preimages $e_j^\pm := (\Psi^\pm)^{-1}(f_j)$ are together for $j = 1, \ldots, n$ flat bases $\underline{e}^\pm = (e_0^{\pm 1}, \ldots, e_0^{\pm n})$ of $H|_{\widehat{I_0^\pm}}$

with

$$P(\underline{e}_0^+(z)^t, \underline{e}_0^-(-z)) = \mathbf{1}_n = P(\underline{e}_0^-(-z)^t, \underline{e}_0^+(z)) \quad \text{for } z \in \widehat{I}_0^+ \tag{17.6}$$

[HS07, Lemma 8.4]. The base change matrix which is defined by (17.7) is upper triangular with 1's on the diagonal and, by (17.6) and (17.8), satisfies

$$\underline{e}_0^-|_{I_0^a} = \underline{e}_0^+|_{I_0^a} T, \tag{17.7}$$

$$\underline{e}_0^-|_{I_0^b} = \underline{e}_0^+|_{I_0^b} T^t \tag{17.8}$$

[HS07, Lemma 8.3]. The sign equivalence class of T is unique.

For any T which is upper triangular with 1's on the diagonal a unique germ $(\mathcal{O}(H)_0, \nabla, P)$ exists [HS07, Lemma 10.1 (1)].

Now we turn to TERP(0) bundles.

Definition 17.6 Let (H, ∇, P, τ) be a semisimple TERP(0) bundle with pairwise distinct eigenvalues $u_1, \ldots, u_n \in \mathbb{C}$ of the pole part at 0 and with a $\xi \in S^1$ such that $\Re(\frac{u_i - u_j}{\xi}) < 0$ for $i < j$.

(a) *The real structure and Stokes structure are compatible if* $\lambda_1, \ldots, \lambda_n \in S^1$ exist such that $(\lambda_1 e_0^{\pm 1}, \ldots, \lambda_n e_0^{\pm n})$ is a flat basis of $H'_{\mathbb{R}}|_{\widehat{I}_0^\pm}$, i.e., if the splitting $\oplus_{j=1}^n \mathbb{C} e_0^{\pm j}$ of the flat bundle on $\widehat{I}_0^\pm$ is compatible with the real structure.

(b) Remark: Then $\lambda_j \in \{\pm 1, \pm i\}$ because by (16.3) for $z \in I_0^+$

$$\overline{\lambda_j^2} = \overline{P(\lambda_j e_0^{+j}(z), \lambda_j e_0^{-j}(-z))} = P(\lambda_j e_0^{+j}(z), \lambda_j e_0^{-j}(-z)) = \lambda_j^2. \tag{17.9}$$

(c) The TERP(0) bundle is a *mixed TERP structure* if in (a) all $\lambda_j \in \{\pm 1\}$.

Remarks 17.7

(i) The Stokes structure is encoded in the splittings $\oplus_{j=1}^n \mathbb{C} e_0^{\pm j}$ of the flat bundle $H|_{\widehat{I}_0^\pm}$ and the Stokes matrix T, which depend on the choice of ξ. One can reformulate Definition 17.6 (a) in a way which does not depend on this choice [Mol1b, 8.1.1]. It says that the *Stokes filtration* and the real structure are compatible.

(ii) An equivalent condition for the Stokes structure and the real structure to be compatible is that the real structure of the TERP(0) bundle induces on the pieces $e^{-u_j/z} \otimes (\mathcal{O}(H_j)_0, \nabla_j, P_j)$ a natural real structure such that these pieces become TERP(0) bundles. Then $\lambda_j f_j$ is a generating flat real section on $\mathbb{C}^*$. The TERP(0) bundle $(H_j, \nabla_j, P_j, \tau_j)$ is automatically pure. It is polarized if $P(\lambda_j f_j, \lambda_j f_j) = 1$, and that holds if $\lambda_j \in \{\pm 1\}$. The alternative is $P(\lambda_j f_j, \lambda_j, f_j) = -1$ and $\lambda_j \in \{\pm i\}$. This motivates the definition of a mixed TERP structure in part (c) above.

(iii) In the non-semisimple case (17.4)–(17.8) hold mutatis mutandis. Then the pieces of the decomposition do not all have rank 1. Part (a) of Definition 17.6

goes through, and if it holds the pieces become TERP(0) bundles. But part (c) has to be replaced by the condition that for each regular singular piece a certain candidate for a sum of two PMHS is indeed a sum of two PMHS. See [HS07] for the definition and explanations.

(iv) Unfortunately, in the proof of [HS07, Lemma 10.1 (2)], the case $\lambda_j \in \{\pm i\}$ and the fact that it is excluded in a mixed TERP structure because of the polarization of the PMHS, had been missed. (Furthermore, in (10.3) and (10.4) $(A^-)^{tr}$ has to be replaced by $(A^-)^{-1}$; they follow from (10.1) instead of (10.2), and in (10.4) the sign $(-1)^w$ has to be deleted.)

(v) The pole at ∞ of a TEP bundle or a TERP(0) bundle can be described analogously to the explanations after Lemma 17.5. We consider only the case of a TERP(0) bundle. Then the eigenvalues of the pole part $[-\nabla_{\partial_z}] : H_\infty \to H_\infty$ are $\overline{u_1}, \ldots, \overline{u_n}$ (Remark 16.2 (vi)). Then for $u_1, \ldots, u_n$ and ξ as above one can define

$$I_\infty^a := I_0^a, \ I_\infty^b := I_0^b, \ I_\infty^+ := I_0^-, \ I_\infty^- := I_0^+. \tag{17.10}$$

Because of the involution τ, the pole at ∞ is a twin of the pole at 0, and the analogously defined flat bases $\underline{e}_\infty^\pm$ on $\widehat{I}_\infty^\pm$ are

$$\underline{e}_\infty^\pm = \tau(\underline{e}_0^\mp) \tag{17.11}$$

with base change matrices

$$\underline{e}_\infty^-|_{I_\infty^a} = \underline{e}_\infty^+|_{I_\infty^a} \ \overline{T}^{-1}, \tag{17.12}$$

$$\underline{e}_\infty^-|_{I_\infty^b} = \underline{e}_\infty^+|_{I_\infty^b} \ (\overline{T}^{-1})^t. \tag{17.13}$$

Define the connection matrix B by

$$\tau(\underline{e}_0^+) = \underline{e}_\infty^- = \underline{e}_0^+ \ B. \tag{17.14}$$

Then Definition 17.6 says that $B = \text{diag}(\varepsilon_1, \ldots, \varepsilon_n)$ with $\varepsilon_1, \ldots, \varepsilon_n \in \{\pm 1\}$ if and only if Stokes structure and real structure are compatible, and that

$$B = \mathbf{1}_n \iff \text{the TERP(0) bundle} \tag{17.15}$$

is a mixed TERP structure.

Corollary 17.8 ([HS07, Lemma 10.1 (2)]) *Fix pairwise distinct values $u_1, \ldots, u_n \in \mathbb{C}$ and $\xi \in S^1$ with $\Re(\frac{u_i - u_j}{\xi}) < 0$ for $i < j$. The 1:1 correspondence in Lemma 17.5 restricts to a 1:1 correspondence between the set of semisimple mixed TERP structures with pole part at 0 having eigenvalues $u_1, \ldots, u_n$, and the set of sign equivalence classes of upper triangular matrices $T \in GL(n, \mathbb{R})$ with diagonal entries equal to 1.*

Proof In the case of a semisimple mixed TERP structure the flat bases $\underline{e}_0^+ = \underline{e}_\infty^-$ and $\underline{e}_0^- = \underline{e}_\infty^+$ of $H|_{\widetilde{I}_0^\pm} = H|_{\widetilde{I}_\infty^\mp}$ are real bases, thus T in (17.7) has real entries.

Conversely, if one starts with the restriction to $\mathbb{C}$ of a semisimple TEP bundle with matrix T with real entries, the real structure with real bases $\underline{e}_0^+$ and $\underline{e}_0^-$ is well defined and compatible with the Stokes structure, and satisfies $B = \mathbf{1}_n$.   □

The following theorem generalizes to all TERP(0) bundles whose formal decomposition at 0 is valid without ramification. This generalization was conjectured in [HS07, conjecture 9.2]. The simpler direction $\Leftarrow$ and the regular singular case of the direction $\Rightarrow$ were proved in [HS07, Theorem 9.3], building on [Mo03]. The general case of the more difficult direction $\Rightarrow$ was proved by T. Mochizuki [Mo11b, Corollary 8.15], building on [Mo11a].

Theorem 17.9 *A semisimple TERP(0) bundle induces a nilpotent orbit $\Longleftrightarrow$ it is a mixed TERP structure.*

Corollary 17.10 *A real solution $f_{mult}(., s, B)|_{\mathbb{R}_{>0}}$ for $(s, B) \in V^{mat,\mathbb{R}}$ (possibly with zeros and/or poles) of $P_{III}(0, 0, 4, -4)$ on $\mathbb{R}_{>0}$ is smooth and positive for large x if and only if $B = \mathbf{1}_2$.*

Proof This follows from Remark 17.4 (i), Theorem 17.9 and the formulae (17.14) and (17.15).   □

The following theorem is a special case of a result in [HS11]. It is relevant for globally smooth real solutions of $P_{III}(0, 0, 4, -4)$ on $\mathbb{R}_{>0}$.

Theorem 17.11 ([HS11, Theorem 5.9]) *Let (H, ∇, P, τ) be a semisimple TERP(0) bundle with pairwise different eigenvalues $u_1, \ldots, u_n \in \mathbb{C}$ of the pole part at 0 and with $\xi \in S^1$ such that $\Re(\frac{u_i - u_j}{\xi}) < 0$ for $i < j$, and suppose that the Stokes matrix T has real entries, so that the TERP(0) bundle is also a mixed TERP structure.*

Then all TERP(0) bundles in its Euler orbit are pure and polarized if $T + T^t$ is positive semidefinite.

Corollary 17.12 *A real solution $f_{mult}(., s, B)|_{\mathbb{R}_{>0}}$ for $(s, B) \in V^{mat,\mathbb{R}}$ of $P_{III}(0, 0, 4, -4)$ on $\mathbb{R}_{>0}$ is everywhere smooth and positive if $|s| \leq 2$ and $B = \mathbf{1}_2$.*

Proof Suppose that $|s| \leq 2$ and $B = \mathbf{1}_2$. Then the TERP(0) bundles in the Euler orbit for the solution $f_{mult}(., s, \mathbf{1}_2)|_{\mathbb{R}_{>0}}$ are mixed TERP structures because of $B = \mathbf{1}_2$, and their Stokes matrix is

$$S = \begin{pmatrix} 1 & s \\ 0 & 1 \end{pmatrix}.$$

Then

$$S + S^t = \begin{pmatrix} 2 & s \\ s & 2 \end{pmatrix}$$

is positive semidefinite. By Theorem 17.11 all TERP(0) bundles are pure and polarized. By Theorem 16.3 (b) (iii) $f_{mult}(x, s, \mathbf{1}_2) > 0$ for all $x \in \mathbb{R}_{>0}$.   □

Remarks 17.13

(i) $B = \mathbf{1}_2$ is necessary for smoothness and positivity for large x (Corollary 17.10), $|s| \leq 2$ is necessary for smoothness near 0 (Theorem 18.2 (c)), therefore $|s| \leq 2$ and $B = \mathbf{1}_2$ are also necessary conditions for global smoothness and positivity (Theorem 18.2 (e)).

(ii) Reference [HS11, Theorem 5.9] is more general in several aspects, but it claims only that the original TERP(0) bundle is pure and polarized. But one can go easily to all TERP(0) bundles in the Euler orbit, because the eigenvalues of their pole parts at 0 are of the form $ru_0^1, \ldots, ru_0^n$ for $r \in \mathbb{R}_{>0}$ and the Stokes matrix T and the connection matrix B do not change (Remark 16.2 (iii)).

(iii) In fact, Theorem 17.11 together with the well known behaviour of the Stokes matrix under the change of the tuple $(u_0^1, \ldots, u_0^n, \xi)$ show that all semisimple TERP structures which are in the isomonodromic family of one TERP structure in Theorem 17.11 are pure and polarized. If the tuple $(u_0^1, \ldots, u_0^n, \xi)$ moves, then in general the Stokes directions change, ξ crosses some Stokes directions, and the values $u_0^1, \ldots, u_0^n$ have to be renumbered. Then the Stokes matrix T changes by some braid group action to a new Stokes matrix $\widetilde{T}$ [Du99, Theorem 4.6]. But $\widetilde{T}$ inherits from T the property that $\widetilde{T} + \widetilde{T}^t$ is positive semidefinite. As we do not need this for the Euler orbits, we shall not give details here.

(iv) The data in [HS11, Theorem 5.9] differ in several ways from the TERP(0) bundles. First, they are more general: instead of a real structure and a $\mathbb{C}$-bilinear pairing one sesquilinear pairing is considered. Second, the structures there do not have weight 0, but weight 1. Therefore the second Stokes matrix and the formal eigenvalues there differ by a sign from the second Stokes matrix and the formal eigenvalues here. Third, not only the semisimple case is considered. Finally, there a *minimality condition* is assumed. But in the semisimple case it is trivial, then $K_c = 0$ holds for all $c \ (= u_0^1, \ldots, u_0^n$ here). $K_c \neq 0$ can arise only if the monodromy of the corresponding piece of the formal decomposition of a TERP(0) bundle has -1 as an eigenvalue. In the semisimple case all pieces have rank 1, and all formal eigenvalues are equal to 1.

Chapter 18
Real Solutions of $P_{III}(0, 0, 4, -4)$ on $\mathbb{R}_{>0}$

The real solutions (possibly with zeros and/or poles) of $P_{III}(0, 0, 4, -4)$ on $\mathbb{R}_{>0}$ are by (15.10) the functions $f_{mult}(., s, B)|_{\mathbb{R}_{>0}}$ for $(s, B) \in V^{mat,\mathbb{R}}$. In this chapter we shall obtain complete results about the sequences of zeros and/or poles of these solutions. These global results will be derived by combining, first, the local behaviour for small x and for large x, and second, the geometry of the spaces $M^{mon}_{3FN,\mathbb{R}}$ and $M^{ini}_{3FN,\mathbb{R}} = M^{reg}_{3FN,\mathbb{R}} \cup M^{sing}_{3FN,\mathbb{R}}$, which was described in Lemma 15.4 (a) and Theorem 15.5 (a).

This kind of argument, from the local behaviour of the solutions near 0 and ∞ and from the geometry of certain moduli spaces to the global behaviour of the solutions, seems to be new in the theory of the Painlevé III equations.

Theorem 18.2 collects the known results on the behaviour of individual solutions near 0 and ∞. Theorem 18.3 develops how the zeros and/or poles behave in certain families of solutions. Theorem 18.4 derives from this the global results on the sequences of zeros and/or poles of all solutions.

The sources for the local results in Theorem 18.2 are: [MTW77], [IN86, ch. 11] and [FIKN06, Ni09, ch. 15] and Chap. 12, and Chap. 17 which builds on [HS07, Mo11b, HS11]. Some of the local results can be derived from several sources. The proof of Theorem 18.2 will explain this. Theorem 18.2 has a qualitative character and does not rewrite the precise asymptotic formulae in [IN86, Ni09, ch. 11] and Chap. 12. It is a feature of the argument leading to Theorem 18.4 that it does not depend on precise asymptotic formulae.

NOTATION 18.1 The following notation allows a concise formulation of the local and global results about sequences of zeros and/or poles of the real solutions of $P_{III}(0, 0, 4, -4)$ on $\mathbb{R}_{>0}$. Recall that there are two types of zeros and two types of poles. If x_0 is a zero (respectively, a pole) of a solution f then $\partial_x f(x_0) = \pm 2$ (respectively, $\partial_x(f^{-1})(x_0) = \pm 2$). A zero x_0 with $\partial_x f(x_0) = \pm 2$ is denoted $[0\pm]$, a pole x_0 with $\partial_x(f^{-1})(x_0) = \pm 2$ is denoted $[\infty\pm]$.

© Springer International Publishing AG 2017
M.A. Guest, C. Hertling, *Painlevé III: A Case Study in the Geometry of Meromorphic Connections*, Lecture Notes in Mathematics 2198, DOI 10.1007/978-3-319-66526-9_18

For a pair of subsequent zeros or poles of a solution f, the value of f is positive between them if they are of one of the types

$$[0+][0-], \ [0+][\infty-], \ [\infty+][0-], \ [\infty+][\infty-]$$

and negative if they are of one of the types

$$[0-][0+], \ [0-][\infty+], \ [\infty-][0+], \ [\infty-][\infty+].$$

Other pairs are not possible.

Let f be a real solution (possibly with zeros and/or poles) of $P_{III}(0,0,4,-4)$ on $\mathbb{R}_{>0}$. Then f is of type $\overleftarrow{>}\,0$ (respectively, $\overleftarrow{<}\,0$) if a $y_0 \in \mathbb{R}_{>0}$ exists such that $f(y) > 0$ (respectively, $f(y) < 0$) for all $y \in (0, y_0)$. Then also $f|_{(0,y_0)}$ is called of type $\overleftarrow{>}\,0$ (respectively, $\overleftarrow{<}\,0$).

f is of type $\overleftarrow{[0+][0-]}$ if a $y_0 \in \mathbb{R}_{>0}$ exists such that $f(y_0) \neq 0$, f has infinitely many zeros or poles in $(0, y_0)$ which are denoted $x_1, x_2, x_3, \dots$ with $x_1 > x_2 > x_3 > \dots$, and x_k is of type $[0-]$ for odd k and of type $[0+]$ for even k. Then also $f|_{(0,y_0]}$ is called of type $\overleftarrow{[0+][0-]}$. The types $\overleftarrow{[0-][0+]}$, $\overleftarrow{[\infty+][\infty-]}$ and $\overleftarrow{[\infty-][\infty+]}$ are defined analogously. Of course f is of type $\overleftarrow{[0+][0-]}$ if it is of type $\overleftarrow{[0-][0+]}$. But for $f|_{(0,y_0]}$ the type of the largest zero is important.

For large x the types $\overrightarrow{>}\,0$, $\overrightarrow{<}\,0$, $\overrightarrow{[0+][\infty-]}$, $\overrightarrow{[\infty-][0+]}$, $\overrightarrow{[0-][\infty+]}$ and $\overrightarrow{[\infty+][0-]}$ are defined analogously. For example $f|_{[y_0,\infty)}$ and f are of type $\overrightarrow{[0+][\infty-]}$ if $f(y_0) \neq 0$, if f has infinitely many zeros and poles in (y_0, ∞) which are denoted $x_1, x_2, x_3, \dots$ with $x_1 < x_2 < x_3 < \dots$ and if x_k is of type $[0+]$ for odd k and of type $[\infty-]$ for even k.

It will turn out that only the types defined here are relevant.

Theorem 18.2 collects known results on the zeros and poles of single real solutions f of $P_{III}(0,0,4,-4)$ on $\mathbb{R}_{>0}$.

Theorem 18.2 *Fix a real solution $f = f_{mult}(., s, B)|_{\mathbb{R}_{>0}}$ of $P_{III}(0,0,4,-4)$ on $\mathbb{R}_{>0}$ and its monodromy data $(s, B) \in V^{mat,\mathbb{R}}$.*

(a) *f has no zeros or poles near $\infty \iff B = \pm 1_2$.*

   More precisely, f is of type $\overrightarrow{>}\,0 \iff B = 1_2$, and f is of type $\overrightarrow{<}\,0 \iff B = -1_2$.

(b) *If $B \neq \pm 1_2$ then for any $y_0 \in \mathbb{R}_{>0}$ $f|_{(y_0,\infty)}$ has infinitely many zeros and/or poles. This case will be described more precisely in Theorem 18.4(a).*

(c) *f has no zeros or poles near $0 \iff |s| \leq 2$.*

   More precisely, f is of type $\overleftarrow{>}\,0 \iff |s| \leq 2$ and $b_1 + \frac{1}{2}sb_2 \geq 1$, and f is of type $\overleftarrow{<}\,0 \iff |s| \leq 2$ and $b_1 + \frac{1}{2}sb_2 \leq -1$.

(d) *If $s > 2$ then f is of type $\overleftarrow{[0+][0-]}$ (thus also of type $\overleftarrow{[0-][0+]}$).*

   If $s < -2$ then f is of type $\overleftarrow{[\infty+][\infty-]}$ (thus also of type $\overleftarrow{[\infty-][\infty+]}$).

(e) If $|s| \leq 2$ and $B = \mathbf{1}_2$ then f has no zeros or poles on $\mathbb{R}_{>0}$ and is positive on $\mathbb{R}_{>0}$.

   If $|s| \leq 2$ and $B = -\mathbf{1}_2$ then f has no zeros or poles on $\mathbb{R}_{>0}$ and is negative on $\mathbb{R}_{>0}$.

Proof (a)+(b) Of course, part (a) implies part (b). Part (a) has at least two different sources, which will be described in the following parts (I) and (II).

(I) One source which gives the complete result is the equivalence between nilpotent orbits of TERP structures and mixed TERP structures from [Mo11b, Corollary 8.15], [HS07, Theorem 9.3] (the relevant special case is reformulated in Theorem 17.11) together with the relation between real solutions of $P_{III}(0,0,4,-4)$ on $\mathbb{R}_{>0}$ and Euler orbits of TERP(0) bundles in Chap. 17 (Remarks 17.2 (vi) and 17.4 (i)). Corollary 17.10 gives the final result

$$f_{mult}(.,s,B) \text{ is of type } \overrightarrow{>0} \iff B = \mathbf{1}_2. \tag{18.1}$$

The symmetry R_2 gives

$$f_{mult}(.,s,B) = -f_{mult}(.,R_2(s,B)) = -f_{mult}(.,s,-B),$$

thus

$$f_{mult}(.,s,B) \text{ is of type } \overrightarrow{<0} \iff B = -\mathbf{1}_2. \tag{18.2}$$

This establishes part (a)

(II) Another source is the combination of [MTW77] and [IN86, ch. 2,11] (or [FIKN06, ch. 15]). However, a priori it gives the result in part (a) only for solutions $f_{mult}(.,s,B)|_{\mathbb{R}_{>0}}$ with $(s,B) \in V^{mat,a \cup c,\mathbb{R}}$ because in [IN86, ch. 11] only such solutions are considered.

In [MTW77], only those real solutions of $P_{III}(0,0,4,-4)$ are studied which are of type $\overrightarrow{>0}$ or type $\overrightarrow{<0}$. In [IN86, ch. 2] isomonodromic families of P_{3D6} bundles are associated to all complex solutions of $P_{III}(0,0,4,-4)$ on $\mathbb{R}_{>0}$. In [IN86, ch. 11] it is stated that for $(s,B) \in V^{mat,a \cup c,\mathbb{R}}$ a solution $f_{mult}(.,s,B)$ is considered in [MTW77] if and only if $B = \pm\mathbf{1}_2$, and the zeros and poles for large x of the other solutions with $(s,B) \in V^{mat,a \cup c,\mathbb{R}}$ are studied. If the arguments in [IN86, ch. 11] work without the restriction to $(s,B) \in V^{mat,a \cup c,\mathbb{R}}$, which they probably do, they give also a proof of part (a).

(c)+(d) For (c) and (d) the best source consists of the asymptotic formulae from [Ni09], which are reformulated in Theorem 12.4 in the formulae (12.21), (12.22), (12.24) and (12.28). They cover $(s,B) \in V^{mat,a \cup b + \cup c+,\mathbb{R}}$. For $(s,B) \in V^{mat,b - \cup c-}$ one needs (12.24) and (12.28) and the symmetry R_1 with

$$f_{mult}^{-1}(.,s,B) = f_{mult}(.,R_1(s,B)) = f_{mult}\left(.,-s,\begin{pmatrix}1&0\\0&-1\end{pmatrix}B\begin{pmatrix}1&0\\0&-1\end{pmatrix}\right).$$

The formulae (12.24) and (12.28) give the type of $f_{mult}(.,s,B)|_{\mathbb{R}_{>0}}$ near 0 immediately. In the case of the formulae (12.21) and (12.22) one has $s \in (-2,2)$ and thus $\alpha_- \in (-\frac{1}{2}, \frac{1}{2})$, so $\kappa_{0,1} > 0$, and for $\delta_2 \in \{\pm 1\}$ one has

$$b_1 + \tfrac{1}{2}sb_2 \in \delta_2 \mathbb{R}_{\geq 1} \iff b_- \in \delta_2 \mathbb{R}_{>0}.$$

This establishes parts (c) and (d).

An alternative would be to apply the equivalence between Sabbah orbits of TERP structures and certain polarized mixed Hodge structures in [HS07, Theorem 7.3], see Remarks 17.4. But this would require quite some additional work. For $|s| \leq 2$ one would have to show that the TERP(0) bundles which are associated to $f_{mult}(.,s,B)|_{\mathbb{R}_{>0}}$ induce a certain PMHS. For $|s| > 2$ they do not because the monodromy Mon does not have eigenvalues in S^1, and thus cannot be an automorphism of a PMHS. But for $|s| > 2$ the finer information that $f_{mult(.,s,B)]}$ is of type $\overleftarrow{[0+][0-]}$ or of type $\overleftarrow{[\infty+][\infty-]}$ does not follow from [HS07, Theorem 7.3]. For these reasons we do not carry out the alternative method.

(e) The necessity of the conditions $|s| \leq 2$ and $B = \pm 1_2$ for the smoothness of a solution $f_{mult}(.,s,B)|_{\mathbb{R}_{>0}}$ follows from (a) and (c). For the sufficiency there are several sources. These will be discussed in the following points (I) to (IV).

(I)   In [MTW77] the smooth solutions $f_{mult}(.s,B)$ for $|s| \leq 2$ and $B = \pm 1_2$ are established, though without identifying (s,B). Another way to establish them is given in [Wi00]. Reference [IN86, ch. 11] gives a criterion in terms of monodromy data which amounts to $|s| \leq 2$ and $B = \pm 1_2$.

(II) Reference [HS11, Theorem 5.9] constructs certain pure and polarized TERP(0) bundles. A special case is reformulated in Theorem 17.11. In particular, the TERP(0) bundles for (s,B) with $|s| \leq 2$ and $B = 1_2$ are pure and polarized. This implies (Corollary 17.12) that for such (s,B) the solution $f_{mult}(.,s,B)$ is smooth and positive on $\mathbb{R}_{>0}$. From the symmetry R_2 one obtains that the solution is smooth and negative in the case $|s| \leq 2$ and $B = -1_2$.

(III) [GL14, theorem 3.1] for the case $w_0 + w_1 = 0$ gives an elementary proof using pde theory.

(IV) Consider the projection $pr_{mat} : M_{3FN,\mathbb{R}} \to V^{mat,\mathbb{R}}$. For $V = V^{mat,a,\mathbb{R}}(B = \pm 1_2)$ (and later also for $V = V^{mat,b,\mathbb{R}}(B = \pm 1_2)$) we want to show that the set $M_{3FN,\mathbb{R}}^{sing} \cap pr_{mat}^{-1}(V)$ is empty. If it were not empty, parts (a) and (c) and the arguments in the proof of Theorem 18.3 (b) would show that the restricted projection

$$pr_{mat} : M_{3FN,\mathbb{R}}^{sing} \cap pr_{mat}^{-1}(V) \to V$$

is a covering with finitely many sheets.

But the existence of the two special smooth solutions $f_{mult}(.,0,\pm 1_2) = \pm 1$ from Remark 10.5 excludes this. Therefore the intersection above is empty.

For $V = V^{mat,b,\mathbb{R}}(B = \pm 1_2)$, $pr_{mat}^{-1}(V)$ is in the closure of $pr_{mat}^{-1}(V^{mat,a,\mathbb{R}}(B = \pm 1_2))$, therefore also $M_{3FN,\mathbb{R}}^{sing} \cap pr_{mat}^{-1}(V)$ is empty.

□

The isomorphism in Theorem 15.5 (a) (iv) between semi-algebraic manifolds is now called

$$\psi_{mat} : M^{mon}_{3FN,\mathbb{R}} \to \mathbb{R}_{>0} \times V^{mat,\mathbb{R}},$$

and the projection to $V^{mat,\mathbb{R}}$ is called

$$pr_{mat} : M^{mon}_{3FN,\mathbb{R}} \to V^{mat,\mathbb{R}}.$$

Part (a) of the following Theorem 18.3 gives a stratification of $V^{mat,\mathbb{R}}$, part (b) states that pr_{mat} restricts above the strata to coverings from $M^{sing}_{3FN,\mathbb{R}}$ and numbers the sheets, part (d) connects different strata, the parts (b)+(c)+(e)+(f) say how the sheets of the coverings glue.

Theorem 18.3

(a) *The decomposition of $V^{mat,\mathbb{R}}$ into the three subsets*

$$V^{mat,\mathbb{R}} = V^{mat,a,\mathbb{R}} \cup V^{mat,b,\mathbb{R}} \cup V^{mat,c,\mathbb{R}}$$

in Theorem 15.5 (a) (ii) is refined by the conditions $B = \pm 1_2$ (respectively, $B \neq \pm 1_2$) into a decomposition into the six subsets

$$V^{mat,\mathbb{R}} = V^{mat,a,\mathbb{R}}(B = \pm 1_2) \cup V^{mat,a,\mathbb{R}}(B \neq \pm 1_2) \qquad (18.3)$$
$$\cup \; V^{mat,b,\mathbb{R}}(B = \pm 1_2) \cup V^{mat,b,\mathbb{R}}(B \neq \pm 1_2)$$
$$\cup \; V^{mat,c,\mathbb{R}}(B = \pm 1_2) \cup V^{mat,c,\mathbb{R}}(B \neq \pm 1_2).$$

They have 2, 4, 4, 4, 4, 8 components respectively (cf. the two pictures in Theorem 15.5 (a) (iii)).

(b) *If V is the set $V^{mat,a,\mathbb{R}}(B = \pm 1_2)$ or the set $V^{mat,b,\mathbb{R}}(B = \pm 1_2)$, then $M^{sing}_{3FN,\mathbb{R}} \cap pr^{-1}_{mat}(V)$ is empty. If V is one of the other four sets in (a), then the restricted projection*

$$pr_{mat} : M^{sing}_{3FN,\mathbb{R}} \cap pr^{-1}_{mat}(V) \to V$$

is a covering.

The sheets of the coverings and the zeros and poles of any solution $f_{mult}(.,s,B)|_{\mathbb{R}_{>0}}$ are indexed by $\mathbb{Z}$ or $\mathbb{Z}_{\geq 1}$ or $\mathbb{Z}_{\leq 0}$ such that $x_k < x_l \iff k < l$ for zeros/poles x_k and x_l. The second column in the following table lists the zeros and poles of a solution $f_{mult}(.,s,B)|_{\mathbb{R}_{>0}}$ for (s,B) in the set in the first column:

$$
\begin{array}{c|c}
V^{mat,a,\mathbb{R}}(B \neq \pm 1_2) & (x_k(s,B))_{k \in \mathbb{Z}_{\geq 1}} \\
V^{mat,c,\mathbb{R}}(B \neq \pm 1_2) & (x_k(s,B))_{k \in \mathbb{Z}_{\geq 1}} \\
V^{mat,b,\mathbb{R}}(B \neq \pm 1_2) & (x_k(s,B))_{k \in \mathbb{Z}} \\
V^{mat,b,\mathbb{R}}(B = \pm 1_2) & (x_k(s,B))_{k \in \mathbb{Z}_{\leq 0}}
\end{array} \qquad (18.4)
$$

The indexing is unique for the three sets with index sets $\mathbb{Z}_{\geq 1}$ or $\mathbb{Z}_{\leq 0}$. It is unique up to a shift for each of the four components of the set $V^{mat,b,\mathbb{R}}(B \neq \pm 1_2)$ with index set $\mathbb{Z}$. There it is fixed by the requirement that the sheets over $V^{mat,b,\mathbb{R}}(B = 1_2)$ with indices $k \in \mathbb{Z}_{\leq 0}$ glue with the sheets with the same indices k over each of the four components of the set $V^{mat,b,\mathbb{R}}(B \neq \pm 1_2)$.

(c) *For $\delta_1, \delta_3 \in \{\pm 1\}$ the sheet with index $k \in \mathbb{Z}_{\leq 0}$ over the component $V^{mat,b\delta_1,\mathbb{R}}(B = -1_2)$ glues to the sheet with index $k - \delta_1\delta_3$ over the component $V^{mat,b\delta_1,\mathbb{R}}(ib_2 \in \delta_3\mathbb{R}_{>0})$.*

(d) *There are continuous maps*

$$\gamma_1 : V^{mat,a,\mathbb{R}} \to \mathbb{R}_{>0},$$

$$\gamma_2 : V^{mat,c,\mathbb{R}} \to \mathbb{R}_{>0},$$

$$\gamma_3 : V^{mat,\mathbb{R}}(B = \pm 1_2) \to \mathbb{R}_{>0},$$

such that

$$\{(x,s,B) \,|\, x < \gamma_1(s,B), (s,B) \in V^{mat,a,\mathbb{R}}\} \cap \psi_{mat}(M^{sing}_{3FN,\mathbb{R}}) = \emptyset, \qquad (18.5)$$

$$\{(x,s,B) \,|\, x < \gamma_2(s,B), (s,B) \in V^{mat,c,\mathbb{R}}\} \cap \psi_{mat}(M^{sing}_{3FN,\mathbb{R}}) = \emptyset, \qquad (18.6)$$

$$\{(x,s,B) \,|\, x > \gamma_3(s,B), B \in \{\pm 1\}\} \cap \psi_{mat}(M^{sing}_{3FN,\mathbb{R}}) = \emptyset. \qquad (18.7)$$

(e) *For $\delta_1, \delta_2, \delta_3 \in \{\pm 1\}$ the sheet with index $k \in \mathbb{Z}_{\geq 1}$ over the component $V^{mat,c\delta_1,\mathbb{R}}(b_1 + \frac{s}{2}b_2 = \delta_2, ib_2 \in \delta_3\mathbb{R}_{>0})$ glues to the sheet with index k over the component $V^{mat,a,\mathbb{R}}(b_1 + \frac{1}{2}sb_2 \in \delta_2\mathbb{R}_{\geq 1}, ib_2 \in \delta_3\mathbb{R}_{>0})$.*

(f) *It glues to the sheet with index $k - \frac{1-\delta_2}{2}\delta_1\delta_3$ over the component $V^{mat,b\delta_1,\mathbb{R}}(ib_2 \in \delta_3\mathbb{R}_{>0})$.*

(g) *Gluing all the sheets over the 26 components of the six sets in (a) gives precisely the four smooth hypersurfaces $M^{sing}_{3FN,\mathbb{R}} = \cup_{k=0}^{3} M^{[k]}_{3FN,\mathbb{R}}$ in $M^{ini}_{3FN,\mathbb{R}}$ of types $[0-], [\infty-], [0+], [\infty+]$ for $k = 0, 1, 2, 3$.*

Proof

(a) This is essentially a definition. The 26 components can be seen in the picture in Theorem 15.5 (a) (iii).

(b) If $V = V^{mat,a,\mathbb{R}}(B = \pm 1_2)$ or $V = V^{mat,c,\mathbb{R}}(B = \pm 1_2)$, then $M^{sing}_{3FN,\mathbb{R}} \cap pr^{-1}_{mat}(V) = \emptyset$ because of Theorem 18.2 (e).

Let V be one of the other four sets in (a). The hypersurface $M^{sing}_{3FN,\mathbb{R}} \subset M^{mon}_{3FN,\mathbb{R}}$ is real analytic and smooth with four components. Because the functions $f_{mult}(.,s,B)$ have only simple zeros and poles, the hypersurface $M^{sing}_{3FN,\mathbb{R}}$ is everywhere transversal to the fibres of the projection pr_{mat}.

Therefore, in order to show that $pr_{mat} : M^{sing}_{3FN,\mathbb{R}} \cap pr^{-1}_{mat}(V) \to V$ is a covering, it is sufficient to show that the following data do not exist:

a C^∞ path $\gamma : [0, 1] \to V$ and a C^∞ lift of $\gamma : |_{[0,1)}$

$$\widetilde{\xi} = (\xi, \gamma) : [0, 1) \to \psi_{mat}(M^{sing}_{3FN,\mathbb{R}}) \cap \mathbb{R}_{>0} \times V \tag{18.8}$$

such that for $r \to 1$, $\xi(r)$ tends to 0 or ∞.

1st case: Suppose that such data exist with $\lim_{r \to 1} \xi(r) = 0$.
1st subcase: $V = V^{mat,b,\mathbb{R}}(B = \pm 1_2)$ or $V = V^{mat,b,\mathbb{R}}(B \neq \pm 1_2)$.

By Theorem 18.2 (d) for any $r \in [0, 1]$ the set $\psi_{mat}(M^{sing}_{3FN,\mathbb{R}}) \cap \mathbb{R}_{>0} \times \{\gamma(r)\}$ is discrete and contains points with x-values arbitrarily close to 0.

Choose any point $(\eta(1), \gamma(1)) \in \psi_{mat}(M^{sing}_{3FN,\mathbb{R}}) \cap \mathbb{R}_{>0} \times \{\gamma(1)\}$. It extends to a C^∞ lift $\widetilde{\eta} = (\eta, \gamma) : [1 - \varepsilon, 1] \to \psi_{mat}(M^{sing}_{3FN,\mathbb{R}}) \cap \mathbb{R}_{>0} \times \gamma([1 - \varepsilon, 1])$ of $\gamma : [1 - \varepsilon, 1] \to V$ for some $\varepsilon > 0$.

For $r \in [1 - \varepsilon, 1)$, there can be only finitely many points in $\psi_{mat}(M^{sing}_{3FN,\mathbb{R}}) \cap \mathbb{R}_{>0} \times \{\gamma(r)\}$ between $\widetilde{\eta}(r)$ and $\widetilde{\xi}(r)$. But only their continuations to $r = 1$ along the path γ can be in $\psi_{mat}(M^{sing}_{3FN,\mathbb{R}}) \cap \mathbb{R}_{>0} \times \{\gamma(1)\}$ and lie with respect to their x-values under $\widetilde{\eta}(1) = (\eta(1), \gamma(1))$. This contradicts Theorem 18.2 (d).

2nd subcase: $V = V^{mat,a,\mathbb{R}}(B \neq \pm 1_2)$ or $V = V^{mat,c,\mathbb{R}}(B \neq \pm 1_2)$.

The asymptotic formulae (12.21), (12.22) and (12.28) in Theorem 12.4 for $f_{mult}(., s, B)|_{\mathbb{R}_{>0}}$ for $x \to 0$ depend real analytically on $(s, B) \in V$. Therefore the data (18.8) do not exist. The 2nd subcase also leads to a contradiction. Hence the 1st case is impossible.

2nd case: Suppose that the data (18.8) exist with $\lim_{r \to 1} \xi(r) = \infty$.
1st subcase: $V = V^{mat,a,\mathbb{R}}(B \neq \pm 1_2)$ or $V = V^{mat,c,\mathbb{R}}(B \neq \pm 1_2)$. or $V = V^{mat,b,\mathbb{R}}(B \neq \pm 1_2)$.

By Theorem 18.2 (b) for any $r \in [0, 1]$ the set $\psi_{mat}(M^{sing}_{3FN,\mathbb{R}}) \cap \mathbb{R}_{>0} \times \{\gamma(r)\}$ is discrete and contains points with arbitrarily large x-values. The same arguments as in the 1st subcase of the 1st case lead to a contradiction.

2nd subcase: $V = V^{mat,b,\mathbb{R}}(B = \pm 1_2)$.

Compare the proof (i) of Theorem 18.2 (a). Choose $(s, B) = (s, 1_2) \in V$ and consider the associated Euler orbit of TERP(0) bundles. By [HS07, Theorem 9.3 (2)] (see also the proof of Corollary 17.10) it is a nilpotent orbit, so there is a lower bound $x_{bound}(s)$ such that the TERP(0) bundles for $x > x_{bound}(s)$ are pure and polarized and thus $f_{mult}(x, s, 1_2) > 0$.

We claim that the bound $x_{bound}(s)$ can be chosen so that it depends continuously on s. This is not explicitly stated in [HS07, Theorem 9.3 (2)], but it follows from its proof. In the semisimple case, which is the only relevant case here, that proof simplifies considerably. In our rank 2 case one just needs that for all $z \in \mathbb{C}^*$ with

$\arg(z) \in (-\varepsilon, \varepsilon)$ for some $\varepsilon > 0$ the matrix

$$\begin{pmatrix} e^{-\frac{x}{z}-xz} & \\ & e^{\frac{x}{z}+xz} \end{pmatrix} \begin{pmatrix} 1 & s \\ 0 & 1 \end{pmatrix} \begin{pmatrix} e^{-\frac{x}{z}-xz} & \\ & e^{\frac{x}{z}+xz} \end{pmatrix} = \begin{pmatrix} 1 & se^{-\frac{2x}{z}-2xz} \\ 0 & 1 \end{pmatrix}$$

is sufficiently close to the unit matrix $\mathbf{1}_2$. For this one can find a lower bound $x_{bound}(s)$ for x which depends continuously on s.

As $f_{mult}(x, s, \mathbf{1}_2) > 0$ for $x > x_{bound}(s)$, data as in (18.8) with $\lim_{r \to 1} \xi(r)_\infty$ cannot exist.

The case $B = -\mathbf{1}_2$ can be reduced to the case $B = \mathbf{1}_2$ with the symmetry R_2. The 2nd subcase also leads to a contradiction. So the 2nd case is impossible.

Therefore the data (18.8) do not exist. Therefore $pr_{mat} : M_{3FN,\mathbb{R}}^{sing} \cap pr_{mat}^{-1}(V) \to V$ is a covering.

(c) Formula (12.25) in Theorem 12.4 gives approximations for the x-values of the sheets above $V^{mat,b\delta_1,\mathbb{R}}$ with indices $k \ll 0$. But beware that the k in (12.25) differs by the sign and a shift from the index k of a sheet.

    If s is fixed and b_- goes once around 0 in the positive direction, then b_- must be replaced by $b_- e^{2\pi i}$, δ^{NI} must be replaced by $\delta^{NI} + 2\pi$, and the sheet with index k turns into the sheet with index $k - 2$. This establishes (c) for $\delta_1 = 1$. For $\delta_1 = -1$ one applies the symmetry R_1 with $R_1(f_{mult}) = f_{mult}^{-1}$, $R_1(s) = -s$, $R_s(b_-) = b_-^{-1}$; see table (12.13).

(d) The existence of γ_1 with (18.5) and of γ_2 with (18.6) follows from the arguments in the 2nd subcase of the 1st case in the proof of (b).

    The existence of γ_3 with (18.7) follows from the arguments in the 2nd subcase of the 2nd case in the proof of (b).

(e) Choose any $x \in \mathbb{R}_{>0}$, consider the point

$$(x, 2\delta_1, \delta_2 \mathbf{1}_2) \in \mathbb{R}_{>0} \times V^{mat,c\delta_1,\mathbb{R}}(B = \delta_2 \mathbf{1}_2)$$

$$\subset \psi_{mat}(M_{3FN,\mathbb{R}}^{reg}) \quad (\Leftarrow \text{Theorem 18.2 (e))},$$

and choose a small open neighbourhood $U \subset \psi_{mat}(M_{3FN,\mathbb{R}}^{reg})$ of this point. Because of $\mathbb{R}_{>0} \times V^{mat,a\cup c\delta_1,\mathbb{R}}(B = \delta_2 \mathbf{1}_2) \subset \psi_{mat}(M_{3FN,\mathbb{R}}^{sing})$ ($\Leftarrow$ Theorem 18.2 (e)) and because of (18.5) and (18.6) all values $x_k(., s, B)$ for $(s, B) \in V^{mat,a\cup c\delta_1,B}(b_1 + \frac{s}{2}b_2 \in \delta_2\mathbb{R}_{\geq 1}, ib_2 \in \delta_3\mathbb{R}_{>0})$ tend along the lines $s = $ constant for $(b_1, b_2) \to (\delta_2, 0)$ to ∞.

    Therefore the parts over $pr_{mat}(U)$ of all sheets over $V^{mat,c\delta_1,\mathbb{R}}(\widetilde{b}_1 = \delta_2, ib_2 \in \delta_3\mathbb{R}_{>0})$ as well as the parts over $pr_{mat}(U)$ of all sheets over $V^{mat,a,\mathbb{R}}(b_1\frac{s}{2}b_2 \in \delta_2\mathbb{R}_{\geq 1}, ib_2 \in \delta_3\mathbb{R}_{>0})$ lie with respect to their x-values above U.

    Therefore the parts over $pr_{mat}(U)$ of the sheets with the same indices glue, therefore the sheets with the same indices glue globally.

(f) By (b)+(c), the sheet over $V^{mat,b\delta_1,\mathbb{R}}(ib_2 \in \delta_3\mathbb{R}_{>0})$ with index $k - \frac{1-\delta_2}{2}\delta_1\delta_3$ for $k \in \mathbb{Z}_{\leq 0}$ glues to the sheet over $V^{mat,b\delta_1,\mathbb{R}}(B = \delta_2\mathbf{1}_2)$ with index k.

As there are no sheets with indices $k \in \mathbb{Z}_{\geq 1}$ over $V^{mat,b\delta_1,\mathbb{R}}(B = \delta_2 \mathbf{1}_2)$, the values $x_k(s,B)$ of the sheets over $V^{mat,b\delta_1,\mathbb{R}}(ib_2 \in \delta_3 \mathbb{R}_{>0})$ with indices $k \geq 1 - \frac{1-\delta_2}{2}\delta_1\delta_3$ tend along all lines $s = $ constant for $(b_1,b_2) \to (\delta_2, 0)$ to ∞.

Therefore the parts over $pr_{mat}(U)$ of all the sheets over $V^{mat,b\delta_1,\mathbb{R}}(ib_2 \in \delta_3 \mathbb{R}_{>0})$ with indices $k \geq 1 - \frac{1-\delta_2}{2}\delta_1\delta_3$ lie with respect to their x-values above U.

On the other hand, because of (18.7) and Theorem 18.2 (e), all values $x_k(s,\delta_2\mathbf{1}_2)$ (automatically $k \leq 0$) for $(s,\delta_2\mathbf{1}_2) \in V^{mat,b\delta_1,\mathbb{R}}(B = \delta_2\mathbf{1}_2)$ tend to 0 for $s \to \delta_2 2$. Therefore the parts over $pr_{mat}(U)$ of all the sheets over $V^{mat,b\delta_1,\mathbb{R}}(B = \delta_2\mathbf{1}_2)$ lie with respect to their x-values below U.

Because the sheets over $V^{mat,b\delta_1,\mathbb{R}}(B = \delta_2\mathbf{1}_2)$ with indices k (automatically $k \leq 0$) glue to the sheets over $V^{mat,b\delta_1,\mathbb{R}}(ib_2 \in \delta_3\mathbb{R}_{>0})$ with indices $k - \frac{1-\delta_2}{2}\delta_1\delta_3$ by (b)+(c), also the parts over $pr_{mat}(U)$ of the sheets over $V^{mat,b\delta_1,\mathbb{R}}(ib_2 \in \delta_3\mathbb{R}_{>0})$ with indices $k - \frac{1-\delta_2}{2}\delta_1\delta_3$ for $k \leq 0$ lie with respect to their x-values below U.

The sheets over $V^{mat,b\delta_1,\mathbb{R}}(ib_2 \in \delta_3\mathbb{R}_{>0})$ and over $V^{mat,c\delta_1,\mathbb{R}}(\widetilde{b}_1 = \delta_2, ib_2 \in \delta_3\mathbb{R}_{>0})$ whose parts over $pr_{mat}(U)$ lie with respect to their x-values above U, must glue pairwise. This establishes (f).

(g) This is clear as the sheets are the components of $M_{3FN,\mathbb{R}}^{sing} \cap pr_{mat}^{-1}(V)$. It follows from (10.13) that $M_{3FN,\mathbb{R}}^{[k]}$ for $k = 0,1,2,3$ are, respectively, of type $[0-], [\infty-], [0+], [\infty+]$. $\square$

Theorem 18.4 *Fix a real solution* $f = f_{mult}(.,s,B)|_{\mathbb{R}_{>0}}$ *of* $P_{III}(0,0,4,-4)$ *on* $\mathbb{R}_{>0}$ *and its monodromy data* $(s,B) \in V^{mat,\mathbb{R}}$.

(a) *(A supplement to Theorem 18.2 (b)) Suppose* $B \neq \pm\mathbf{1}_2$.

 If $ib_2 < 0$ *then* f *is of type* $\overrightarrow{[0-][\infty+]}$ *(thus also of type* $\overrightarrow{[\infty+][0-]}$*).*

 If $ib_2 > 0$ *then* f *is of type* $\overrightarrow{[0+][\infty-]}$ *(thus also of type* $\overrightarrow{[\infty-][0+]}$*).*

(b) *(Global results) A* $y_0 \in \mathbb{R}_{>0}$ *exists such that* $f(y_0) \neq 0$ *and such that* $f|_{(0,y_0]}$ *and* $f|_{[y_0,\infty)}$ *are each of the type determined by the behaviour near* 0 *and* ∞. *This means there is no intermediate mixed zone.*

The following two tables list the 18 possible combinations of

$$\text{type of } f|_{(0,y_0]} \text{ \& type of } f|_{[y_0,\infty)}.$$

In each of the four cases with $|s| > 2$ *and* $B \neq \pm\mathbf{1}_2$ *there are two ways to position* y_0 *and to split the sequence of zeros and poles into a type near* 0 *and a type near* ∞. *In the other 10 cases there is only one way to position* y_0.

It follows that there are 14 possible sequences of zeros and poles. Each is realized above one of the 14 components of

$$V^{mat,a\cup c,\mathbb{R}}(B = \pm\mathbf{1}_2) \quad \text{(2 components)},$$

$$V^{mat,a\cup c,\mathbb{R}}(B \neq \pm\mathbf{1}_2) \quad \text{(4 components)},$$

$$V^{mat,b,\mathbb{R}}(B = \pm\mathbf{1}_2) \quad \text{(4 components)},$$

$$V^{mat,b,\mathbb{R}}(B \neq \pm 1_2) \quad \text{(4 components)}.$$

$|s| \leq 2$	
$B = 1_2$	$\overleftarrow{>0}$ & $\overrightarrow{>0}$
$B = -1_2$	$\overleftarrow{<0}$ & $\overrightarrow{<0}$
$b_1 + \frac{1}{2}sb_2 \geq 1, ib_2 < 0$	$\overleftarrow{>0}$ & $\overrightarrow{[\infty+][0-]}$
$b_1 + \frac{1}{2}sb_2 \geq 1, ib_2 > 0$	$\overleftarrow{>0}$ & $\overrightarrow{[0+][\infty-]}$
$b_1 + \frac{1}{2}sb_2 \leq -1, ib_2 < 0$	$\overleftarrow{<0}$ & $\overrightarrow{[0-][\infty+]}$
$b_1 + \frac{1}{2}sb_2 \leq -1, ib_2 > 0$	$\overleftarrow{<0}$ & $\overrightarrow{[\infty-][0+]}$

(18.9)

	$s > 2$	$s < -2$
$B = 1_2$	$\overleftarrow{[0-][0+]}$ & $\overrightarrow{>0}$	$\overleftarrow{[\infty-][\infty+]}$ & $\overrightarrow{>0}$
$B = -1_2$	$\overleftarrow{[0+][0-]}$ & $\overrightarrow{<0}$	$\overleftarrow{[\infty+][\infty-]}$ & $\overrightarrow{<0}$
$ib_2 < 0$	$\overleftarrow{[0-][0+]}$ & $\overrightarrow{[0-][\infty+]}$	$\overleftarrow{[\infty-][\infty+]}$ & $\overrightarrow{[0-][\infty+]}$
	or $\overleftarrow{[0+][0-]}$ & $\overrightarrow{[\infty+][0-]}$	or $\overleftarrow{[\infty+][\infty-]}$ & $\overrightarrow{[\infty+][0-]}$
$ib_2 > 0$	$\overleftarrow{[0-][0+]}$ & $\overrightarrow{[\infty-][0+]}$	$\overleftarrow{[\infty-][\infty+]}$ & $\overrightarrow{[\infty-][0+]}$
	or $\overleftarrow{[0+][0-]}$ & $\overrightarrow{[0+][\infty-]}$	or $\overleftarrow{[\infty+][\infty-]}$ & $\overrightarrow{[0+][\infty-]}$

(18.10)

Proof For each of the sheets over the 26 components of the six sets in Theorem 18.3 (a), it is crucial to determine to which of the four hypersurfaces $M^{[k]}_{3FN,\mathbb{R}}$ it belongs. First we read off from the gluing information in Theorem 18.3 (b)+(c)+(e)+(f) which sheets glue together to one of these hypersurfaces, then we will find its type.

Theorem 18.3 (e) allows one to join for $\delta_2, \delta_3 \in \{\pm 1\}$ the sheets over $V^{mat,a,\mathbb{R}}(b_1 + \frac{s}{2}b_2 \in \delta_2\mathbb{R}_{\geq 1}, ib_2 \in \delta_3\mathbb{R}_{>0})$ and $V^{mat,c,\mathbb{R}}(b_1 = \delta_2, ib_2 \in \delta_3\mathbb{R}_{>0})$ to sheets over $V^{mat,a\cup c,\mathbb{R}}(b_1 + \frac{s}{2}b_2 \in \delta_2\mathbb{R}_{\geq 1}, ib_2 \in \delta_3\mathbb{R}_{>0})$. The sheets over $V^{mat,b,\mathbb{R}}(B = \pm 1_2)$ will not be considered separately, as the sheets on both sides of them will be glued via them. The sheets are denoted as follows:

$u_k^{\delta_2}$ sheet with index k over $V^{mat,b+,\mathbb{R}}(ib_2 \in \delta_3\mathbb{R}_{>0})$, here $k \in \mathbb{Z}$,

$d_k^{\delta_2}$ sheet with index k over $V^{mat,b-,\mathbb{R}}(ib_2 \in \delta_3\mathbb{R}_{>0})$, here $k \in \mathbb{Z}$,

$l_k^{\delta_2}$ sheet with index k over $V^{mat,a\cup c,\mathbb{R}}(b_1 + \frac{s}{2}b_2 \in \mathbb{R}_{\geq 1}, ib_2 \in \delta_3\mathbb{R}_{>0})$, here $k \in \mathbb{Z}_{\leq 0}$,

$r_k^{\delta_2}$ sheet with index k over $V^{mat,a\cup c,\mathbb{R}}(b_1 + \frac{s}{2}b_2 \in \mathbb{R}_{\leq -1}, ib_2 \in \delta_3\mathbb{R}_{>0})$, here $k \in \mathbb{Z}_{\leq 0}$.

The gluing information in Theorem 18.3 (b)+(c)+(f) is rewritten for $-3 \leq k \leq 4$ in the following four chains of sheets. They extend in the obvious way to all k:

$$..u_{-3}^- \, u_{-3}^+ \, u_{-1}^- \, u_{-1}^+ \, u_1^- \, r_1^- \, d_1^- \, l_2^- \, u_3^- \, r_3^- \, d_3^- \, l_4^- ..$$
$$..u_{-2}^- \, u_{-2}^+ \, u_0^- \, u_0^+ \, l_1^+ \, d_2^+ \, r_2^+ \, u_2^+ \, l_3^+ \, d_4^+ \, r_4^+ \, u_4^+ ..$$
$$..d_{-3}^+ \, d_{-3}^- \, d_{-1}^+ \, d_{-1}^- \, d_1^+ \, r_1^+ \, u_1^+ \, l_2^+ \, d_3^+ \, r_3^+ \, u_3^+ \, l_4^+ ..$$
$$..d_{-2}^+ \, d_{-2}^- \, d_0^+ \, d_0^- \, l_1^- \, u_2^- \, r_2^- \, d_2^- \, l_3^- \, u_4^- \, r_4^- \, d_4^- ..$$

$$(18.11)$$

Each chain is one of the four hypersurfaces.

By Theorem 18.2 (d), the first two chains are the hypersurfaces of types $[0+]$ and $[0-]$, and the last two chains are the hypersurfaces of types $[\infty+]$ and $[\infty-]$. But it remains to determine which is which.

The sheets u_0^+ and u_0^- are glued via the highest sheet (it has index 0) over $V^{mat,b+,\mathbb{R}}(B = 1_2)$. Above this sheet $f_{mult}(.,s,B)$ is of type $\overrightarrow{> 0}$. Thus $f_{mult}(.,s,B)$ is positive right above u_0^+ and u_0^-. Therefore the hypersurface which contains u_0^+ and u_0^- is of type $[0+]$, and the hypersurface which contains u_1^- and u_{-1}^+ is of type $[0-]$. Analogously, the hypersurface which contains d_0^+ and d_0^- is of type $[\infty+]$, and the hypersurface which contains d_{-1}^- and d_1^+ is of type $[\infty-]$.

The following table lists for each of the 14 components in Theorem 18.4 (b), above (or under) which sheet the value y_0 has to be positioned so that $f|_{(0,y_0]}$ and $f|_{[y_0,\infty)}$ are of one of the types in the table. For the first 10 components there is only one possibility, for the last four components there are two possibilities.

$V^{mat,a\cup c,\mathbb{R}}(B = \pm 1_2)$ (2 components): y_0 arbitrary in $(0,\infty)$
$V^{mat,a\cup c,\mathbb{R}}(B \neq \pm 1_2)$ (4 components): y_0 under the lowest sheet
$V^{mat,b\delta_1,\mathbb{R}}(B = \delta_2 1_2)$ (4 components): y_0 above the highest sheet
$V^{mat,b\delta_1,\mathbb{R}}(ib_2 \in \delta_3\mathbb{R}_{>0})$ (4 components): y_0 above the sheet

u_0^+	u_0^-	d_0^+	d_0^-
or u_{-1}^+	or u_1^-	or d_1^+	or d_{-1}^-
for $(\delta_1, \delta_3) = (1, 1)$	$(1, -1)$	$(-1, 1)$	$(-1, -1)$

This proves Theorem 18.4 (b). It also proves Theorem 18.4 (a).   □

Remarks 18.5

(i) The following three pictures give an idea of the gluing of the sheets and of
the sequences of zeros and/or poles of the real solutions of $P_{III}(0,0,4,-4)$ on
$\mathbb{R}_{>0}$.

  The first picture is a sketch of the restriction to some fixed value $s_0 > 2$ of
$M^{mon}_{3FN,\mathbb{R}}$, that is

$$M^{mon}_{3FN,\mathbb{R}}|_{s=s_0} \cong \mathbb{R}_{>0} \times V^{mat,\mathbb{R}}|_{s=s_0} \cong \mathbb{R}_{>0} \times S^1 \quad \text{(cf. (15.8)).}$$

Here $V^{mat,\mathbb{R}}|_{s=s_0} \cong S^1$ is cut open at $b_- = i$ (i.e. $b_1 + \frac{1}{2}s_0 b_2 = 0$, $b_2 = i/\sqrt{\frac{1}{4}s^2 - 1}$—see (5.24)). In the white region f is positive, in the shaded region f is negative. The lines are the intersections of the glued sheets, i.e. of the four hypersurfaces, with $M^{mon}_{3FN,\mathbb{R}}|_{s=s_0}$.

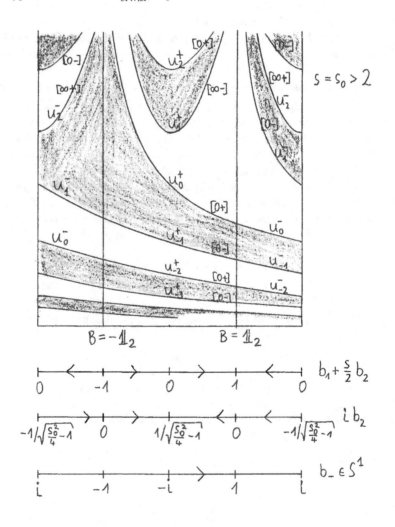

The second picture is a sketch of the restriction to some value $s_1 \in [-2,2]$ of $M_{3FN,\mathbb{R}}^{mon}$, that is

$$M_{3FN,\mathbb{R}}^{mon}|_{s=s_1} \cong \mathbb{R}_{>0} \times V^{mat,\mathbb{R}}|_{s=s_1} \cong \mathbb{R}_{>0} \times \mathbb{R}^* \quad \text{(cf. (15.7))}.$$

It has two components. The symmetry R_2 is a horizontal shift—it maps the two components to one another and exchanges white and shaded regions and the types $[0+] \leftrightarrow [0-]$ and $[\infty+] \leftrightarrow [\infty-]$.

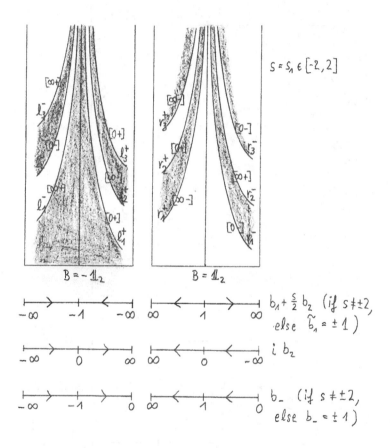

The third picture is a sketch of the restriction to the value $s = s_2 = -s_0 < -2$ of M_{3FN}^{mon}. It is obtained from the first picture using the symmetry R_1, which acts as reflection along the line $B = 1_2$. It preserves white and shaded regions, but exchanges the types $[0+] \leftrightarrow [\infty+]$ and $[0-] \leftrightarrow [\infty-]$.

The symmetry R_2 is visible in the first and the third pictures as a horizontal shift of half the length of the S^1, which exchanges white and shaded regions and the types $[0+] \leftrightarrow [0-]$ and $[\infty+] \leftrightarrow [\infty-]$.

(ii) The pictures are only sketches. All the lines are drawn as convex graphs, which is what we expect. Conjecture 18.6 (a) makes this expectation more precise and extends it.

(iii) The two regions in $M_{3FN,\mathbb{R}}^{mon}$ where f_{mult} is positive (white region) or negative (shaded region) are the two components of $M_{3FN,\mathbb{R}}^{reg} \cong \mathbb{R}_{>0} \times \mathbb{R}^* \times \mathbb{R}$, so they are connected and contractible.

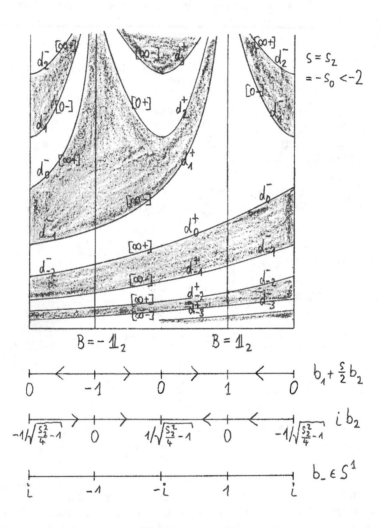

In the picture for $s = s_0 > 2$ the lower white and shaded strips, that is, the strips between lines of types [0+] and [0−], are glued by gluing the two vertical bounding lines where $b_- = i$. The movement to lower and lower strips in $M^{mon}_{3FN,\mathbb{R}}|_{s=s_0}$ within one region projects in $V^{mat,\mathbb{R}}$ to a positive winding around the hole with $s = \infty$. The same holds for the lower strips in the picture for $s = s_2 = -s_0 < -2$ and the hole with $s = -\infty$. Observe that, for $s = s_2$, $b_- \in S^1$ winds in the negative direction around the hole $s = -\infty$.

But for the upper strips one needs all three pictures. It is necessary to alternate two types of moves, moving x_0 appropriately so that one stays in the white or the shaded region:

(α) for fixed $ib_2 \in (-1/\sqrt{\frac{1}{4}s_0^2 - 1}, 1/\sqrt{\frac{1}{4}s_0^2 - 1})$ move between $s = s_0$ and $s = s_2 = -s_0$,

(β) for fixed $s = \pm s_0$ move between points near $B = 1_2$ and points near $B = -1_2$.

Together this gives two movements in $M^{mon}_{3FN,\mathbb{R}}$ whose projections to $V^{mat,\mathbb{R}}$ are windings in the positive direction around the hole with $ib_2 = \infty$ or the hole with $ib_2 = -\infty$.

Parts (b) and (c) of the following conjecture would allow us to make all these movements within $\{x_0\} \times V^{mat,\mathbb{R}}$ for any fixed $x_0 \in \mathbb{R}_{>0}$.

Conjecture 18.6

(a) The components of $\psi_{mat}(M^{sing}_{3FN,\mathbb{R}}) \cap \mathbb{R}_{>0} \times V^{mat,a\cup c,\mathbb{R}}$ and of $\psi_{mat}(M^{sing}_{3FN,\mathbb{R}}) \cap \mathbb{R}_{>0} \times V^{mat,b,\mathbb{R}}$ are convex surfaces, where $V^{mat,a\cup c,\mathbb{R}}$ is reparametrized by $(s, \delta_2, r) \in [-2, 2] \times \{\pm 1\} \times (-1, 1)$ with $b_-/|b_-| = \delta_2$, $ib_2 = r/(1 - r^2)$, and $V^{mat,b,\mathbb{R}}$ is reparametrized by $(s, r) \in (\mathbb{R}_{<-2} \cup \mathbb{R}_{>2}) \times (\frac{-3\pi}{2}, \frac{\pi}{2}]$ with $b_- = e^{ir}$.

(b) For fixed $b_- \in S^1$, and any $k \in \mathbb{Z}$ if $b_- \neq \pm 1$, or any $k \in \mathbb{Z}_{\leq 0}$ if $b_- = \pm 1$, then $x_k(s, B)$ tends to ∞ in both limits $s \to \infty$ and $s \to -\infty$.

(c) For fixed $s_1 \in [-2, 2]$ and any $k \in \mathbb{Z}_{\geq 1}$, then $x_k(s_1, B)$ tends to 0 in any of the four limits $b_- \to -\infty$, $b_- \nearrow 0$, $b_- \searrow 0$, $b_- \to +\infty$.

Remarks 18.7

(i) s In Conjecture 18.6 (c), a priori some $x_k(s_1, B)$ could also have a finite limit or even tend to $+\infty$. Neither the asymptotic formulae in [IN86, ch. 11] nor the results on nilpotent orbits of TERP structures in [Mo11b] are formulated in a way which would control the behaviour of $x_k(s_1, B)$ for the four limits of b_-. Nor can Conjecture 18.6 (b) be answered from the asymptotic formulae in [IN86, ch. 11].

(ii) If Conjectures 18.6 (b) and (c) are correct, then also the following picture is essentially correct.

It is a conjectural sketch of the restriction to some $x_0 \in \mathbb{R}_{>0}$ of $M^{mon}_{3FN,\mathbb{R}}$, that is

$$M^{mon}_{3FN,\mathbb{R}}|_{x=x_0} \cong \{x_0\} \times V^{mat,\mathbb{R}},$$

together with the lines which are the intersection of $M^{mon}_{3FN,\mathbb{R}}|_{x=x_0}$ with $M^{sing}_{3FN,\mathbb{R}}$. The regions where f is positive or negative are again white or shaded. $V^{mat,\mathbb{R}}$ is a sphere with four holes. In the picture all four holes are shifted to the front. The back of the sphere is not visible. It is part of the shaded region.

Conjecture 18.6 (b) gives the spirals around the two holes $s = \pm\infty$. For large x_0 they are small, for small x_0 they are large, and the spiral around $s = \varepsilon\infty$ for $\varepsilon \in \{\pm 1\}$ approaches the line $s = 2\varepsilon$ near $B = \mathbf{1}_2$ and near $B = -\mathbf{1}_2$ as $x_0 \to 0$.

Conjecture 18.6 (c) gives the spirals around the two holes $ib_2 = \pm\infty$. For small x_0 they are small, for large x_0 they are large, and the spirals around both holes approach both lines $B = \pm\mathbf{1}_2$ as $x_0 \to \infty$.

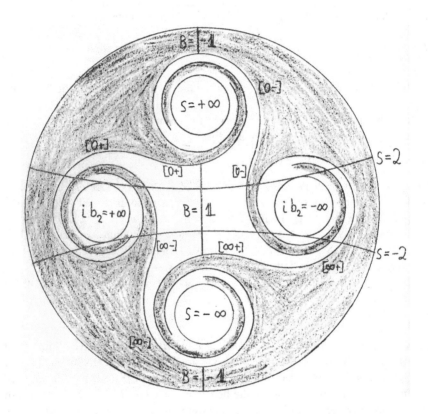

References

[BV89] Babbitt, D.G., Varadarajan, V.S.: Local moduli for meromorphic differential equations. Asterisque **169–170**, 1–217 (1989)

[BJL79] Balser, W., Jurkat, W.B., Lutz, D.A.: Birkhoff invariants and Stokes' multipliers for meromorphic linear differential equations. J. Math. Anal. Appl. **71**, 48–94 (1979)

[BS76] Bănică, C., Stănăşilă, O.: Algebraic Methods in the Global Theory of Complex Spaces. Editura Academiei, Buchurest (1976)

[Bo01] Boalch, P.: Symplectic manifolds and isomonodromic deformations. Adv. Math. **163**, 137–205 (2001)

[Bo05] Boalch, P.: Quasi-hamiltonian geometry of meromorphic connections. Duke Math. J. **139**, 369–404 (2007)

[BI95] Bobenko, A.I., Its, A.R.: The Painlevé III equation and the Iwasawa decomposition. Manuscripta Math. **87**, 369–377 (1995)

[BIK04] Bolibruch, A., Its, A.R., Kapaev, A.A.: On the Riemann-Hilbert-Birkhoff inverse monodromy problem and the Painlevé equations. Algebra i Analiz **16**, 121–162 (2004)

[CKS86] Cattani, E., Kaplan, A., Schmid, W.: Degeneration of Hodge structures. Ann. Math. (2) **123**(3), 457–535 (1986)

[CV91] Cecotti, S., Vafa, C.: Topological–anti-topological fusion. Nuclear Phys. B **367**(2), 359–461 (1991)

[CV93] Cecotti, S., Vafa, C.: On classification of $N = 2$ supersymmetric theories. Comm. Math. Phys. **158**(3), 569–644 (1993)

[Co99] Conte, R. (ed.): The Painlevé property: one century later. CRM Series in Mathematical Physics. Springer, New York (1999).

[Do08] Dorfmeister, J.: Generalized Weierstraß representations of surfaces. In: Surveys on Geometry and Integrable Systems. Adv. Stud. Pure Math. **51**, 55–111 (2008)

[DGR10] Dorfmeister, J., Guest, M.A., Rossman, W.: The tt^* structure of the quantum cohomology of $\mathbb{C}P^1$ from the viewpoint of differential geometry. Asian J. Math. **14**, 417–438 (2010)

[DPW98] Dorfmeister, J., Pedit, F., Wu, H.: Weierstrass type representations of harmonic maps into symmetric spaces. Commun. Anal. Geom. **6**, 633–668 (1998)

[Du93] Dubrovin, B.: Geometry and integrability of topological-antitopological fusion. Comm. Math. Phys. **152**(3), 539–564 (1993)

[Du99] Dubrovin, B.: Painlevé equations in 2D topological field theories. In: [Co99], pp. 287–412

© Springer International Publishing AG 2017

M.A. Guest, C. Hertling, *Painlevé III: A Case Study in the Geometry of Meromorphic Connections*, Lecture Notes in Mathematics 2198, DOI 10.1007/978-3-319-66526-9

[FLPP01] Ferus, D., Leschke, K., Pedit, F., Pinkall, U.: Quaternionic holomorphic geometry: Plücker formula, Dirac eigenvalue estimates and energy estimates of harmonic 2-tori. Invent. Math. **146**, 507–593 (2001)

[FPT94] Ferus, D., Pinkall, U., Timmreck, M.: Constant mean curvature planes with inner rotational symmetry in Euclidean 3-space. Math. Z. **215**, 561–568 (1994)

[FN80] Flaschka, H., Newell, A.C.: Monodromy- and spectrum-preserving deformations I. Comm. Math. Phys. **76**, 65–116 (1980)

[FA82] Fokas, A.S., Ablowitz, M.J.: On a unified approach to transformations and elementary solutions of Painlevé equations. J. Math. Phys. **23**, 2033–2042 (1982)

[FMZ92] Fokas, A.S., Mugau, U., Zhou, X.: On the solvability of Painlevé I, III, and IV. Inverse Prob. **8**, 757–785 (1992)

[FIKN06] Fokas, A.S., Its, A.R., Kapaev, A.A., Novokshenov, V.Yu.: Painlevé Transcendents: The Riemann-Hilbert Approach. Mathematical Surveys and Monographs, vol. 128. Amer. Math. Soc., Providence (2006)

[GJP01] Gordoa, P.R., Joshi, N., Pickering, A.: Mappings preserving locations of movable poles: II. The third and fifth Painlevé equations. Nonlinearity **14**, 567–582 (2001)

[Gr73] Gromak, V.I.: The solutions of Painlevé's third equation. Differ. Equ. **9**, 1599–1600 (1973)

[Gr79] Gromak, V.I.: Algebraic solutions of the third Painlevé equation. Dokl. Akad. Nauk BSSR **23**, 499–502 (1979)

[GL79] Gromak, V.I., Lukashevich, N.A.: Special classes of solutions of Painlevé equations. Differ. Equ. **18**, 317–326 (1980)

[GLSh02] Gromak, V.I., Laine, I., Shimomura, S.: Painlevé Differential Equations in the Complex Plane. Walter de Gruyter, Berlin (2002)

[Gu08] Guest, M.A.: From Quantum Cohomology to Integrable Systems. Oxford Univ. Press, Oxford (2008)

[GL14] Guest, M.A., Lin, C.-S.: Nonlinear PDE aspects of the tt^* equations of Cecotti and Vafa. J. Reine Angew. Math. **689**, 1–32 (2014)

[He03] Hertling, C.: tt^* geometry, Frobenius manifolds, their connections, and the construction for singularities. J. Reine Angew. Math. **555**, 77–161 (2003)

[HS11] Hertling, C., Sabbah, C.: Examples of non-commutative Hodge structures. J. Inst. Math. Jussieu **10**(3), 635–674 (2011)

[HS07] Hertling, C., Sevenheck, Ch.: Nilpotent orbits of a generalization of Hodge structures. J. Reine Angew. Math. **609**, 23–80 (2007)

[HS10] Hertling, C., Sevenheck, Ch.: Limits of families of Brieskorn lattices and compactified classifying spaces. Adv. Math. **223**, 1155–1224 (2010)

[Heu09] Heu, V.: Stability of rank 2 vector bundles along isomonodromic deformations. Math. Ann. **344**, 463–490 (2009)

[Heu10] Heu, V.: Universal isomonodromic deformations of meromorphic rank 2 connections on curves. Ann. Inst. Fourier **60**, 515–549 (2010)

[HL01] Hinkkanen, A., Laine, I.: Solutions of modified third Painlevé equation are meromorphic. J. Anal. Math. **85**, 323–337 (2001)

[IS12] Inaba, M., Saito, M.-H.: Moduli of unramified irregular singular parabolic connections on a smooth projective curve. Kyoto J. Math. **53**, 433–482 (2013)

[IN11] Its, A.R., Niles, D.: On the Riemann-Hilbert-Birkhoff inverse monodromy problem associated with the third Painlevé equation. Lett. Math. Phys. **96**, 85–108 (2011)

[IN86] Its, A.R., Novokshenov, V.Yu.: The isomonodromic deformation method in the theory of Painlevé equations. Lecture Notes in Mathematics, vol. 1191. Springer, Berlin (1986)

[IN88] Its, A.R., Novokshenov, V.Yu.: On the effective sufficient conditions for solvability of the inverse monodromy problem for the systems of linear ordinary differential equations. Funct. Anal. i Prilozhen. **22**(3), 25–36 (1988)

[IKShY91] Iwasaki, K., Kimura, H., Shimomura, S., Yoshida, M.: From Gauss to Painlevé. A modern theory of special functions. Aspects of Mathematics, vol. E 16. Vieweg, Braunschweig (1991)

[JM81] Jimbo, M., Miwa, T.: Monodromy preserving deformation of linear ordinary differential equations with rational coefficients. II. Physica D **2**, 407–448 (1981)

[Ki88] Kitaev, A.V.: The method of isomonodromy deformations and the asymptotics of solution of the "complete" third Painlevé equation. Math. USSR Sb. **62**, 421–444 (1988)

[Ko11] Kobayashi, Sh.: Real forms of complex surfaces of constant mean curvature. Trans. AMS **363**(4), 1765–1788 (2011)

[Lu67] Lukashevich, N.A.: On the theory of Painlevé's third equation. Differ. Equ. **3**, 994–999 (1967)

[Ma83a] Malgrange, B.: La classification des connexions irrégulières à une variable. In: Séminaire de'l ENS, Mathématique et Physique, 1979–1982. Progress in Mathematics, vol. 37, pp. 353–379. Birkhäuser, Boston (1983)

[Ma83b] Malgrange, B.: Sur les déformations isomonodromiques, I, II. In: Séminaire de'l ENS, Mathématique et Physique, 1979–1982. Progress in Mathematics, vol. 37, pp. 401–438. Birkhäuser, Boston (1983)

[MW98] Mansfield, E.L., Webster, H.N.: On one-parameter families of solutions of Painlevé III. Stud. Appl. Math. **101**, 321–341 (1998)

[MMT99] Matano, T., Matumiya, A., Takano, K.: On some Hamiltonian structures of Painlevé systems, II. J. Math. Soc. Jpn. **51**, 843–866 (1999)

[Mat97] Matumiya, A.: On some Hamiltonian structure of Painlevé systems, III. Kumamoto J. Math. **10**, 45–73 (1997)

[MTW77] McCoy, B.M., Tracy, C.A., Wu, T.T.: Painlevé functions of the third kind. J. Math. Phys. **18**(5), 1058–1092 (1977)

[MCB97] Milne, A.E., Clarkson, P.A., Bassom, A.P.: Bäcklund transformations and solution hierarchies for the third Painlevé equation. Stud. Math. Appl. **98**, 139–194 (1997)

[Mi81] Miwa, T.: Painlevé property of monodromy preserving deformation equations and the analyticity of τ-functions. Publ. RIMS, Kyoto Univ. **17**, 703–712 (1981)

[Mo03] Mochizuki, T.: Asymptotic behaviour of tame harmonic bundles and an application to pure twistor $\mathcal{D}$-modules, Part 1. Mem. Am. Math. Soc. **185**(869), xi+324 p. (2007)

[Mo11a] Mochizuki, T.: Wild harmonic bundles and wild pure twistor D-modules. Astérisque **340**, x+607 p. (2011)

[Mo11b] Mochizuki, T.: Asymptotic behaviour of variation of pure polarized TERP structures. Publ. RIMS Kyoto Univ. **47**(2), 419–534 (2011)

[Mu95] Murata, Y.: Classical solutions of the third Painlevé equation. Nagoya Math. J. **139**, 37–65 (1995)

[Ni09] Niles, D.: The Riemann-Hilbert-Birkhoff inverse monodromy problem and connection formulae for the third Painlevé transcendents, vii+76 p. Dissertation, Purdue University, Indiana, 2009

[No04] Noumi, M.: Painlevé equations through symmetry. Translations of Math. Monographs, vol. 223. Amer. Math. Soc., Providence (2004)

[NTY02] Noumi, M., Takano, K., Yamada, Y.: Bäcklund transformations and the manifolds of Painlevé systems. Funkcial. Ekvac. **45**(2), 237–258 (2002)

[NY04] Noumi, M., Yamada, Y.: Symmetries in Painlevé equations. Sugaku Expositions **17**, 203–218 (2004)

[OKSK06] Ohyama, Y., Kawamuko, H., Sakai, H., Okamoto, K.: Studies on the Painlevé equations, V, third Painlevé equations of special type $P_{III}(D_7)$ and $P_{III}(D_8)$. J. Math. Sci. Univ. Tokyo **13**, 145–204 (2006)

[OO06] Ohyama, Y., Okumura, S.: A coalescent diagram of the Painlevé equations from the viewpoint of isomonodromic deformations. J. Phys. A **39**, 12129–12151 (2006)

[Ok79] Okamoto, K.: Sur les feuilletages associés aux equations du second ordre à points critiques fixes de P. Painlevé. Jpn. J. Math. **5**, 1–79 (1979)

[Ok86] Okamoto, K.: Isomonodromic deformation and Painlevé equations and the Garnier system. J. Fac. Sci. Univ. Tokyo Sect. IA Math. **33**, 575–618 (1986)

[Ok87] Okamoto, K.: Studies on the Painlevé equations. IV. Third Painlevé equation P_{III}. Funkcial. Ekvac. **30**, 305–332 (1987)

[PS03] van der Put, M., Singer, M.F.: Galois theory of linear differential equations. Grundlehren der mathematischen Wissenschaften, vol. 328. Springer, New York (2003)

[PS09] van der Put, M., Saito, M.-H.: Moduli spaces for linear differential equations and the Painlevé equations. Ann. Inst. Fourier (Grenoble) **59**(7), 2611–2667 (2009)

[PT14] van der Put, M., Top, J.: Geometric aspects of the Painlevé equations $PIII(D_6)$ and $PIII(D_7)$. SIGMA **10**, 24 p. (2014)

[Sa02] Sabbah, C.: Déformations isomonodromiques et variétés de Frobenius. Savoirs Actuels, EDP Sciences, Les Ulis, 2002, Mathématiques. English translation: Isomonodromic deformations and Frobenius manifolds. Universitext, Springer and EDP Sciences, 2007

[Sa05a] Sabbah, C.: Fourier-Laplace transform of a variation of polarized complex Hodge structure. J. Reine Angew. Math. **621**, 123–158 (2008)

[Sa05b] Sabbah, C.: Polarizable twistor $\mathcal{D}$-modules. Astérisque **300**, vi+208 p. (2005)

[Sa13] Sabbah, C.: Introduction to Stokes structures. Lecture Notes in Mathematics, vol. 2060, xiv + 249 p. Springer, New York (2013)

[Sa01] Sakai, H.: Rational surfaces associated with affine root systems and geometry of the Painlevé equations. Comm. Math. Phys. **220**, 165–229 (2001)

[Sch73] Schmid, W.: Variation of Hodge structure: the singularities of the period mapping. Invent. Math. **22**, 211–319 (1973)

[ShT97] Shioda, T., Takano, K.: On some Hamiltonian structures of Painlevé systems, I. Funkcial. Ekvac. **40**, 271–291 (1997)

[Si67] Sibuya, Y.: Perturbation of linear ordinary differential equations at irregular singular points. Funkcial. Ekvac. **11**, 235–246 (1968)

[Si90] Sibuya, Y.: Linear differential equations in the complex domain: problems of analytic continuation. Translations of Math. Monographs, vol. 82. Amer. Math. Soc., Providence (1990)

[Ta07] Takei, Y. (ed.): Algebraic, analytic and geometric aspects of complex differential equations. Painlevé hierarchies. RIMS Kôkyûroku Bessatsu B2. RIMS, Kyoto University (2007).

[Te96] Temme, N.M.: Special Functions. An Introduction to the Classical Functions of Mathematical Physics. Wiley, New York (1996)

[Te07] Terajima, H.: Families of Okamoto-Painlevé pairs and Painlevé equations. Ann. Mat. Pura Appl. (4) **186**(1), 99–146 (2007)

[TOS05] Tsuda, T., Okamoto, K., Sakai, H.: Folding transformation of the Painlevé equations. Math. Annalen **331**, 713–738 (2005)

[Um98] Umemura, H.: Painlevé equations and classical functions. Sugaku Expositions **11**, 77–100 (1998)

[UW98] Umemura, H., Watanabe, H.: Solutions of the third Painlevé equation. I. Nagoya Math. J. **151**, 1–24 (1998)

[Um01] Umemura, H.: Painlevé equations in the past 100 years. Am. Math. Soc. Transl. (2) **204**, 81–110 (2001)

[LW92] Levi, D., Winternitz, P. (eds.): Painlevé transcendents. In: Their Asymptotics and Physical Applications. Nato ASI Series, Series B: Physics, vol. 278 (1992)

[Wi00] Widom, H.: On the solution of a Painlevé III equation. Math. Phys. Ana. Geom. (4) **3**, 375–384 (2000)

[Wi04] Witte, N.S.: New transformations for Painlevé's third transcendent. Proc. Am. Math. Soc. **132**(6), 1649–1658 (2004)

Index of Notation

© Springer International Publishing AG 2017
M.A. Guest, C. Hertling, *Painlevé III: A Case Study in the Geometry of Meromorphic Connections*, Lecture Notes in Mathematics 2198, DOI 10.1007/978-3-319-66526-9

LECTURE NOTES IN MATHEMATICS   🐴 Springer

Editors in Chief: J.-M. Morel, B. Teissier;

Editorial Policy

1. Lecture Notes aim to report new developments in all areas of mathematics and their applications – quickly, informally and at a high level. Mathematical texts analysing new developments in modelling and numerical simulation are welcome.

   Manuscripts should be reasonably self-contained and rounded off. Thus they may, and often will, present not only results of the author but also related work by other people. They may be based on specialised lecture courses. Furthermore, the manuscripts should provide sufficient motivation, examples and applications. This clearly distinguishes Lecture Notes from journal articles or technical reports which normally are very concise. Articles intended for a journal but too long to be accepted by most journals, usually do not have this "lecture notes" character. For similar reasons it is unusual for doctoral theses to be accepted for the Lecture Notes series, though habilitation theses may be appropriate.

2. Besides monographs, multi-author manuscripts resulting from SUMMER SCHOOLS or similar INTENSIVE COURSES are welcome, provided their objective was held to present an active mathematical topic to an audience at the beginning or intermediate graduate level (a list of participants should be provided).

   The resulting manuscript should not be just a collection of course notes, but should require advance planning and coordination among the main lecturers. The subject matter should dictate the structure of the book. This structure should be motivated and explained in a scientific introduction, and the notation, references, index and formulation of results should be, if possible, unified by the editors. Each contribution should have an abstract and an introduction referring to the other contributions. In other words, more preparatory work must go into a multi-authored volume than simply assembling a disparate collection of papers, communicated at the event.

3. Manuscripts should be submitted either online at www.editorialmanager.com/lnm to Springer's mathematics editorial in Heidelberg, or electronically to one of the series editors. Authors should be aware that incomplete or insufficiently close-to-final manuscripts almost always result in longer refereeing times and nevertheless unclear referees' recommendations, making further refereeing of a final draft necessary. The strict minimum amount of material that will be considered should include a detailed outline describing the planned contents of each chapter, a bibliography and several sample chapters. Parallel submission of a manuscript to another publisher while under consideration for LNM is not acceptable and can lead to rejection.

4. In general, **monographs** will be sent out to at least 2 external referees for evaluation.

   A final decision to publish can be made only on the basis of the complete manuscript, however a refereeing process leading to a preliminary decision can be based on a pre-final or incomplete manuscript.

   Volume Editors of **multi-author works** are expected to arrange for the refereeing, to the usual scientific standards, of the individual contributions. If the resulting reports can be forwarded to the LNM Editorial Board, this is very helpful. If no reports are forwarded or if other questions remain unclear in respect of homogeneity etc, the series editors may wish to consult external referees for an overall evaluation of the volume.

5. Manuscripts should in general be submitted in English. Final manuscripts should contain at least 100 pages of mathematical text and should always include

   - a table of contents;
   - an informative introduction, with adequate motivation and perhaps some historical remarks: it should be accessible to a reader not intimately familiar with the topic treated;
   - a subject index: as a rule this is genuinely helpful for the reader.
   - For evaluation purposes, manuscripts should be submitted as pdf files.

6. Careful preparation of the manuscripts will help keep production time short besides ensuring satisfactory appearance of the finished book in print and online. After acceptance of the manuscript authors will be asked to prepare the final LaTeX source files (see LaTeX templates online: https://www.springer.com/gb/authors-editors/book-authors-editors/manuscriptpreparation/5636) plus the corresponding pdf- or zipped ps-file. The LaTeX source files are essential for producing the full-text online version of the book, see http://link.springer.com/bookseries/304 for the existing online volumes of LNM). The technical production of a Lecture Notes volume takes approximately 12 weeks. Additional instructions, if necessary, are available on request from lnm@springer.com.

7. Authors receive a total of 30 free copies of their volume and free access to their book on SpringerLink, but no royalties. They are entitled to a discount of 33.3 % on the price of Springer books purchased for their personal use, if ordering directly from Springer.

8. Commitment to publish is made by a *Publishing Agreement*; contributing authors of multiauthor books are requested to sign a *Consent to Publish form.* Springer-Verlag registers the copyright for each volume. Authors are free to reuse material contained in their LNM volumes in later publications: a brief written (or e-mail) request for formal permission is sufficient.

Addresses:
Professor Jean-Michel Morel, CMLA, École Normale Supérieure de Cachan, France
E-mail: moreljeanmichel@gmail.com

Professor Bernard Teissier, Equipe Géométrie et Dynamique,
Institut de Mathématiques de Jussieu – Paris Rive Gauche, Paris, France
E-mail: bernard.teissier@imj-prg.fr

Springer: Ute McCrory, Mathematics, Heidelberg, Germany,
E-mail: lnm@springer.com

Printed in the United States
By Bookmasters